T0134327

Parasites and Infectious Disease
Discovery by Serendipity, and Otherwise

This series of entertaining essays provides a unique insight into some of the key discoveries that have shaped the field of parasitology. Based on interviews with eighteen of the world's leading parasitologists and infectious disease epidemiologists, the stories of their contributions to discovery in contemporary parasitology and infectious disease are told. Taken together, the essays represent a beautifully written account of the development of the field and provide a real insight into the thought processes and approaches taken in generating breakthrough scientific discoveries, ranging from immunology to ecology, and from malaria and trypanosomiasis to schistosomiasis and Lyme disease. Some of these discoveries were made serendipitously and others only after relentless effort pointed to a specific solution. This engaging and lively introduction to discovery in parasitology will be of interest to all those currently working in the field and will also serve to set the scene for future generations of parasitologists.

GERALD W. ESCH is the Charles M. Allen Professor of Biology at Wake Forest University, Winston-Salem, North Carolina, U.S.A. He is Editor of the *Journal of Parasitology*, the author of *Parasites, People and Places: Essays on Field Parasitology* (2004) and coauthor of the textbook *Parasitism* (2001). He is a recipient of the Louis T. Benezet Distinguished Alumnus Award from his undergraduate alma mater, Colorado College, in 1992, and of the Clark P. Read Mentor Award from the American Society of Parasitologists in 1999.

Parasites and Infectious Disease

Discovery by Serendipity, and Otherwise

GERALD W. ESCH
*Wake Forest University,
Winston-Salem, North Carolina*

CAMBRIDGE UNIVERSITY PRESS
Cambridge, New York, Melbourne, Madrid, Cape Town, Singapore, São Paulo

Cambridge University Press
The Edinburgh Building, Cambridge, CB2 8RU, UK

Published in the United States of America by Cambridge University Press, New York

www.cambridge.org
Information on this title: www.cambridge.org/9780521858823

First published 2007

Printed in the United Kingdom at the University Press, Cambridge

A catalog record for this publication is available from the British Library

ISBN 978-0-521-85882-3 hardback
ISBN 978-0-521-67539-0 paperback

Contents

Preface   *page* vii

Prologue   1
   Introduction   1
   Dick Seed   3
   Keith Vickerman   6
   Bob Desowitz   9
   K. Darwin Murrell   12
   Bill Campbell   18
   Richard Tinsley   25
   Sidney and Margaret Ewing   28
   Don Bundy   37
   Peter Hotez   42
   David Rollinson   48
   John Hawdon   53
   Mark Honigsbaum   61
   Roy Anderson   65
   Steve Nadler   77
   Jim Oliver   86
   Pat Lord   95
   J. P. Dubey   100

1 African trypanosomes and their VSGs   108

2 Malaria: the real killer   128

3 The HIV–AIDS vaccine and the disadvantage of natural
selection: the yellow fever vaccine and the advantage of
artificial selection   150

4   Lyme disease: a classic emerging disease   164

5   The discovery of ivermectin: a 'crapshoot', or not?   175

6   "You came a long way to see a tree"   188

7   Infectious disease and modern epidemiology   203

8   The 'unholy trinity' and the geohelminths: an
    intractable problem?   219

9   Hookworm disease: insidious, stealthily treacherous   236

10 The spadefoot toad and *Pseudodiplorchis americanus*: an
    amazing story of two very aquatic species in a very dry
    land   254

11 The schistosomes: split-bodied flukes   265

12 *Dicrocoelium dendriticum* and *Halipegus occidualis*: their
    life cycles and a genius at work   282

13 Trichinosis and *Trichinella* spp. (all eight of them, or is
    it nine?)   299

14 Phylogenetics: a contentious discipline   315

15 *Toxoplasma gondii*, *Sarcocystis neurona*, and *Neospora
    caninum*: the worst of the coccidians?   328

    Summary   345

    Index   348

Preface

Any book of the present sort requires a large number of sources, and I sought as many as I could. First, there were those folks who took the time to sit and talk with me. Without exception, they all opened up and answered every question I asked. But far more than this, they helped lead me into areas that I did not know about or might not have otherwise probed. Second, I want to thank them for all of the reprints they generously provided so that I could 'bone up' on their areas of interest before the interview was consummated. Third, there are several books that describe the fascinating history of parasitology. I am confident I have given appropriate credit for all that I used as I went along. There are, however, several special authors and books that I want to emphasize and from which I drew invaluable information. These of course include Bob Desowitz and all of his popular tomes, but especially those that dealt with the history of malaria and its treatment, primarily *The Malaria Capers*. Mark Honigsbaum's *The Fever Trail* was an extraordinary account of the history of quinine and the cinchona tree from which this herbal remedy comes. It should be read by anyone with an interest in malaria. I also read a large number of general historical accounts dealing with our discipline. The best was *A History of Human Helminthology*, by D. I. Grove. This is a really excellent encyclopedia of human helminthogy and should be in any university library.

I want to thank several folks who read various sections of the book along the way. This included a 'bunch' of my general parasitology students who were involuntarily cajoled into reading several of the essays as I wrote them. My good and treasured friend, Herman Eure, read several for me. Dan Johnson, a nonparasitologist colleague here at Wake Forest consented to read several of the essays and I appreciate his interest. Ralph Amen, my personal 'editor', read the entire book and offered invaluable input, in his own inimical way. One of my graduate students, Nick Negovetich, read several of the essays and parts of the Prologue. He had some very good suggestions. Another old friend, Ron Hathaway, at Colorado College, read the entire book as well. I sort of 'conned' him into doing it so that he could be my 'shill' when I gave a presentation to the Rocky Mountain Conference of Parasitologists meeting in September 2006.

My friend Ward Cooper was my original Commissioning Editor at Cambridge University Press. He helped get things started at the outset, but then moved to Blackwell, where he is now a Senior Commissioning Editor. Katrina Halliday stepped in to take Ward's place and actually got the contract through the 'dons' at Cambridge. I really owe her a huge thank you! There were four reviewers who took the time to provide some excellent comments on the proposal. I know two of them and have thanked them personally. I don't know the other two, but I thank them now. I also want to thank Mrs. Vickie Hennings for helping in the *Journal of Parasitology* office. It made writing a whole lot easier! I also had the great pleasure of working with Clare Georgy, Assistant Editor at Cambridge University Press; she was most helpful while I was trying to get the final version completed. Janice Robertson was my copy-editor and she was very supportive as well – I might add, she was as thorough as any copy-editor with whom I have worked during my career. I also want to thank Jeanette Alföldi for her help in guiding me through the new indexing process.

When I was thinking about a cover for *Parasites, People, and Places* I recalled a photograph of Slapton Ley I had taken back in 1987 during my second visit to the University of Exeter and Clive Kennedy. Of course, this is the site where Clive spent 35 years doing research. I persuaded my daughter Lisa to paint it and my good friend, the late Charlie Allen, to photograph the painting so I could use it as the cover for the book. That cover has received some really nice compliments from a wide range of readers. For the cover of the present book, I searched and searched for an idea. I had thought about a photo from the Chelsea Physic Garden, or the front of the Natural History Museum, both in London. Then, I recalled a couple of photographs I had taken from a moving train as we passed through the Midlands along the western side of England in May 2004. My wife, Ann, and I were returning to London from Glasgow where we had gone so I could interview Keith Vickerman for the present book. An old, nineteenth-century train trestle that stands in the middle of a really beautiful green valley grabbed my attention and, as the train passed by, I snapped off a couple of what turned out to be really good shots. But, you say, what is the connection between an old train trestle and discovery in my discipline of parasitology? I know it is a 'stretch', but throughout the book, I have attempted to link the past and the present. This train trestle represents the bridge I am trying to capture in the new book. As I said, this is a 'stretch', but I thought the photographs were really quite striking. So, I asked Lisa if she would paint me a new cover and she agreed. She has a real talent for capturing things like this on canvas. I think she has done a very good job again, and I thank her for her contribution.

Finally, I thank Ann for sticking with me while I was doing my thing over all these years, 47 to be exact. By the time this book is in print, I trust it will be 48!

Prologue

INTRODUCTION

In the summer of 2003, I finished work on a book entitled *Parasites, People and Places: Essays on Field Parasitology*. My wife, Ann, and I were in our cabin in Green Mountain Falls, Colorado, and I was trying to tell her the story from the book that had to do with the discovery by William Walter Cort of the cause of swimmer's itch back in 1927. At the same time, she knew I was sort of lamenting the absence of a new project. She must have been impressed by my tale, because out of the blue, she said, "Why don't you write a book about discovery in parasitology?"

This started me thinking about the possibility of doing something along that line. Gradually, over the next several months, I put together an idea. Stories regarding the discovery of the transmission of malaria or sleeping sickness have been told many times over the years, so they are sort of 'old hat'. But, then I thought, are they really?

I recalled the way I teach my own general parasitology course to undergraduates. I know that I mention Ronald Ross and David Bruce, among others, but I really do not get into much detail about how Ross and Bruce did their work regarding malaria or African sleeping sickness, respectively. Then, I began thinking about some of the new discoveries regarding malaria (*Plasmodium* spp.) and sleeping sickness (*Trypanosoma* spp.) that have been made since their life cycles, and those of other parasites, first were resolved. For example, consider the variant surface glycoproteins of the trypanosomes. Who did this research, and what led them to do it? Since this is definitely not my area of expertise, I honestly did not know for certain who did what until after I began researching the background information necessary to write this series of essays. So, I thought, why not briefly retell some of the old stories

1

regarding discovery in the nineteenth and early twentieth centuries, and link these historical accounts, where relevant, to some of the newer work that has been done in the last 50 years. In a few essays, I have simply gone to a prominent parasitologist and asked them to tell me about their lifetime of research.

Another idea emerged when I began thinking about this approach. I became intrigued by the possible role of serendipity in all of these parasitological discoveries. My *Random House* dictionary defines serendipity as "the faculty for making desirable discovery by accident." Then, I began to wonder, do some folks have a faculty for making "desirable discovery by accident"? Was Louis Pasteur correct when he said, "In the field of observation, chance only favors the mind which is prepared"? What about the life cycles of *Plasmodium* spp. and *Trypanosoma* spp.? A number of investigators were looking for the way in which the malarial parasite was transmitted, not just Ross. Was Ross endowed with a 'special' faculty for discovery? Why did he 'hit the jackpot' sooner than the others? In the case of sleeping sickness, Aldo Castellani first saw tryps in the cerebrospinal fluid of humans, but he thought initially the disease was caused by a streptococcus infection of the heart. It was David Bruce, however, who is generally given credit for identifying the etiological agent of sleeping sickness. Why not Castellani?

I also began thinking about something else. Ross, Bruce, and the other giants of their era have been dead for many years. But, what about those individuals who made important discoveries in the past fifty to sixty years? It dawned on me that many of these parasitologists were either retired, or were close to it, and some have even died. They have important stories to tell, but they are generally not being told. I think these stories need to be out there as well, in full view. The new parasitology students need to know who these people were/are and they need to know why these folks, or what I call the 'middle generation' of parasitologists, did what they did.

Who would I choose to talk about and why would I select certain ones to focus upon? I realized early that this proposition could become rather 'dicey'. Whenever anyone makes a selection of this kind, some folks will be annoyed because one of their favorite people was left out. On the other hand, since I am the one doing the writing, it must be my choice. So, that's what I did, I made some hard choices. I considered a long list of ideas and possibilities. I then proceeded to choose my favorite 'discoveries', my favorite parasites, and my favorite people (at least some of them). As it turned out, it is a very eclectic group, in all three categories. In a few cases, I could not conduct an interview because the person is dead. So, I had to rely on someone else to tell the story for them, but I was lucky because I had some very good sources.

Another point, I am writing these 'stories' as essays because I think this format gives me freedom to roam. I am not bound by a particular style, and can pretty much go where I want and take a particular topic as far as I want. I also decided to place in a Prologue at least some of the biographical information regarding each person that I have interviewed, or that I talk about extensively in the specific essay. Each of the interviews begins the same way. Where did you go to school, first as an undergraduate, and then as a graduate student? How did you get into parasitology? When did you graduate? Then, I wanted to use fairly standard questions in an effort to get some sort of idea about how they think, where they might have traveled to do their research, why their work went the way it did, etc. In obtaining information like this, I was able to generate some useful comparisons of some very good parasitologists.

It was fun!

DICK SEED

Figure 1. J. Richard Seed, retired Professor, Department of Epidemiology, School of Public Health, University of North Carolina-Chapel Hill, Chapel Hill, North Carolina

While thinking about discoveries in parasitology over the past fifty years, it occurred to me that the research on variant surface gly-coproteins (VSG) would be of interest in connection with the story of how the life cycle of the African trypanosomes *Trypanosoma brucei brucei* was worked out by David Bruce. After doing some 'snooping' around in the appropriate literature, two names, J. Richard Seed and Keith Vicker-man, F.R.S., recurred with some frequency. I decided to interview both of them and ask about their contributions to the VSG tale. Dick Seed was just down the road in Chapel Hill, North Carolina, an easy drive from Wake Forest and Winston-Salem, and Keith Vickerman was at the Glasgow University in Scotland, so I went there to see him as well, an easy trip by plane and train via London.

In May of 2004, I drove to Chapel Hill and spent the morning drinking tea and visiting with my friend Dick Seed. I have known Dick for nearly thirty years, having met him the first time at an American Society of Parasitologists meeting in San Antonio, Texas, in 1976. That was the year he received the H. B. Ward Medal for his contributions to the immunology and biology of the African trypanosomes.

Dick's undergraduate days in the mid 1950s were spent at Lafayette College in Pennnsylvania. He said that his interest in biology was already present when he entered Lafayette. While there, his enthu-siasm for microbiology was 'tweaked' by Professor Willis ("Bugsy") Hunt, a 'Yaley', who then encouraged him to follow his lead and also head for Yale and his Ph.D., which he completed in just three years, working in the lab of David Weinman. According to Dick, "The latter was an M.D., and had spent much of his life . . . doing parasitology all over the world." Weinman was working on the African trypanosomes and this is where Dick focused his graduate research as well. While Dick's experience at Yale was a good one, he lamented that it was too short, just three years, and that there were many techniques, etc., he felt he should have learned, but did not have the time. For any young person reading this book, this is very good advice. Along the way, someone may advise you to skip the Master's degree and go straight for the Ph.D. I disagree. I strongly believe the Master's degree is excellent prepara-tion for pursuing the Ph.D. degree. It presents you with an element of practice for what is to come.

I was interested to learn that, as a youngster, Dick had read Paul de Kruif's (1926) *The Microbe Hunters*, and that this is what stim-ulated his interest in infectious disease and, subsequently, in para-sitology and trypanosomiasis. Another book, Geoffrey Beale's (1954)

The Genetics of Paramecium aurelia, had also caught his attention. Among other things, the latter book detailed the phenomenon of antigenic variation, which ultimately led him to a postdoc with Irving Finger at Haverford College in Pennsylvania. His thinking was that "antigenic variation among the African trypanosomes was similar to what occurred in the mobilization antigen on *Paramecium*. If I learned about the latter, then I could use the tools of the *Paramecium* geneticist to understand the African tryps." (When you read on, you will discover that the same two books had a huge influence on the career of Keith Vickerman too, another veteran of the antigenic variation effort, plus several others that I interviewed for the present book.)

As I will detail in my essay on African trypanosomiasis and VSGs, Dick did not really follow up on any of his Ph.D. research after he finished at Yale. In 1960, Jacob and Monod won a Nobel Prize for their discoveries regarding the regulatory processes associated with β-galactosidase in *Escherichia coli*. Dick's mission for most of his post-Ph.D. professional life was to search for what turned out to be an elusive regulatory process in trypanosomes. In graduate school and throughout his career, he made several important discoveries, but not the one for which he was ultimately searching. As will be seen from the trypanosome essay, however, the absence of 'ultimate' success was not for a lack of trying.

Dick's stay at Haverford was brief since 'Uncle Sam' decided to 'hire' him as a Medical Service Officer for the U.S. army. He was assigned to Fort Baker, just north of San Francisco, where he spent the next two years running a parasitology diagnostic laboratory. It was not a wasted time as he had two very good technicians from whom he learned a "lot about parasitology and diagnostic procedures." Following Fort Baker, Dick traveled to New Orleans where he became an Assistant Professor at Tulane University, advancing to the rank of Professor over the next eight years. After that came five years at College Station, Texas, and Texas A&M University, where he served as Chairman of the Biology Department. It was then on to Chapel Hill, North Carolina, where he became Head of the Department of Parasitology in their School of Public Health. He serves there as Professor of Epidemiology, although this (2006) is his last year before retirement.

I really enjoyed that morning in Chapel Hill with Dick. My stay was greatly enhanced by the presence of Dick's lovely wife, Judy, who provided tea and pleasant conversation during our breaks.

KEITH VICKERMAN

Figure 2. Keith Vickerman, FRS, retired Professor, University of Glasgow, Glasgow, Scotland

In late May 2004, my wife, Ann, and I drove to Charlotte, North Carolina, and boarded a flight for Gatwick Airport, near London. After a couple of days of getting our body clocks reset, we took a train to Glasgow where I was to interview Professor Emeritus Keith R. Vickerman, F.R.S. We arrived at Central Station around 2:30 P.M. and checked into our hotel. Ann suggested we take the underground out to the campus of Glasgow University and scout out Keith's office, so I wouldn't be late for my appointment the next morning. We not only found it, but also actually met and spoke with Keith for a few minutes that afternoon. Ann then wanted to visit the wonderful Hunterian Museum on the University's campus, which we did. We did not see much because they ran us out at 5:00. As we were leaving, we were caught in a classic thunderstorm. By the time we made it back to the tube station, we were drenched (no umbrellas – great planning!). When we emerged from the tube station at the other end, it was still pouring. So, she suggested we make at dash for a nearby 'Boots' (a chain drugstore in the U.K.) and buy a couple of cheap umbrellas. We came out with our new protection and walked back to the hotel, 'singing and dancing', in the rain. (By the way, as I could have predicted, we did not use the umbrellas over the next two weeks of our stay in the U.K.)

The interview the next morning went really well. Keith had even prepared a written autobiography and a full CV for me to take home.

He said it was part of an 'obituary' he was required to write when he was elected Fellow in the Royal Society. I don't know if he was 'pulling my leg' or not, but the written information did come in handy while preparing the essay on trypanosomiasis.

He told me that his entry into the biological realm was triggered by a serendipitous event, one that occurred early in his life. As a twelve-year-old in grammar school, he was enrolled in a second year science course devoted to the history of microbiology. With the sudden departure of his instructor, a mathematician with absolutely no training in biology was recruited as a replacement. Not knowing what to teach, the teacher, Keith said, "read to us from Paul de Kruif's *The Microbe Hunters*. I was spellbound," and a brilliant career began.

Keith went on to say that Elie Metchnikoff, the legendary Russian zoologist, was to become his idol. It was Metchnikoff who deduced that phagocytic cells in mammals may be involved as a defense against intruding pathogens while he watched similar-type cells attack thorns experimentally introduced into the bodies of larval starfish. Keith said he realized "later that it [Metchnikoff's idea] represents a perfect example of Popper's view of the scientific method – that a single observation inspires a flash of intuition that leads to a fashioned hypothesis that can be tested by further observation and experiment with a view to falsification or corroboration." What a great description of the scientific method!

Keith's undergraduate academic career began at University College London (UCL) in 1952, under the tutelage of Peter Medawar (who was to become a Nobel Laureate in 1960 for his work on skin graft rejection). While he enjoyed his relationship with Medawar, he considered him to be somewhat 'baronial'. When Keith announced his interest in parasitology for graduate work, Medawar was rather scornful, referring to parasitology, "as a somewhat philistine pursuit, far too long cut off from the advances of mainstream fundamental biology." In our interview, he said that Medawar considered the discipline as "impenetratively deaf to all the advances in biology for the past 50 years." (Keith told me he thought at the time, "Well, there must be an awful lot of catching up to do – all the more reason for getting into it.") However, for Medawar, protozoology was another matter. He was very impressed with the great American ciliatologist and geneticist, Tracy Sonneborne, and said to Keith, "The Research Councils are very concerned about the dwindling population of protozoologists, so why not become a protozoologist," and Keith responded, "I will." Keith actually became a protozoan parasitologist, with an early, and then a long-term, research focus on the African trypanosomes.

In fact, because he wanted to work on trypanosomes, doing his Ph.D. presented some difficulty because he could not settle on someone with whom to study or who would take him on as a student. He finally ended up at the University College of the South West, in Exeter, with R. S. J. Hawes, working with protozoans of soil and soil-dwelling insects: "not exactly what I had envisaged, but I did manage to find a trypanosomatid in tipulid (crane fly) larvae." He was not entirely pleased with his experience there, although he did manage to spend a "term" working at "Edinburgh University in Michael Swann's thriving cell biology group." It wasn't totally bad at Exeter because he also met F. E. G. (Frank) Cox, a fortuitous event. Keith said that, "Frank was in a class I had demonstrated to and he had been a technician with P. C. C. Garnham" at the London School of Hygiene and Tropical Medicine. Keith and Frank were to become life-long friends and colleagues.

When Keith finished at Exeter, he went back to UCL and Peter Medawar's immunology lab, where he was offered the opportunity to work on any protozoan he wanted. He naturally went to the African trypanosomes. He also chose to focus his efforts on antigenic variation because of a book by Geoffrey Beale, *The Genetics of* Paramecium aurelia (the same one that Dick Seed had read). When I returned from London/Glasgow, I phoned Seed to tell him about my interview with Keith and then sent him a copy of the information that Keith had prepared for me. A few days later, Dick returned my phone call and excitedly described for me how many parallels there were between his early biological experiences and those of Vickerman. Both were to have huge successes in their work on antigenic variation in the African trypanosomes.

Keith's "fascination with life cycle changes and their relation to survival in changing environments" began as a student and stayed with him throughout his career. His research on the African trypanosomes actually focused on two areas, both of which were tied to life cycle changes and changing environments. Thus, for example, he spent a great deal of time attempting to understand the energy metabolism of the organisms, discovering in the process that in the tsetse fly gut, the amino acid proline is the main energy source. In the vertebrate, the trypanosomes switch to an aerobic glycerophosphate oxidase system. He was able to correlate these metabolic differences with significant morphological changes in the mitochondria as the parasite moves from the fly to the vertebrate host and back. It was while working on these changes using electron microscopy that he made his second really important discovery. That was the physical presence of an antigen coat, the VSG, on the surface of the metacyclic form of the trypomastigote in the salivary glands of the fly. This was a significant event and one that was serendipitous – he was not looking for it. But, when he saw it, he knew about the significance of the coat, and he went after it. He made a huge discovery and a really momentous research contribution as a result.

Keith had a marvelous career at Glasgow University, retiring in 1993. He continues to write and do research in spite of a serious back injury recently suffered in a fall at his home.

BOB DESOWITZ

Figure 3. Bob Desowitz, retired Professor, University of Hawaii, Southern Pines, North Carolina

Bob Desowitz completed his undergraduate work at the University of Buffalo, in New York, with a two-year interruption by service in the U.S. Army right after WWII. He had become interested in microbiology and parasitology and decided to pursue the topics at the next level, graduate school. He asked his advisor at Buffalo where he should go and his mentor suggested the London School of Hygiene and Tropical Medicine. So, he applied and was accepted.

In the fall of 1948, Bob arrived at the London School of Hygiene and Tropical Medicine (LSHTM), ready to pursue his Ph.D. degree in protozoology with Henry Shortt, a former Colonel in the British Colonial Service. Fortuitously, Bob was in place to witness one of the great dramas of discovery in parasitology. Henry Shortt and P. C. C. Garnham were about to take the final step in pursuit of the *Plasmodium* spp. life cycle and solve a mystery that had been around for nearly 50 years. It seems that Fritz Schaudinn had reported in 1903 that the sporozoites of *Plasmodium* spp., on being inoculated into the blood of their vertebrate hosts by mosquito vectors, disappeared after about 30 minutes – correct! But, he also said that the sporozoites then penetrated red blood cells directly – incorrect! When Bob arrived, Shortt and Garnham had just finished the first effort to purge this assertion using *P. cynomolgi* and monkeys. I'll write more about this huge discovery later.

During my interview with Desowitz, he described Henry Shortt as a "truly wonderful man, and was marvelous to work for." Bob also explained, "Shortt was big on lineages. He [Shortt] was taught by Sir Rickard Christophers, and Christophers was taught by Sir Ronald Ross, so I'm a direct descendant of Ronald Ross!" Bob's Ph.D. research was on *Histomonas meleagridis*, the causative agent of 'turkey blackhead'. He told me that he kept all of his turkeys up on the roof of the building housing the LSHTM. It was soon after the war and there were still food shortages in the U.K. Each year, at Christmas time, Bob said, "Shortt and Garnham would show up at his lab space and inquire as to the availability of a control turkey."

When he completed his dissertation research, he was searching for a place to publish his work. Shortt persuaded him to submit it to *Nature* where it appeared as his first publication ("not a bad place to start," he proudly remarked during our interview). With his Ph.D. in hand, he was about to take a position in the local poultry industry when a monacled Englishman showed up in his office. He introduced himself as Colonel ("another one!") Hugh Mulligan of the British Colonial Service and said to Desowitz, "You come highly recommended by Henry Shortt. Would you like to go to work for me? I am setting up a new research institute in northern Nigeria." Bob responded immediately, "Yes, but I'm an American." "Don't worry about that, I'll fix it," said Mulligan, "Let's go to the pub and discuss it." "So," Bob continued, "we headed for the Duke of Wellington Arms to talk about it", adding, "I think all recruiting for the British Colonial Service takes place in pubs!" He was off to Africa and nine years of research on trypanosomes as an American in the British Colonial Service.

Toward the end of his stay in Nigeria, the Provost at the University of Singapore, an old friend from his graduate school days at the London School of Hygiene and Tropical Medicine, contacted him. He was invited to take the position of Head of their Department of Parasitology. He accepted the challenge and spent the next five years there. He also switched his research into the diagnostic area of malaria. Then, during his last year in Singapore, he was persuaded by Elvio Sadun to come to the SEATO (Southeast Asia Treaty Organization) research lab in Bangkok, Thailand, as Chief of the Department of Medical Parasitology, and he accepted. The U.S. Army operated the lab as an activity of the Walter Reed Army Institute of Research and, although a Department of Defense civil service civilian, he was given the courtesy/substantive rank of "bird colonel". He laughed about the service appointment because immediately after WWII he had been drafted into the army as a lowly private. He considered his rapid rise in rank rather amusing.

Several years later, he was approached by the University of Hawaii's administration. After fourteen years of administrative responsibilities he told me it was a happy relief to accept a professorship with a more simple charge to run the parasitology component within the Department of Tropical Medicine and Medical Microbiology. He was tired of administration by this time, and accepted the offer with a quid pro quo that he would have an administrative assistant to handle some of the chores. After thirty years in Hawaii, he is now retired and living comfortably in Southern Pines, North Carolina. He has since emerged as a successful and widely read 'popular' author dealing with tropical disease and parasitology, mostly based on his wonderful experiences in Africa and southeastern Asia.

K. DARWIN MURRELL

Figure 4. K. Darwin Murrell, retired scientist, Agricultural Research Service, U.S. Department of Agriculture, Beltsville, Maryland

As is the case with many of us who wind up in academia, industry, or in government labs to do research, we were drawn into our professions by a teacher, usually in a college or university when we take our first biology or zoology course. This was true in the case of Darwin Murrell. On graduating from high school in northern California, his intention was to join the Marine Corps, but one of his buddies 'chickened out'. So, when the rest of his friends went to Chico State College, Darwin trailed along, not really knowing what he wanted to do. His first courses were in agriculture, but this direction was quickly aborted due to a colossal case of monotony. He recalled that in the spring of his first year, he signed up for a beginning course in zoology, taught by Don Wootton. Don always took his entire class for a week of marine biology at Bedoga Bay and, in Darwin's words, "That blew my mind! Wow," he continued, "I never imagined such stuff. Then he taught us how to cook our own things from the sea." This reminds me of Mick Burt, the great Canadian parasitologist, who required his students to learn how to cook the hosts of whatever parasite project they might choose for his parasitology course – my favorite was garlic sea cucumbers (see p. 177 in Esch, 2004).

After being hooked by Wootton and his course in general parasitology, Uncle Sam sent him a letter indicating the Army needed him for a spell. With help from Don, however, he managed to secure a deferment. This allowed him to travel to the University of Michigan Biological Station (UMBS) at Douglas Lake where he took the helminth parasitology course from Jim Hendricks. Throughout his undergraduate career, each summer, he and usually a couple of other students would travel with Wooton to UMBS, camping along the way and visiting the great natural areas of the west on their way to and from Michigan. While he was at UMBS in the summers, he was talked into heading south to University of North Carolina-Chapel Hill and Hendricks' home base where he was to begin pursuit of his Ph.D. degree in the early 1960s. From his new Chapel Hill base, he returned to UMBS for one more summer. I asked him if Will Cort was still going up to the Station at that time, and he said no, at least Cort was not physically present. But, he told me that "Will Cort and Paul Beaver were still there in spirit, and their names and vials of materials were everywhere."

Before he could complete his degree at UNC, Uncle Sam decided it was time to end the deferment and he was reclassified 1-A. All of us who experienced those days are well aware that a 1-A classification meant a draft notice was on its way, and Darwin was informed in December that he would be called up in March. Rather than having his draft board make the decision on how he would serve, he headed for the offices of the local recruiters. Darwin was exceedingly lucky (serendipity) because it was from one of these recruiters he learned about the Naval Medical Research Institute (NAMRI) in Bethesda, Maryland. After filling out all the appropriate papers and interviewing with personnel at NAMRI, he received an ensign's commission in the Medical Service Corps as a parasitologist. He was initially based in Bethesda where he spent his initial four months in the service, before being transferred to Naval Medical Research Unit-2 in Taipei, Taiwan. It was there he had another stroke of luck. John Cross, a civilian, had just been named the new head of the parasitology section at NAMRU-2. John had lived in China for many years, making him a perfect fit for the position in Taiwan.

Their workload was heavy. Darwin told me about his "parasitological responsibilities in developing diagnostic tests for the Marine Hospital in DaNang, Vietnam, plague in Indonesia, capillariasis in the Philippines, parasitic diseases in Taiwan – we were on the road a lot." Cross knew that Darwin needed to do some more research to complete his dissertation. He told Darwin that anytime he was not busy, he could work on his dissertation research, and that he would even buy the equipment for him. His Ph.D. research involved a study of the effects of antibody and complement on the tegument of larval cestodes, specifically cysticerci of *Taenia crassiceps*. One of the instruments purchased for Darwin was the old Warburg manometer. He reminded me, "I was taught how to use the Warburg from a young postdoc in Chapel Hill, by the name of Jerry Esch." I had almost forgotten, probably because I hated the apparatus so much. Anyone who has ever used it would wholeheartedly agree.

A female missionary in Taiwan improved his German to the point he was eventually able to pass the language exam, then required for most of us seeking a Ph.D. in the old days. At night, he would write his dissertation and in the day, his wife, Joyce, would type what he had written the previous evening. It was then sent to Norman Weatherly, his dissertation advisor in Chapel Hill, who would work through it and then return it to Darwin for further revision. After his tour in Taiwan was completed, he returned to Bethesda. His Ph.D. defense was in 1969, four years after entering the Navy. He was tempted to stay in the Navy with NAMRI because of the opportunities afforded him by their research facilities in Africa and Asia. But, Darwin believed, the Navy had a policy of transferring personnel with some regularity and he felt he could not get a good research program started at any of these venues before he would be transferred again. Moreover, he felt all the moving and travel would be tough for a growing family.

So, after his tour of duty in the Navy, he headed for the University of Chicago and the legendary Bob Lewert, with whom he worked for the next two years. Darwin had, by then, begun thinking about a job and his long-term future. There was a particularly serious problem at that time though, since positions in academia were terribly scarce. Even with his wonderful résumé and overseas experience, his job search was not successful. By then, the University of Chicago had instituted a two-year M.D. program for folks with the kind of background possessed by Darwin and he was sorely tempted to pursue it. In fact, he applied and was admitted. However, just before entering the program, he received a phone call from Dick Beaudoin (head of malaria research at the NAMRI).

Dick informed him that Peg Stirewalt, who was leading the schistosome research program at NAMRI, was about to retire and encouraged him to apply for the position. So, he told Dick to pass the word to the appropriate people at NAMRI that he was interested. A few days later, he received a phone call with an invitation to come to Bethesda for an interview, which he did. The job was offered and he accepted on the spot. The folks in the medical school at Chicago made a considerable effort to keep him, but his mind was made up: he was going to Bethesda.

Life at NAMRI was good. He was there for seven years. Then, as he related, "The Navy's Head of Research and Development showed up, asking for a report on their research. After the meeting, he said, 'Well, you are doing nice research here, but I have to be honest with you, schisto research is of no interest. They don't even have it in Brazil'." Darwin was shocked, "I couldn't believe he said that!" He consulted with a friend in the R&D Command of the Navy who told him he thought schisto research was going to be chopped. He believed malaria research would continue, but no more with the schistosomes.

Then came the next move. He knew the NAMRI job was coming to an end. He said that not long after learning his position with the Navy was going to be terminated, he ran into Ron Fayer. Ron said he had heard of Darwin's predicament and that there was a lab chief's opening in helminthology over at Beltsville (USDA). It would be in veterinary helminthology and a change in research emphasis, but he needed a job, so he interviewed and was hired. The interesting thing about the offer was that he was given *carte blanche*. He was told "to look around and see what you think is of importance," another lucky break in his research career. He "decided that strongyloidiasis was a big problem in the swine industry. So I began working on *Strongyloides ransomi* and *S. ratti* as a complementing lab model."

Things went well for about two and a half years when he received a phone call one day from an official in the Agriculture Research Services (ARS), the R&D agency of the Department of Agriculture. He asked, "Are you doing anything on *Trichinella*?" Darwin responded, "No, don't you remember a couple of years ago when I came here that we decided to phase that out?" The ARS person said, "Well, the pork industry raised a hell of a lot of money for the ARS a few years back and they want to know what we did with it. Can't you come up with something to do with regard to *Trichinella*?" Darwin was then told by his ARS colleague to "think about it."

"Well," as Darwin put it, "it was one of those coincidences in life that a scientist over in the Meat Institute called me up about that time and said, 'you know I have been working on a grant from the fast food industry and the pork industry to look at ways to cook pork very fast for fast food outlets.' They want these methods evaluated for their effectiveness in killing trichina and I don't know anything about how to do it." He invited Darwin to collaborate, and he agreed. He immediately set up animal bioassays for the rapidly cooked pork.

Irony in their research was subsequently to come in several unusual ways. For example, when they completed their work and wrote up the paper, it was submitted to the ARS for clearance prior to being sent to a journal for publication. It had to go through this sort of process because *T. spiralis* was on the list of 'sensitive' parasites, even though no research sponsored by the government had been done on these worms for several years. The conclusion in their manuscript was that fast cooking with microwave technology could not guarantee that trichinae in pork would be killed, due to uneven cooking by microwave ovens. The next day, he received a call from his colleague in the ARS saying, "Jesus Christ, what are you guys trying to do, get me fired? If we let you publish this before we inform the pork and microwave industry, all hell will break loose." Darwin said that they went ahead and published the report anyway. As predicted, "All hell did break loose! Newspapers, radio, television, everyone, wanted to know about our *Trichinella* results! Why?", Darwin asked rhetorically, "Because microwave cooking won't protect you from *Trichinella*." On the other hand, he remarked, again rhetorically, "Have you ever known anyone infected with *Trichinella*?" Their results ultimately created such an in-house ruckus at ARS that further work on *T. spiralis* suddenly became an absolute necessity, and the next appropriation from Congress included a sizable amount of money for *T. spiralis* research. That is how he developed a "lifelong and intimate relationship with this exotic parasite". A further irony was that research in the Department of Parasitology at UNC-Chapel Hill, where he obtained his Ph.D. some thirty years earlier, was almost entirely focused on *T. spiralis* immunity. As Darwin said, "What goes around comes around."

In our conversation, Darwin remarked that after nine years he took a particular fork in his career path that, in retrospect, was not altogether wise because it led him away from the lab for the next thirteen years, and into administration. Although life as an administrator had its fine moments, the nagging regret of not being 'hands on' in research never left him. The only way he could be involved was through the occasional review or book chapter he was asked to contribute.

Then, a stunning loss to the worldwide veterinary parasitology community came with the untimely passing of Peter Nansen, who was Director of Denmark's Centre for Experimental Parasitology, where their focus was on basic research on parasites, mostly helminths, of veterinary importance. Darwin had spent six months in the Centre as a Fulbright Fellow in 1996 while on sabbatical from ARS in 1996. Knowing that he had retired from the USDA, the Foundation's Board called on Darwin to become the new Head, replacing Nansen. He signed a three-year contract, and he and Joyce were off to Copenhagen. I could tell by our conversation that he really enjoyed his tour of duty in Europe. Their research was focused on any number of parasitic diseases, ranging from *Ascaris* and *Trichinella* to neurocysticercosis, and he was an active participant in much of it. In fact, one of the projects focused on neuro-cysticercosis in collaboration with a group in India. Their target was to develop an immunological diagnosis for the single cyst granuloma form of the disease, which, apparently, was behaving differently than it was in many other parts of the world. In India, the preponderance of infections are apparently single cysts, which are very difficult to diagnose in contrast to other areas where the easier to diagnose multiple cyst infections predominate. In India, the epidemiology of neurocysticercosis is not well enough understood to be successful in the disease's control.

Another particularly rewarding project he undertook was to conduct epidemiological studies with colleagues in Serbia on the resurgence of trichinellosis following the outbreak of civil war there in 1989. In addition to identifying some key social and agricultural factors that greatly increased the risk of *Trichinella* transmission, he guided a project that produced the first experimental data on horse infections, which have been responsible for the majority of the recent outbreaks of trichinellosis in western Europe. Sadly, it was also an opportunity of seeing what devastation a senseless war could bring to a society and a country.

After the three years in Denmark, he decided to break free of administration, so he and Joyce returned to their home in Rockville, Maryland. However, this does not mean he has stopped working; he has just gone off in a new direction. One of his current projects involves fish-borne parasites in Vietnam, funded by the Danish International Development Agency (Danida). Apparently, there is still a huge problem with 'honey buckets', raw sewage, raw fish, and fish-borne flukes in many areas of southeastern Asia. Another includes a study with an Australian collaborator on canine zoonoses in Thailand. He is also advisor to an FAO research project on trichinellosis in Argentina, and

frequently serves as a consultant to WHO and FAO, particularly in editing new guidelines for cysticercosis and trichinellosis. For Darwin, as
for so many colleagues, there is no retirement in the classic sense. He
is now enjoying every minute of his newly discovered occupation, that
of a freelance parasitologist.

BILL CAMPBELL

Figure 5. Bill Campbell, retired scientist, NAS, Merck, Rahway, New
Jersey

We transition now to another *Trichinella* expert, although that is
not the reason I wanted to interview Bill. Among other things, I consider William C. Campbell to be among a handful of true 'Renaissance
Men' in the American Society of Parasitologists. Whether one agrees
with my judgment on this notion or not, something one cannot disagree about was his justifiable election to the National Academy of

Sciences, U.S.A. The primary reason for this esteemed and high honor was his contribution at Merck to the discovery of ivermectin, a truly remarkable drug for the treatment of parasitic helminths and certain ectoparasitic arthropods in both humans and their domesticated animals. In addition to treating a broad spectrum of diseases, the drug is now being used to save the eyesight of thousands of people in developing countries where *Onchocerca volvulus* is still a dread organism.

Bill's journey from his native Donegal in Ireland to Rahway in New Jersey was a long, but fascinating, trip. After boarding school in Belfast in Northern Ireland, he headed for Dublin, the Republic of Ireland, and Trinity College, the charter of which was granted by Queen Elizabeth I in 1592. His headmaster had thought he should pursue medicine, but his biology teacher at boarding school had such a great influence on him that he ended up in natural (biological) science, a very wise choice as it turned out. I asked if he had made contact with the great parasitologist, J. Desmond Smyth, in his first year at Trinity and he responded in the affirmative. At the time, Smyth was not *the* professor in the department, although he did acquire a professorship subsequently. One should know that in those days it was typical in the British system to have but a single professor in an academic department. Faculty members without professorships had to wait for death, retirement, or transfer of their senior faculty member before a younger faculty member could rise to the highest rank in the British system. More often than not, the professor was also head of the department as well, making this person, in many cases, almost 'god-like'. Fortunately, the system in Britain has changed a great deal since then.

As Bill said, "Yes, I had contact with him. Desmond Smyth became a hero of mine." This was not surprising to me, because several years after Desmond had influenced Bill, I was on sabbatical at Imperial College of the University of London, and was able to work with Smyth for nine months. He became a hero of mine as well. I am absolutely confident that Bill and I represent just the 'tip of the iceberg' in this regard, because I know this wonderful fellow had a huge influence on a lot of young parasitologists of that era. He was not only respected for his science, he was cherished as a man and a friend by all fortunate enough to know him.

Bill continued, "You know, there was something strange about that. At the time, I was rather diffident, totally lacking in self confidence, coming from Donegal into a situation like that at Trinity." He told me that there was fairly high attrition among the biology students who came in with him. There were 48 at the beginning and "something like 14 who went ahead with their fourth year." This was the honor's year, when each student could pick a topic and pursue research on their own, but always under the tutorial of one of the faculty.

"As the third year came to an end, we were told we should pick a special subject for our final year. There were two guys in our small group . . . that had grown up in Kenya, part of the British colonial system in Africa. They were tremendously cosmopolitan and were 'way up there' in terms of sophistication. They had arranged to do their projects with Desmond Smyth in parasitology. It was part of the culture there that one does not try and 'horn in' on somebody who had already staked out a piece of the action." Bill said that he was greatly, shall we say, chagrined that he was unable to pick Desmond as a person with whom to work. In retrospect, he said his feelings "were totally irrational, but," he continued, "I didn't have the nerve to say that I wanted to do that because Desmond had already committed to these two boys." So, he thought he would work with someone else. "But, when I came back from the summer holiday to start that final year, there on the table holding the assignments, etc., were three folders, including a notebook and suggested projects by Desmond for parasitology." I asked, "Then he must have picked you?" Bill responded, "Yes, he picked me. I don't know whether he sensed I wanted to do it. But that changed my life!" he said excitedly. "The fact that he had made three folders and put me into the group was just amazing." He went on to say that early in his third year, he had been asked by Desmond to stay on one afternoon to watch the master remove a large plerocercoid of *Schistocephalus* from a stickleback abdomen to use in an in vitro culture experiment, and that this was the moment of his 'capture' by Smyth and by parasitology.

"The year passed quickly, but in the spring, I recall that we three honors students were called together and told of a letter Desmond had received from Arlie Todd at the University of Wisconsin. Even though Smyth had never met Todd, the latter had written to Desmond asking him if he had any graduate students who would like to come to Wisconsin to work on their Ph.D.s." Bill admitted to not having given much thought to beyond the fourth year, but that he became intrigued with the idea when Smyth encouraged all three to apply and they did – and all three were accepted. The only trouble was that the other two dropped out, leaving 'diffident' and shy Bill Campbell as the only one of the three to go across the ocean. Bill recalled one of his two friends writing him a letter just before he was to leave and adding a postscript that said, "For God's sake, don't panic when you get off the boat in New York City!"

Getting ready for the great adventure was rather humorous. First, he had to get a visa for entry, so he headed down to Dublin and the U.S. Consulate. On arrival, he had to answer a bunch of questions from an official. One of them was, "How much income will you have to live on?" Bill responded, "$1280 per year as a research assistant." The official replied, "That's not enough." So, he was instructed to go back home and write a letter to Todd. The official wanted to see a statement from the new mentor in Wisconsin that he could make it okay on $1280. Todd responded quickly that it was enough money. The Consulate official muttered on reading the letter, "Well, it wouldn't be enough in New Hampshire, where I come from." Bill thought to himself, "I don't even know where New Hampshire is!"

The next problem was how to pay for the trip. Bill said, "At that time, most European countries were strict on how much cash one could carry when traveling abroad. In the Republic of Ireland's case, it was only $15," a real problem indeed. So, he applied for a Fulbright travel grant through Trinity College, the first student from there to ever apply for such a grant, but he received it, enough to pay his travel costs. He knew, however, that $15 would fall far short of covering his living expenses (including rent) for the month, and he would have to wait before getting his first check as a research assistant.

By a very strange quirk of fate, his father was to receive a letter from a man named Cavanaugh in Chicago, Illinois. It seems that Mr. Cavanaugh was a native of Donegal and had left Ireland as a boy to travel to the U.S.A., where he was to grow up and became a successful lawyer. He offered to Bill, through his father, the princely sum of $200, as a loan to be paid back as time would allow. The funny thing to Bill over the years was how this man found out that he was going to the U.S.A. and he still does not know. He said that Mr. Cavanaugh had also included newspaper clippings describing the University of Wisconsin's upcoming trip to the Rose Bowl – he hadn't the faintest idea of what the Rose Bowl was, but he was now ready to leave.

On arrival in New York City, he was met at the dock by an 'angel', Mrs. Minucci ("At least that's the way I would describe her," said Bill), from the Institute of International Education. She was a volunteer whose job it was to get him settled into a hostel in New York City where he would stay until he could pick up the $200 from Mr. Cavanaugh's friend. Bill described the first day as one in which he was actually too afraid to speak. He even found a 'deli' where he could point to the dish he wanted rather than ordering it out loud in a restaurant. This made things rather difficult because after taking a very long tourist walk in the city, he had to remember how to get back to the 'deli' where he could eat another meal without speaking. In a letter to me subsequently, Bill explained his reluctance to talk out loud was because of his lack of confidence, his Irish accent, and because "I didn't know the names of the food and the dishes that were on display or listed on menus, or how to pronounce them, or even what they were. Words like ravioli, hash browns, hot dog, and so on, were to be mastered gradually, over several months. Even after being settled in Madison, I had breakfast every morning at the same little place, not only because it was cheap, but also because I had discovered the words 'short stack' would magically produce a plate of pancakes!"

When he finally arrived in Madison, he was hit by a couple of serious culture shocks. First, he discovered he had to take more courses. This was unheard of in the European system, for in Europe the Ph.D. was strictly a research degree. Second, he also had to secure housing. Despite the $200 advance from Mr. Cavanaugh, he found himself in the hole financially, almost immediately. He said that Todd got out his checkbook and wrote him a check for $50 and he was covered. He remarked, "It wasn't that long before I had it all [his debt] taken care of."

His experience at Wisconsin was evidently a good one as he spoke very highly of Arlie Todd, who was a faculty member in the Department of Veterinary Science, and of Chester Herrick in the Department of Zoology, from which he also obtained a degree. When I first contacted Bill about submitting to an interview, I mentioned in passing something about Wendell Krull and *Dicrocoelium dendriticum*. In his response to my request, he told me an interesting story about a seminar he had given at Wisconsin that had to do with the discovery of this parasite's life cycle, Wendell Krull, and the venerable T. W. M. Cameron. During our interview in Philadelphia, I reminded Bill of our correspondence and he roared with laughter.

He told me that he was "required to give two seminars as a graduate student that I can remember. Maybe I gave more than that, but I can't remember. One was on the newly discovered life cycle of *Dicrocoelium dendriticum*. I was really intrigued by it and wanted to learn more." He went on, "You know, of course, that at that time at least, seminars and debating societies in the British Isles were a big thing, more so than here, in terms of being effective. So, you wanted to include some sort of surprise or humor as a way of advancing your presentation. There was always a high standard and I had never given a seminar before, so I wanted to do well. And for some reason, I was struck by the fact that T. W. M. Cameron had published an erroneous account [in 1930] of this parasite's life cycle, so I wrote to him about it. What possessed me was that I was giving this seminar and I wanted to do a good job of making a presentation, and I felt this would be really neat to find out what his response would be" to the new accounts so recently published by Krull and Mapes and their new discovery. "So, I wrote to him pointing out that he was wrong, and wondered what had happened.

"Now then, one must understand the whole situation. T. W. M. Cameron was a huge persona in international parasitology during those days and had been for many years. For example, he was the first Director of the Institute of Parasitology at McGill University in Montreal, a cornerstone of Canadian parasitology. At that time, he was probably at the pinnacle of his great career. He also was of the 'old school' and, I am certain, was unquestionably unprepared for a letter that came 'out of the blue', from a lowly graduate student that I am sure he would not know from 'Adam'. Moreover, the implication of the letter, though written very carefully and with great respect, was clear – how in the world did you 'blow it' when you published that paper in 1930 where you said the life cycle of *D. dendriticum* was identical to that of *Fasciola hepatica*?" Bill said, "Even though he was absolutely wrong, he wrote me the nicest letter. It was on blue stationery, and handwritten. He could have torn up my letter, or slapped me down in some way, because he had a reputation for being a very tough sort of person." We agreed that he "had rank and was aware of his position," but he certainly didn't act that way. "So," Bill said, "I gave the seminar and, at just the right time, I pulled Cameron's letter out of my pocket and read it to my colleagues." He recalled, "that Cameron could not account for his findings, except it was obvious he was not working in an ant-proof barn!"

On completing his degree at Wisconsin, he said he was certain he would go into academics. However, chance (serendipity?) was to play a huge role in his actual career decision. In fact, he said, "I was just beginning to look at the names of colleges, and colleges that might be looking for new, young faculty." However, he recalled that Desmond Smyth received a letter from Arlie Todd asking if he had any students who might want to pursue a graduate degree, and the same thing happened again. This time it was a letter from Ashton Cuckler at Merck who wrote Todd asking him if he had anyone who was about to finish up the Ph.D. "Todd showed the letter to me," but he said his reaction was the same as when he was shown Todd's letter to Smyth at Trinity, "I didn't really want to go into industry." But, he said, "Todd was very pro-industry and was very cagey about it. He said, 'why don't you at least go back and talk to them about it? What do you have to lose?'" So, he decided to go.

The first thing he did was to look at a map, and, he said, "That didn't look very promising at all." But, he thought, "I'll go back; at least I will get to see New York again. So, I went at Merck's expense, and one of the first things I did was to go see Charles Laughton (for $8) in George Bernard Shaw's *Major Barbara*." Since I knew of Bill's great interest and success in amateur theater over the years, I asked if he had had any previous interest in the theater. He said he hadn't and I asked why he went. He responded, "Oh, well, I went for the same reason you would go see a play in London's West End. But, there was a name I could recognize from the movies."

During his interview at Merck, he spent time with Ash Cuckler and, "He was very nice, and the work was much more interesting than I thought it was going to be." He sort of hesitated at this point in our conversation and said, "You know I really had an interest in chemotherapy," like it was some sort of memory that he was conjuring up from his past. Then he remarked, "I think I had an interest in chemotherapy for a long time. I remember going to an agricultural show . . . when I was in Ireland and the only thing I recall about it was picking up a brochure at an ICI booth dealing with a drug used to treat *Fasciola hepatica* in sheep. It was hexachloroethane. I devoured that leaflet. And you know, that only came back to me recently when I was looking up something having to do with *Fasciola* or something like that." It was then he told me that at Wisconsin he had also dabbled in chemotherapy in trying to treat cattle or deer that were infected with *Fascioloides magna*. He said, "There was something fascinating about the idea of curing a disease. There was something there early." Even so, he said, "I really didn't take the interview at Merck very seriously."

As soon as he returned to Wisconsin, he received a letter from Ash Cuckler offering him a job at Merck. He said, "I was just at the point of writing colleges, and here was an offer in hand." It was, needless to say, a real quandary, "I didn't know what to do, a salary of something like $9080." So, he consulted with one of his zoology professors and asked, "If I take this job for a year or so, it won't taint me, will it?" The professor laughed and said, "'Oh no, it isn't that bad.' He said he wouldn't hesitate. He went on to say, 'You're not doing it for life'." Bill said that Todd was quietly implying the same thing. "He did not have an anti-industry bias, as I did, and I certainly had an anti-industry bias, at least at that time, but I accepted the offer."

While at Merck, Bill's contribution to parasitology was to be enormous because, as will be seen in a later essay, he was to play a central role in the development of ivermectin, one of the most important drugs in modern veterinary medicine. The story of its discovery is an absolutely fascinating tale.

RICHARD TINSLEY

Figure 6. Richard Tinsley, Professor, University of Bristol, Bristol, England

The first time I met Richard was when he was teaching during the late 1980s at Westfield College, part of the University of London system. My wife, Ann, and I were on one of my U.K. junkets. I had made arrangements to see Richard for tea and we ended up spending the entire morning talking about his research in the southwestern part of the United States. After that, we would see each other at the occasional British Society for Parasitology meetings I might attend. When he moved over to the University of Bristol as Head of the Department of Zoology, we visited him there a couple of times as well. Then, in early June of 2004, Ann and I traveled again to Bristol so that I could interview him for the essay I wanted to write about his fieldwork and research.

Prior to our visit, he had sent me a number of reprints, so I was prepared to face him in his office. He was ready for me as well. He had set up a fascinating slide presentation featuring his study sites, the field station in Arizona, some of the interesting students with whom he collaborated, and the very fascinating hosts and parasites he encountered in his many summers in the U.S.A. I'll tell you what, if anyone who reads this book is looking for a really good seminar speaker, Richard would be an excellent choice. I was absolutely impressed by the specifics of his research, as well as his breadth of knowledge regarding the general biology of the parasite and the amphibian hosts. I believe you will see that his strong suit is directly related to the breadth of coursework in which he was involved as an undergraduate student at the University of Leeds.

When his slide show was over, I began my biographic 'interrogation'. His undergraduate days (1963–67) were spent at the University of Leeds where he came under the influence of Robert Wynne Owen. Richard described his mentor "As not a big player in research, but he was a very stimulating teacher. His interests were mainly in fish parasites." He also related, however, "I chose Leeds because it was one of the few universities in the U.K. at the time that would allow me the opportunity of doing a degree in both botany and zoology." I asked if it took him longer to complete, but he replied that it did not and that he was the first student to graduate under this new plan. He said that he could not do a complete program in either zoology or botany, however, but that he managed to accomplish about three-quarters of each. He said, "It was exactly what I wanted because it gave me this broad base." Because he was a sort of double major in a new program, the Department did not know how to deal with him in his last year, the honors year in British universities. Most students would do a single research project. Richard did three! One of these research projects involved research with Owen on a monogene in *Xenopus*, his introduction to a host–parasite system with which he would spend a great deal of time in years to come. He said, "That hooked me. As I look back on it, it was a fairly prophetic project. I really enjoyed it because I could see all sorts of potential."

He enjoyed the setting at Leeds and decided to remain there to do his Ph.D. with Owen. As Richard put it, Owen produced quite a "stable" of Ph.D. students, including, among others, Chris Arme, Les Chappell, Roger Sweeting, John Riley, Geoff Boxshall, plus several who went on to Canada as faculty members. He then told me that Owen was a student of Gwendolyn Rees, a Fellow in the Royal Society, and one of the very bright stars in British parasitology in the 1940s and 50s. Richard is obviously proud of his professional heritage.

His Ph.D. project began with an effort to resolve the life cycle of a polystomatid monogenean that cycled through *Xenopus*. These studies were based in the laboratory and Richard soon wanted to explore the system in the natural environment in Africa. So, he applied for a grant from the African Studies Unit in order to go to Uganda and study it first hand. At that time, Uganda was both independent and stable (it was pre-Amin). His experience there was a very positive one. In fact, he was able to take side trips into Kenya and South Africa during the same foray to Africa. While there, he shipped back some live *Xenopus* so that he could have the appropriate host for some experimental infections with his proterodiplistomatid when he returned home.

As it turned out, the cross infections he attempted when he returned to Leeds would not work. When I asked why, he responded, "I had been told while in Uganda what species [of *Xenopus*] it was. However, the parasites were not taking in these hosts, which supposedly was the right species. But I soon discovered that what I was working with was the wrong host species. In fact, it was one that had not even been described yet. So, the first development of my doctoral project was being able to describe a new species of vertebrate and I felt quite daunted by it!" He may have felt daunted at that time, but his subsequent work with toads and toad parasites was to make him one of the world's experts on both groups.

On completing his Ph.D. at Leeds, he applied for a postdoctoral fellowship. He was successful in making the case that he should stay at Leeds to do his postdoc because it would be "so much more efficient to do so." In reality, it also enabled his wife, Heather, the opportunity of completing her Ph.D. in geography, which she was able to do over the next two years of his postdoc. They then had even more good luck when they were both able to secure academic appointments at the University of Keele, he in biology and she in geography. He noted that at that time, Keele was rather like a liberal arts college in the United States, with most of the emphasis on teaching and very little on research, "very laid back", as he described it. However, "We really enjoyed that period because we could do almost anything we wanted, and the Department was very happy because I put on courses in parasitology, which they never had before. So, I could develop my parasitology teaching there in complete freedom. I really enjoyed that. However, it was a temporary post."

From there, he moved on to a letctureship at Westfield College, part of London University. "In those days," Richard explained, "London University was made up of a large number of small colleges. Alongside Imperial College, King's, and University College, there was this constellation of small schools, including Westfield, which now do not exist." Ten years into his career at Westfield, a decision was made to merge these small colleges. Westfield was moved to Queen Mary's College in east London. It was, according to Richard, "a fairly tough environment, but it was during this time that I developed my work on desert parasites, working most summers in the field in Arizona and then studying the host–parasite system in the lab on toads that I shipped back."

Then, in 1993, he moved to the University of Bristol as Head of Zoology. And this is where we traveled in May of 2004 so that I could sit for several hours and talk parasitology with this very knowledgeable and amiable man. It was a hugely delightful experience.

SIDNEY AND MARGARET EWING

Figure 7. Margaret and Sidney Ewing, retired Professors, Oklahoma State University, Stillwater, Oklahoma

In 2003, I was putting the final touches on a book called *Parasites, People, and Places*, when I received *Wendell Krull: Trematodes and Naturalists*, written by Sidney Ewing at the College of Veterinary Medicine, Oklahoma State University. Sidney sent it to me for review in the *Journal of Parasitology*. At the time, I had no idea about doing another book – i.e., the present one – this urge was to come later. However, I was nonetheless intrigued by what Sidney had written about Wendell Krull because of my long interest in *Halipegus occidualis*, a hemiurid trematode, the life cycle of which Krull had brilliantly resolved and published in 1935. Then, when I began to think about writing the present book, I had to make some decisions about what to include. The resolution of the *H. occidualis* life cycle and a biographic sketch of Wendell Krull written by Sidney were 'no brainers' for the new book. Since Sidney Ewing is the 'biographer' of Krull, this meant that he had to be interviewed. Subsequently, I discovered that Sidney's delightful wife, Margaret, had actually taken Krull's veterinary parasitology course while pursing her Ph.D. in zoology at Oklahoma State University, so she became a 'must' for an interview as well.

Although I grew up on the Great Plains of North America, I do not go back very often. Despite the difficulty in getting to Stillwater from Winston-Salem, I had no qualms about returning to the wide-open spaces in that part of the country. However, Sidney and Margaret bailed me out by offering to come and see me instead. It seems that the Ewings' daughter lives in Durham, North Carolina, only about ninety minutes from Winston-Salem, and they volunteered to drive over for the interview. In return, I promised them a barbeque lunch at a local diner named 'Little Richard's'. Everything came off without a hitch and they arrived about 10 o'clock on a Thursday morning in March of 2005 ready to answer questions about themselves and Wendell Krull. It was a fun interview.

I discovered that Sidney is a native southerner, having grown up in rural Georgia. Actually, after listening to him for about thirty seconds, I easily decided about his roots – he has a great southern accent. His college career included stints at Oxford College (a branch of Emory University in Atlanta), then the University of Georgia, and finally the University of Georgia's School of Veterinary Medicine, obtaining his D.V.M. in 1958.

His attraction to parasitology was initially piqued by Frank Hayes and Helen Jordan, both then faculty members of the veterinary school at the University of Georgia. The former had also started the vet school's wildlife disease unit, which is still fully operational nearly 45 years later. Hayes' training in parasitology came at Auburn University with Will

Bailey who, coincidentally, had been Wendell Krull's teaching assistant during Krull's first year as a vet student and instructor in parasitology at Auburn in 1942. Jordan had a D.V.M. degree from Georgia and was pursuing her Ph.D. with the legendary Elon Byrd, also at the University of Georgia, but in the Zoology Department. In the summer after completing the D.V.M. at Georgia, Sidney returned to their School of Agriculture where he completed his undergraduate degree. As he said, it was an odd sequence, "a D.V.M. in June and a B.S. in August."

He recalled that he was required to write a term paper while taking his parasitology course at Georgia, and "the phenomenon of parasitism captured my imagination." His topic was nematodes since he had discovered a roundworm that was parasitic on the roots of pine trees in nurseries while working a previous summer as an assistant in the U.S. Forest Service. As it turned out, it was a new species and it ended up being named (by another person) *Tylenchorhynchus ewingi*. Hayes and Jordan obtained some fish from the Pacific Northwest, "and I, as a veterinary student, took care of the dog that developed salmon poisoning disease for a demonstration to my classmates. We also managed to recover the flukes from the dog."

After receiving his D.V.M. and B.S. from Georgia, he headed for the University of Wisconsin and Arlie Todd (also the mentor of Bill Campbell) to work on his Master's degree. At this point in our interview, he somewhat sheepishly admitted to being very naïve at the time. He said, "I thought that I had to go off and teach for a while to earn enough money so I could go to graduate school. At that time, Georgia's veterinary faculty included mostly D.V.M.s, with very few who also had graduate degrees. But I happened to ask one of them to write me a letter of recommendation to Emory University so I could go teach and earn enough money to allow me to go to graduate school. He said, 'Sure, I'll help, but you don't need to do that. You can get a teaching assistantship.' Well, I had never heard of an assistantship until that minute. So, that's how I ended up at Wisconsin, and with an assistantship." He had originally thought he would work on dairy cattle parasites (having been raised on a dairy farm), but when "I got there, I was informed that I would be working on *Metastrongylus* in the lungs of hogs since they had just received an NIH [National Institutes for Health] grant to work on hog influenza. Well, I didn't know anything about hogs because all the hogs in Georgia are in the south and my experience was with dairy cattle in north Georgia. But, I learned!"

His stay at Wisconsin was a great success. The *Metastrongylus* spp. research to which he was assigned was to produce some rather interesting results regarding the interactions of three species of these lungworms. Anyone who has attempted to work with parasite communities will understand immediately that they can be complex and frequently involve some rather subtle, but powerful, interactions. The three species of *Metastrongylus* in the lungs of hogs illustrate this phenomenon quite well. For example, Spindler (1934) reported the prevalence of infections of *M. apri* was 69%; for *M. pudendotectus* it was 50%, and for *M. salmi* it was 12%. Ewing found essentially the same levels of infection nearly thirty years later (Ewing and Todd, 1961a). These numbers are not remarkable in and of themselves. However, the combinations of different species were of real interest. For example, *M. apri* and *M. pudendotectus* co-occurred 76% of the time and to the nearly complete exclusion of *M. salmi*. Thus, the three species were together in only 11% of the hosts examined. Moreover, experimental infections (Ewing and Todd, 1961b) showed that *M. pudendotectus* was less able than *M. apri* to successfully infect pigs when present alone. Ewing and Todd (1962b) characterized the association between *M. apri* and *M. pudendotectus* as a 'mutualistic' one, with the former species referred to as a facultative mutual and the latter an obligate mutual. The latter study ended with the proposal that "*M. pudendotectus* may be an obligate mutualist that must encounter *M. apri* in the lung to mature sexually." Whatever the explanation for this association, it can easily be agreed that a most unusual relationship exists between metastrongyles in the lungs of swine.

After his Master's degree, Sidney decided to continue his graduate work and obtain a Ph.D. But, he wanted to study with someone who had both veterinary and Ph.D. degrees, and Arlie Todd had just the latter. Sidney heard via the 'grapevine' that "Wendell Krull might have a position, if you would be willing to teach as well as work on the degree. At Oklahoma State University at that time, an Instructor could work on a degree. Actually, as it turned out, an Assistant Professor could work on a degree as well. So, I went down and visited with Krull, and he decided to hire me." Although Todd had a wonderful reputation, Krull was even better known and better suited for Sidney's interests.

The decision to enter Krull's graduate program was unquestionably one of the best of his life, both professionally and personally. In 1960, when Sidney arrived at Oklahoma State, Krull was definitely *the* helminth person in his department (Veterinary Parasitology). Everett Besch was another graduate student in Krull's department and he was *the* arthropod specialist. Krull had actually sent Besch off to a special institute one summer to learn more about this group of organisms so that he could teach about them to the veterinary students. This left the protozoans 'uncovered' by a specialist. Krull had decided his new graduate student should develop expertise in parasitic protozoans and, moreover, the best place to do it would be at the University of Michigan Biological Station on Douglas Lake, near Pellston, Michigan. According to Ewing, "He also thought I needed the experience of a biological station per se," and that Douglas Lake would be a great place to get it, especially since he himself had spent so much time up there while he was a graduate student of George R. LaRue back in the 1920s. James H. (Jim) Barrow, a faculty member at Hiram College in Ohio, was teaching a couple of courses in protozoology at the Station during that summer of 1961 and the opportunity for Sidney to learn about parasitic protozoans was the perfect setup in Krull's mind. It was to be a great summer for parasitology, but an even greater summer on a personal note because Sidney was to encounter Margaret (Steffens), who was to become his wife, the mother of his children, and his friend, companion, and colleague of some 42 years (at the time this essay was being written).

It was at this point in our conversation that Margaret inserted an interesting comment regarding Krull. She reminded me that he was a veterinarian. Sidney was also a veterinarian. However, she emphasized emphatically that Krull had the insight to want Sidney to work at a biological field station as a way of broadening his intellectual and biological experience. Margaret continued, "The probability of a faculty member in a *veterinary school* [her emphasis] saying to a graduate student, 'you have to go to Douglas Lake next summer', is *very* [her emphasis] small, yet that is exactly what Krull did." One must conclude that Krull was more than a veterinarian, or even a parasitologist. He was first a biologist, and, as we shall see later, a very good one.

Sidney said that he had a lot of fun that summer. Even though he was considerably older than most of the other students, he was able to interact in a positive way with all of them. In fact, he made several lifelong friends, including Dick Kocan and Darwin Murrell, both of whom were to become my friends just a few years later. Indeed, Dick Kocan was to be my teaching assistant in my first year of teaching field parasitology at Michigan State University's W. K. Kellogg Biological Station on Gull Lake in southwest lower Michigan, and Darwin's interview is included elsewhere in this book.

Margaret said that Dick and Sidney also became her friends that summer. Both of them actually dated her, in fact, not long after they all arrived at the Station. It seems that not far from Pellston, in the sleepy village of Petoski, Michigan, there was a summer theater where they performed plays and musicals for the locals. She said that early in the session, Dick invited her to go to a play. Sidney chimed in, "She had a car. Dick didn't!" Margaret then picked up on the story, "Unbeknownst to me, and for reasons unclear to me, Dick had borrowed a suit to wear, from Sidney. It seems that Sidney arrived at the Station wearing a suit, having just come from some sort of veterinary medicine meeting. I think he was surely the only student in the history of the Station, or at least in that decade, to have arrived at the field station in a suit." Sidney added at this point, "My suitcase was lost on my way up, and so there I was, until it was found, still dressed up!" Margaret continued, "Then, Sidney asked me to go to the same play the next weekend. I had one dress and that was probably unusual too. So, the next weekend, I went to the same play, wearing the same dress, in the same car [Sidney didn't have a car either!], but with a different date in the same suit." Sidney continued courting Margaret and they were married a couple of years later, in 1963, after she graduated from Oberlin College in Ohio and had completed her first year of graduate work at Oklahoma State University (OSU).

I then asked Sidney why he switched from helminths to ticks during his work toward the Ph.D. He reminded me that Krull had sent him to UMBS to learn about protozoans. Then, he explained, "You know, each veterinary school has a teaching hospital, and a teaching hospital is a laboratory of naturally occurring disease. I was looking around for something to do in terms of research and, as I said earlier, Krull really wanted me to work with protozoans. That's why I went to Douglas Lake. It happens that we had a case of babesiosis in a dog, which wasn't very common. So, I started fooling around with *Babesia*, and I was able to passage the parasite by blood transfusion. I didn't do any tick work at this point. In the process, I came to realize that I was working with two different organisms, rather than one, and that the second was *Ehrlichia*, a rickettsial pathogen. They both will cause disease in dogs, but by completely different mechanisms. I had written an NIH grant proposal to support my dissertation research and it was funded. It was, however, based on what turned out to be a totally false hypothesis. The original idea was that a big, purple 'thing' that I was seeing in stained white blood cells was an undescribed schizogonous phase in the *Babesia* life cycle. Why such a hypothesis? Well, the Babesiidae were separated from the Theileriidae on the basis that there was no schizogonous phase in the Babesiidae cycle, whereas the theileriids had these things called Koch's blue bodies, which are multiplying forms in leukocytes. There is some resemblance between Koch's blue bodies, which are very well known in animals with East Coast fever and ehrlichial morulae. And, I was thinking that these blue things I am seeing look a whole lot like those Koch's blue bodies. Maybe, I thought, the Babesiidae and Theileriidae are separated on a false basis. So, I'm going along merrily working on this problem and then one day, one of my dogs shows up with only the blue bodies. The dog is very sick and there were no *Babesia* present. This is when it became clear that I was working not with one disease agent, but two! The separation was accidental. I still do not know how it happened. It is likely that a tick that had nailed a dog in the teaching hospital passed this thing transtadially and a later tick stage got on one of my control dogs and produced ehrlichiosis. This is how it's recorded in the literature. It was an accidental separation of these two agents. It was a piece of great good fortune because ehrlichiosis had never been seen in North America up to that time. It was known in the Old World for many years, since the time I was a boy."

I interjected at this point, "In other words, this was a clear case of serendipity?" He responded, "It was. But I was prepared. And I was a very meticulous kind of worker. I mean I saw my dogs multiple times a day. I bled them every day and I read the slides myself every day. I was at the right place at the right time. But I was not looking for it." I then asked him how long it took to figure out what was going on. He answered, "It was a while, because I actually took these pictures of the dog's white blood cells to several scientific meetings, but nobody knew what they were. Nobody else had ever seen them. As it turns out, in Algeria, about the time I was born, some French veterinarians had described *Ehrlichia canis*. It was originally described as *Rickettsia canis*, but was transferred to *Ehrlichia* when that genus was erected. So," he chortled, "that was my Ph.D. dissertation, all based on a false hypothesis."

Sidney continued working on ehrlichiosis for a number of years, culminating with a 1995 paper in which he and several other colleagues demonstrated the experimental transmission of the newly described *Ehrlichia chaffeensis* among white-tailed deer by *Amblyomma americanum*. An important conclusion of this paper was that white-tailed deer were probably important reservoir hosts for the passage of ehrlichiosis to humans in North America.

When Margaret arrived in Stillwater she began her graduate work in the laboratory of Troy Dorris, a limnologist. She immediately was also urged to write a grant proposal to support her research, which she did and was successful. This outside funding allowed her to finish her Master's and Ph.D. degrees in just four years. She said, "I never had to do anything for anyone else. The ironic thing is that I never taught while I was a graduate student, yet teaching became my passion."

In February of 1965, they left OSU for Kansas State University's School of Veterinary Medicine and an academic position for Sidney. This was after Sidney completed his Ph.D., but before Margaret had time to do her dissertation research. Fortunately, she had completed her preliminary examinations at OSU and was able to work at Kansas State University (KSU) in the laboratory of Dick Marzolf, a well-known aquatic biologist. Her research involved an analysis of the effects of nutrient loading on algal populations in local fish ponds. Even though Sidney went to Kansas State for a teaching position, he also left OSU in part because he was angry that Wendell Krull had been fired, even after having been promised that he could work until he was 70. I asked the two of them why the Board of Regents had done it, especially after promising him otherwise. Sidney responded, "The story we heard eventually was that there was a faculty member they wanted to remove and they could not do it unless they applied the new rule across the board. So, they just summarily fired everyone over the age of 65."

Sidney said he went to Kansas State expecting to be there for forty years. However, after just a few years, Mississippi State University contacted him and invited him to come as Chairman of their Department of Veterinary Sciences. At the time, "they were ready to construct a new building and it just seemed like an exciting opportunity. I was involved with teaching and advising pre-vet students since Mississippi did not have a veterinary school yet. All of their students would go either to Auburn or Texas A&M University. I was at Mississippi State long enough to get the building fully funded and partially built when the opportunity came up to go back to Oklahoma State University. So, we were at Mississippi State for just two years. When Krull was fired by OSU, Everett Besch was hired as Chairman of the Department of Veterinary Parasitology to take his place. After two years, Besch was lured away by [Louisiana State University] to be their founding Dean of the new School of Veterinary Medicine and the job at OSU came open again. It was Krull's old position and they offered it to me."

Sidney continued, "Interestingly, when Krull was axed at OSU, he and his wife Nellie packed up everything and moved to New York City. Of course, when he got to his new job, they would not allow him to infect his dogs with heartworm like he wanted to, and he was devastated. I had stayed in touch with him and learned about his predicament. Fortuitously, the Associate Dean at KSU had been the founding Chairman of the Department of Pathology at OSU the same time that Krull had come on board as the founding Head of Veterinary Parasitology, so obviously he knew Krull. One day I happened to mention Krull's situation to Associate Dean Trotter and asked if KSU would be interested in hiring him to help teach veterinary parasitology. To make a long story short, he was soon a faculty member at KSU."

They stayed at OSU for four years before leaving for the University of Minnesota where Sidney became Dean of their School of Veterinary Medicine. As an 'old Dean' myself (of the Graduate School here at Wake Forest), I asked him if he liked it. He responded, "I did. I enjoyed it. I even enjoyed certain aspects of working with the legislature. However, I knew that I would never like to be Dean at Oklahoma State because the legislature in Oklahoma is the worst scenario that I can think of!" Margaret said, "Minnesota was a great place to try out new ideas. And we loved being there." Then, in 1978, he was offered his old job at OSU again, after six years of being Dean at Minnesota. He decided he had had enough of 'Deaning', so he accepted the offer and they returned to Stillwater for his third stint.

Sidney and Margaret have three children. Although research was restricted somewhat because of family constraints, Margaret managed to do some teaching along the way, at both Mississippi State and at Oklahoma State University. At Mississippi State, for example, she taught ecology in the fall and principles of adaptation (really evolution) in the spring. Between semesters of their first year there, she even had their second child! When they returned to Oklahoma State the second time, she started teaching invertebrate biology. As she said, "at first, it was part-time professional work." Teaching openings were scarce in Minnesota, so she opted to do some freelance editing for the University of Minnesota Press, a job that she liked quite well.

Three years after they returned to Oklahoma State, Margaret finally obtained a tenure track position and a full-time appointment in the Department of Zoology. She was finally able to teach parasitology and pursue research more readily. She and Sidney "applied for a grant from the USDA to look at the effects of a stressor (copper) on the susceptibility of catfish to *Ichthyophthirius multifiliis*, also known as 'Ich'. The wonderful thing about 'Ich', I thought, was that it would be an easy thing to quantify infection because all you had to do was count the white spots. Someone who had some experience with this parasite once said when he began working on it, 'he thought he had found a bird nest on the ground', and so did I. I discovered though that it's not quite that simple. I did not realize how hard it was to keep it going in the lab. Nonetheless, I was doing research and teaching, and finally was in a tenure track position. And here I am."

Sidney's last years of research were spent working on *Anaplasma marginale*, a rickettsial organism in cattle, and *Hepatozoon americanum*, which, along with other species in the same genus, infects a wide range of vertebrate hosts. A letter Sidney sent me after our interview here at Wake Forest particularly fascinated me. In it, he wrote, "So . . . , that's a long-winded way to say that my natural history outlook, fostered by Krull and the Douglas Lake experience, has been important to my work with tick-borne agents, prokaryotic and eukaryotic alike. Understanding that 'some things didn't quite fit' – and having a generous measure of good luck – certainly made for an interesting life in parasitology for me. In spite of my excursions into administration on several occasions – like you, who was similarly afflicted – I never lost my enthusiasm for parasites or for interacting with students, veterinary and graduate alike. I am grateful that so many natural history puzzles presented themselves in [veterinary] teaching hospitals. One could not have asked for better opportunities. So, let's hear it for time and support to study naturally occurring parasitic diseases!"

Sidney and Margaret are now retired as faculty members at OSU. I had never met either one before that day in March 2005, but by the time we finished talking and then eating lunch at 'Little Richard's', I felt as though we had known each other all our lives. It was a very enriching experience for me. I learned a lot about both of them and, from them, about Wendell Krull. It was a very good day!

DON BUNDY

Figure 8. Don Bundy, Senior Scientist, World Bank, Washington, D.C.

I met Don Bundy for the first time when I attended a fall meeting of the British Society for Parasitology in London about fifteen years ago. The topic of the meeting was the geohelminths and, although I have not done any research with parasites in this group, I was very keen on learning much more about them. I was not disappointed. The last speaker in the afternoon was Don and it was fascinating to hear him talk. He is now with the World Bank in Washington, D.C. I called and asked if he would be willing to sit down with me for a few hours and subject himself to an interview. He agreed and, again, I was not disappointed with what he had to say.

So, in early April of 2005, I flew to Dulles and took a taxi to my hotel in Arlington across the grand Potomac River from Washington. The next morning, I caught another taxi to the World Bank. As I was obtaining my security badge, Don was just coming into work and spotted me. He greeted me enthusiastically and after a quick ride to the 8th floor and walk to his office, we began a conversation that was to last through lunch and into the afternoon. It was what I hoped for.

The discussion was free ranging, but initially we talked about him and his career. I began by asking him how he got into parasitology. He responded by asking a rhetorical question, "Why would a Brit be interested in worms?" He continued, "Let me tell you a little story. I was actually born in Singapore." He said that his "father was a power engineer and went there in the 1920s, staying until the 1960s. In fact, he built the first power station in Singapore." I asked if he had been there at the outbreak of WWII and Don replied that he was, and "that he spent the war in a camp, a guest of the Japanese. My father was Welsh, being born in a tiny village in south Wales. After the war, he married my mother, a wonderful Scottish woman. My brother and I grew up in Singapore, and we did get worms. That was what happened to all of us children at the time. They eradicated malaria in the 1950s, but Singapore is in the tropics, and so are worms. My mother tells me that one night when my brother and I were dutifully saying our prayers at bedtime, I petitioned, 'God bless auntie Nan, God bless auntie Dorothy, and God bless auntie Par.' Of course I was quite young and did not realize that my 'auntie Par' [i.e., anti-PAR] was the ICI brand name for piperazine in those days. 'Auntie Par' was definitely part of my life at the time."

After growing up in the Far East, Don headed back to the U.K. and started his college schooling at Plymouth Polytechnic where he pursued a Higher National Diploma, as it was called then, in applied biology. Most who followed that line would end up in a technical career.

However, he said, "while at Plymouth, I took a course in parasitology from the Mathews, a husband and wife team of lecturers, and I was hooked." His newly acquired interest in parasitology caused him to apply to work for an undergraduate degree with Phil Whitfield at King's College in London. It was a good choice because Frank Cox was also at King's and, a year or so later, Roy Anderson came as a young faculty member. He and Roy were to ultimately develop an important and lasting relationship.

He told me that when he first went to King's, he "was like any other kid. I was interested in rowing, rugby, and other sports. Academics were something I was interested in, but it wasn't part of my core. It was really Phil Whitfield who made me understand the importance of science in my life. Looking back on it, I can see the value of being a tutor, like Phil, and I've tried to reflect that in my own life. He really changed the way I thought about things because he had this view of parasitology, in particular, that parasites are a fine example of ecology, of the web of life, and of how everything is connected to everything else. I think the term is symbiology, the idea that all organisms have to interact. He really engaged my thinking on all of this."

He continued, "It was really a very small department [of biology] when I was there, which was brilliant for a student. This meant we would often have tutorials with just two people in them. It was terrific for the kids who were there. Phil has this wonderfully agile mind. It was a great lesson for me growing up. I hadn't had an academic background and I hadn't run with an academic crowd. He grounded me in parasitology and not just in one area, but also in the whole thing. He helped me do work on systematics, taxonomy, ecology, and physiology. It was an extraordinarily broad education while I was an undergraduate. I ended up writing my honors thesis in two volumes because of this. And I published in as many different kinds of journals, i.e., physiology, taxonomy, systematics, epidemiology . . . People now may sort of frown on such an approach, but I don't. You know, people use this phrase nowadays, 'thinking outside the box', but who defines the box is my question? I think that is the problem, that we have boxes in the first place. As parasitologists, defining boundaries is not the way we have been brought up to think. This is crucial in making connections."

Don stayed on at King's to do his Ph.D. with Phil. His dissertation research focused on the population biology and ecology of a *Transversotrema patielense*, a digenetic trematode that lives under the scales of tropical fishes. According to Don, "It was a wonderful model because you didn't have to kill the host to count the parasites. There was the host and there was the parasite as well. But what was fun was doing whatever I wanted. I was engaged in all sorts of things. I did the whole life cycle and the physiology of the parasite. We even did some high-speed photography to see if we could understand the hydrodynamics of swimming behavior by cercariae. Looking back, it was far more work than anyone would do now."

He was fortunate because it was about this time that parasite epidemiology and mathematical modeling began to blossom, and one of those responsible for stimulating work in this area was Roy Anderson. Don told me that Roy had been thinking about some of these ideas, translating them from the purely ecological side of things to the epidemiological side of infectious disease. He said, "Roy had put in a proposal to the College to buy a Hewlett-Packard calculator, which was something like £2000, or $4000, and at a time when a hand-held calculator was £400, as I recall. They turned him down, and asked why would anyone need this kind of technology? At that time, we had a big mainframe, probably an IBM, you know what I mean, the one that still required punch cards. What he was asking for was the precursor of what we've all got on our desks right now. The work he was doing at that time was on intestinal worms," and, according to Don, "the data were somewhat insecure at that point." He went on to note that there was a "huge history of work on these intestinal worms, but the idea of recording the intensity of infection, the number of worms that people had, the pattern that it might be, and how that might influence treatment, was not something that parasitologists and epidemiologists had come to grips with yet. It was translating what was going on more generally in ecology to parasitology and epidemiology. To me, it was striking to realize, for example, that you could, hypothetically, treat just some of the people in a village, but have a huge effect on the population as a whole. The basis for this idea was that the most intense infections were in a few individuals, or a particular age group", a direct application of the overdispersion concept published in the early 1970s by Harry Crofton. It was also based on the ideas developed by "Roy Anderson and Bob May that if you could take the intensity of parasitism below some threshold in a given village, and keep it low, that you could drive the parasite to local extinction."

I asked if he had known Crofton. He said that he had and, in fact, was using his textbook in one of his courses, reminding me of the genius of the man because the book dealt with the physiology of nematodes, and the two papers that he had published were absolutely seminal in this area of parasite ecology/epidemiology. I went on to tell Don one of my favorite stories regarding my first experience with Crofton's two papers in parasite ecology. I had taken them to an ecology friend and asked him to look at them and that he had come back to me and said, "So what? Everything is overdispersed spatially in nature." My colleague could not see the significance of just a few people having most of the parasites in a given population and the significance of this point in the possible treatment and control of parasitic disease. Don responded with a chortle, "And that was a boundary and crossing issue, wasn't it? Any ecologist would call that normal." We agreed, however, that it wasn't until Roy Anderson and Bob May's papers in the *Journal of Animal Ecology* and then in *Nature* not long after that the full application of Crofton's ideas was to emerge.

Around that time, Neil Croll, another of the young British epidemiologists, was to help to revolutionize the way in which the epidemiology of the geohelminths, indeed a vast array of eukaryotic parasites, is now done. It was Croll and Gahdarian who, while working in Iran, initially "bridged the gap between what Anderson and May were saying about what could be true and what was true in the real world. It would provide an empirical basis for those theoretical constructs."

When Don finished at King's College, he transitioned immediately into epidemiology of the geohelminths and has spent the rest of his professional career pursuing these parasites, quite literally, all over the world. After his stint at the University of the West Indies, he spent time at Imperial College in London, then at Oxford and, most recently, at the World Bank in Washington, D.C. The essay that comes later details his experiences and provides some real insights into how a modern epidemiologist is trying to free humanity of the geohelminth scourges.

My interview with Don opened up an entirely new vista for me, one that relates to geopolitics, as well as to geohelminths. The approach that he is taking at the World Bank requires not only intimate knowledge of the enteric helminths and their treatment, but a strategy for coping with the enormous scale of treatment. He is obviously concerned about health care delivery in an isolated village and the problems attendant with such a process. However, at the same time, his concern also extends across the landscape of entire countries and geographic regions, and to people with different cultures, mores, languages, traditions, and political schemes. In thinking about it, his job also involves discovery, both of a political nature, and a scientific one. Not many of us are skilled, or even schooled, in these sorts of things. It appears as though Don has acquired the experience and knowledge to pull it off, and I wish him well in this very exciting and positive endeavor!

PETER HOTEZ

Figure 9. Peter Hotez, Professor and Administrator, George Washington University School of Medicine, Washington, D.C.

In 2001, Peter Hotez received the Henry Baldwin Ward Medal at the annual meeting of the American Society of Parasitologists in Monterrey, California. That was my first encounter with Peter. A couple of years ago, I asked him to provide me with some stories regarding his field experiences in China. I was writing a book at the time dealing with field parasitology and he was generous in giving me the information that I wanted – some very interesting tales too.

In thinking about topics for the present book, I decided that I wanted to deal with hookworm immunity, as well as the geohelminths from an epidemiological perspective. I determined that a very good choice for the latter topic would be Don Bundy. Hookworm immunity was easy too. The right person to talk with was Peter Hotez and he willingly obliged. So, Don and Peter agreed to sit with me for a few hours in Washington, D.C. in April of 2005. Both were great interviews.

Peter was born and raised in Hartford, Connecticut. His grandparents were part of the Jewish Diaspora. I am embarrassed to say that this last word I had to look up in my 'trusty' Random House dictionary. It refers to the scattering of Jews outside of Palestine into other countries following their liberation from the Babylonians. His grand-father, Morris Goldberg, with whom he was very close while growing up, was raised in poverty in the Jewish quarter (*Le Marais*) in Paris. His mother and stepfather were actors in the Yiddish theater, and did not have the time or wherewithal to support him. He was very much a Paris street urchin.

Later, Morris' biologic father, who was living in Montreal, sent for Peter's grandfather about the turn of the twentieth century. Even-tually, his grandfather then made his way down to New Jersey, then up to Connecticut where he met and married Peter's grandmother, Rose, who lived in New Britain, which is just outside Hartford. There was a long line of scholars and Jewish rabbis on his mother's side of the fam-ily. In fact, his grandmother's brother, David Krech, at one time was a professor at the University of Chicago. He was, however, ironically fired from his position there by the great immunoparasitologist, William Taliaferro, who was then Dean of their School of Medicine. It seems that his uncle was rather 'pink' politically and somewhat extreme. He ended up on the faculty at the University of California-Berkeley, an appropriate association! He later went on to become one of the found-ing fathers of modern psychology in North America. Other influential people in Peter's formative years include his Uncle, Irv Goldberg, cur-rently a distinguished biochemist at Harvard Medical School, and his cousin, Daniel Goldberg, a Howard Hughes Medical Institute Fellow at Washington University in St. Louis, Missouri.

Peter's mother was a homemaker. His father had been in the Navy during WWII before returning to Hartford and a career with Pratt and Whitney, the manufacturer of airplane engines. Peter became motivated toward biology at a very early age. One of the reasons was that he had read *The Microbe Hunters* by Paul de Kruif (1926). I should note here that Peter is the third parasitologist of the several I have interviewed so far who have been hugely influenced by this book. I must admit that I had not read it myself, so finally I checked out a copy from our library and read it. I can easily see why a young person would be affected. The writing style of de Kruif is highly 'romantic', and would be quite appealing to the open mind of a youngster.

Peter's parents rewarded his passion for science when they pur-chased a microscope for him. He told me that, "In fact I still have that microscope today and I use it when I go into schools and give demon-strations for the classes where my kids are students." He described a small brook near his house where he would go and collect 'critters' to examine with his new microscope. He was particularly interested in water fleas, "and I used to buy them and other things from Carolina Biological Supply. Then, I would to go to the local library and pick up any new science book they might have. That is where I came across Asa Chandler's [1949] *Introduction to Parasitology*. I was struck to find that water fleas served as the definitive host for a parasitic nematode, and I have stuck with the nematodes ever since." He added, "Years later, I went back to the same library and the book was still there!"

He thus was hooked on studying parasitology and tropical medicine while still in junior high school. So, when it was time to go off to college, he began looking at schools that would afford him the opportunity of study and research in these areas. "There was something fascinating about these diseases in humans. By the time I finished high school, I had finished reading Craig and Faust's [1949] *Clinical Parasitology* from cover to cover. I ended up going to Yale on a partial scholarship from United Technologies because Curtis Patton and Frank Richards were there and I wanted to work with them." At Yale, he majored in biochemistry. I asked why he chose that direction instead of ecology or epidemiology? He responded, "Because I had taken a lot of AP [Advanced Placement] chemistry and physics in high school and it seemed like a natural way to go. I also had an influential high school teacher named Daniel Hoyt. This was in the late 1970s and molecular biology was just taking off. This was something new and exciting. By that time, I had also met Dick Seed and George Cross, both of whom were working on antigen surface coats of trypanosomes. In college, I wound up spending all of my free time working in the laboratory of Frank Richards, a Cambridge-trained physician and immunochemist who had come to Yale and had switched to trypanosomes as well. So, by the time I was a junior, I already knew a lot of the people working on the molecular side of parasitology. It was Curtis and Frank who then told me about Rockefeller University and how great a place it would be to study medical parasitology."

I then asked Peter when had he decided to pursue both the Ph.D. and M.D. degrees. He responded, "I originally thought I wanted to do just a Ph.D. in parasitology. But I was talking to my Dad about that and he became very upset. Part of the reason my Dad balked was that he had wanted to go to medical school. He even had a promise from the Navy that they would send him, but instead he ended up serving as an officer on an LST [landing ship, tanks] in the South Pacific in WWII. When he came back, he married, had some kids, and had to support a family, which was why he was working for Pratt and Whitney. So, instead, he made damned sure that all three of his sons went to medical school and that's what happened. For me, the M.D./Ph.D. combination was sort of a compromise. The combination of medicine and molecular parasitology came because of my growing interest in drugs and vaccines, and how they work against parasites."

He began looking into M.D./Ph.D. programs. There were not that many, but the one at Rockefeller was clearly superior. He also interviewed at Johns Hopkins, but he had this 'thing' about going to New York. He said it was sort of ironic that he "wanted to go to school there when my father had spent so much time and effort trying to escape from the place.

"Before I finished at Yale, I had done enough work to be a coauthor on several papers that were published in *Experimental Parasitology*. By the time I finished my interview at Rockefeller, I was told that I would be admitted to school." He originally wanted to do his Ph.D. with Bill Trager, "But the Deans were very discouraging about working with Trager, claiming his best years were behind him, etc. The person I did my Ph.D. with was Tony Cerami, a fascinating guy who called his working space the Laboratory of Medical Biochemistry. He would take disease problems and try to identify the mechanism associated with the disease. He had people working, for example, on diabetes. He had discovered hemoglobin A1C, and was one of the discoverers of tumor necrosis factor as well. At the time I was at Rockefeller, there were a lot of great molecular biologists, many of whom were working on the molecular aspects of host/parasite systems. So I started looking around for a problem. One of the things I noticed was that the trypanosome field was already getting sort of crowded, and so was malaria. The interesting thing was that as a new M.D./Ph.D. student at Rockefeller, you were told that you were handpicked, so you need to do something that no one else is doing. It was a wonderful, although at times pretentious, place to be in, but I bought into it."

He continued, "So, I wanted to pick a parasite of great public health importance. I was in the library one day and came across Norman Stoll's [1947] paper in *Experimental Parasitology*, the one entitled 'Endemic hookworm: where do we stand today?'. In this paper, Stoll had written, 'As it was when I first saw it, so it is now, one of the most evil of infections, not like filariasis or schistosomiasis, but with damage silent and insidious. Now that malaria is being pushed back, hookworm remains the great infection of mankind. In my view it out-ranks all other worm infections of mankind . . .' I looked in *Index Medicus* and found that no one was doing anything with hookworm." So, hook-worm became Peter's parasite.

He explained to me that when the M.D./Ph.D. program starts, the first two years are spent in medical school at Cornell Medical Center, followed by the Ph.D. at Rockefeller, and then the last two years back at Cornell. For the first two years, Cerami would meet Peter every Fri-day morning for breakfast and talk about what he wanted to do. Peter said, "For me, that was the best part of my entire education, to be able to meet with this extraordinary scientist who would take time to have breakfast to discuss health problems in developing countries." It was at one of these meetings, soon after reading Stoll's paper, when he told Tony of his interest in hookworm. He said that Cerami was sort of surprised and asked what had he wanted to do with hookworm. Tony had said, "You can't do much in the way of an in vitro system, and where are you going to get hookworms?" The only person working with hookworm at that time was Gerry Schad, who had by then left Johns Hopkins and moved to the University of Pennsylvania. Peter said that he used to take the train once a month from New York to Philadelphia and bring hookworms back. He said that at first he maintained a single immunosuppressed dog infected with *Ancylostoma duodenale*, but later switched to *A. caninum*. Another nice thing about Rockefeller was that it had so much money that it was easy for Cerami to keep a dog in the animal facility. In the end, Peter became the first person to run a polyacrylamide gel on a hookworm, the first to try and clone a gene from hookworm, etc. At some point during the discussion, Peter admit-ted to actually having two persons for a professor. The first was Tony Cerami for the biochemistry and the second was Gerry Schad for the parasitology. Yet another person who was very instrumental in Peter's work with hookworm was Ken Warren.

It seems, at the time, that Cerami was getting funding from the Rockefeller Foundation via Ken Warren. The Rockefeller Foundation had set up "this incredible network of research laboratories called the GND, the Great Neglected Diseases. They had these fantastic annual meetings where everyone funded by the Foundation would come to Woods Hole. We would go to these meetings every year and I would get the chance to meet all of these extraordinary people. This meant a lot because today I am still in touch with a lot of these same people who have gone on to have distinguished careers in tropical medicine, including Gerald Keusch, James Kazura, Dyann Wirth, and Richard Guerrant. That was one of the great things that Ken Warren did, setting up this network. It was Ken who actually convinced Tony to let me work on hookworm. I can remember at one of these Wood's Hole Conferences sitting with Tony, Ken, and John David, who was the Chair of the Department of Tropical Health at Harvard, and John David saying to Tony, 'If he wants to work on hookworm, let him work on hookworm'!"

So, Peter worked on hookworm for his Ph.D. and he continues with it up to this day. A few years ago, he left Yale for George Washington University where he presently serves as Chairman of the Department of Microbiology, Immunology, and Tropical Medicine. He has been successful as a researcher in many ways. As I mentioned at the outset, the American Society of Parasitologists recognized his work with the prestigious Henry Baldwin Ward Medal.

Peter presently has funding from the Gates Foundation and has begun field trials in Brazil with a hookworm vaccine that he and several colleagues have been trying to develop over the past several years. It took time to procure the antigens they had discovered in order to produce them as a vaccine under a regulatory umbrella. This required a team of fermentation and process development experts (led by Goddom Goud), quality control experts (led by Aaron Miles and Jordan Plieskatt), and strong program management (led by Maria Elena Bottazzi, Kari Stoever, and Ami Shah Brown). Once they had the vaccine, serious attention had to be given in setting up these field trials. As Peter said, "It's not like the clinical team swoops in on a helicopter. They work with the local institute, a branch of the highly regarded Instituto Oswaldo Cruz, as well as the Oswaldo Cruz Foundation. I also have a faculty member who lives in Brazil. His name is Jeff Bethony, who was trained by Phil LoVerde. Jeff is in the field on a regular basis, along with David Diemert, our clinical principal investigator. The personnel in the Institute are also in the field on a regular basis and have been for a long time. They know the local customs. There is a local review board. It's all done with the approval and permission from the local and federal government. We have largely stopped working elsewhere because we had such strong connections down there."

I have known Peter for several years now and followed his career with interest, even though he is molecular and I am ecological. This vaccine he has been pursuing is almost like his 'Holy Grail'. An interesting thing about Peter is that he has been consciously, and unconsciously, chasing this goal since the first time he looked into a microscope and the first time he opened de Kruif's *The Microbe Hunters*. I hope he has now succeeded. We will see.

DAVID ROLLINSON

Figure 10. David Rollinson, Senior Scientist, The Natural History
Museum, London, England

The purpose of our trip to London in May of 2005 was, among
other things, so I could interview David Rollinson at the Natural History Museum where he has spent his entire professional career. I do not
know how many of you have ever visited this hugely important facility, but if you have not, I strongly recommend it. It is so very impressive. Ascending into the enormous anteroom, you are confronted by
the skeleton of a giant dinosaur. At the far end of the room, there
is a stairway that splits after a number of steps, creating a landing
several feet above the floor. On the landing is a statue of Sir Richard
Owen, a provocative Director of the Museum in the middle part of
the nineteenth century. For those of you with some historical perspective, you will recognize Owen as the scientist who described *Trichinella
spiralis* (after having virtually stolen the parasites from James Paget, a
medical student who had discovered them during an autopsy at St.
Bartholomew's Hospital in London). Owen was also a very strong adver-
sary of Charles Darwin and a highly vocal opponent of Darwin's Theory.
I have often thought how strange his dedicated presence is in a museum
now so strongly committed to understanding the significance of natural selection and evolution. Moreover, it has always been ironic to
me that Queen Victoria knighted Owen, but not Darwin. I have often

speculated about whether Darwin ever thought about the very obvious snub. I seriously doubt it considering how self-effacing Darwin was during his lifetime. On the other hand, it is Darwin, not Owen, who rests in Westminster Abbey next to Sir Isaac Newton. On considering the enormity of each honor, I think Darwin got the better of the two deals.

Almost each time my wife and I go to London, a highlight is our visit to the Museum, during which we usually take tea with many of the parasitologists in residence, followed by a pub lunch with David and his colleagues. The May 2005 visit, however, was strictly one-on-one with David in his office (although after the interview, we did manage our pub lunch). The time spent was well worth it.

I learned from my first question that David was born in the rolling hills of Yorkshire, in the north of England. As a young teenager, he moved south with his family to Hertfordshire, and a small town named Ware. There, he attended Waring Grammar School. He was quick to identify Alfred Russell Wallace as another alumnus of his small school. He told me that he was turned on to the life sciences by two inspiring teachers, one in zoology and another in botany. From Ware, he went on to University College Cardiff. He had considered Liverpool and Exeter, but he said he just felt that Cardiff was a better fit. He said he was quite interested in marine biology at that time and there were good opportunities for its study in Cardiff.

His focus on marine biology was to change early in his stay in Cardiff. He told me that, "I was particularly fortunate in coming under the influence of David Erasmus, one of the very good parasitologists of that era. He was just getting into electron microscopy, and he hired me to work for him one summer. It was a great experience and helped shape my career, even though I was unaware of it at the time. My honors project involved circadian rhythms and egg excretion patterns in *Schistosoma mansoni*. An evolutionary ecologist, Mike Claridge, who was an insect man in the Zoology Department at Cardiff, also had a great effect on me. In fact, Mike was probably the spark responsible for my interest in evolutionary biology.

"Toward the end of my stay at Cardiff, I began writing around, looking for a place to continue my studies. By that time, I knew that I was hooked on research and wanted to do a Ph.D. So I spoke with Dave and we came up with a few names and addresses. One of the people I had considered working with was Desmond Smyth at Imperial College in London and I wrote to him about the possibility. While finishing my honors program at Cardiff, I went into speak with him and, there he was, correcting proofs for his new introductory parasitology book. I had a good chat with him. Fortunately, after returning to Cardiff, Desmond wrote to me and said there was a good chance for an MRC [Medical Research Council] studentship with Elizabeth Canning. I didn't know too much about Elizabeth Canning at the time, so I thought I had better go back and have a chat with her. When I arrived at the Ascot train station, she picked me up in her green Triumph sport car and whisked me off to Silwood Park, in Ascot, where a lot of Imperial's parasitologists were located and where her work was centered. She talked to me about *Eimeria* and possibly looking at relationships between different species. My evolutionary interest came into play at this point. Isozyme work had just started, and people were beginning to do enzyme electrophoresis for the first time. Between the two of us, we hatched a project to actually work on the *Eimeria* of chickens. So, I ended up at Imperial College, beginning in the fall of 1973."

He was particular impressed with the parasitologists in residence at Silwood at the time, i.e., Elizabeth Canning, Neil Croll, Bob Killick-Kendrick, Bob Sinden, and P. C. C. Garnham. According to David, "It was really a very good 'working' laboratory. Everyone was totally focused on what they were doing, but not so much so that they didn't take the time to interact with one another and with the students who were out there. So, I worked with her toward my Ph.D. She was an excellent supervisor in the sense that she was there when she was needed. She's very meticulous and serious about her research." He continued, "At about the same time I was starting my Ph.D. research, comparable work had actually begun in a government facility, called Houghton Poultry Research Station. It was led by a fellow named Peter Long, so I had the opportunity of collaboration with them as well."

Noting that he does not do protozoan work now, I asked, "How did you get into schistosome research?" He responded, "Well, that was an interesting situation. I had been hooked, of course, on *Eimeria* while doing my Ph.D., but I was still fascinated by the work I had done with David Erasmus on circadian rhythms and the release of eggs from *Schistosoma mansoni*. As it happened, I had met a person at Imperial College who knew someone out in Tanzania, a man called John MacMahon. He was a medic doing field epidemiology. He too had been looking at egg excretion patterns in humans, and we began an exchange of letters. Then, during the second year of my Ph.D., I saw this job advertised for a population geneticist of schistosomes at the Natural History Museum. I thought that would be really good because I had just checked out the literature as it applied to an enzyme electrophoresis approach to schistosomes, and there were lots of interesting research ideas that could be followed up. So, I wrote to Chris Wright who was Head of what was then called Experimental Taxonomy, and put my case on the table. Not surprisingly, I received a charming, handwritten letter back from Chris saying that they were very interested in my application, but they wanted to appoint someone now. Moreover, they really did not want a parasitologist. They wanted a population geneticist. So, I continued my work with *Eimeria* and thought no more about it. In the third year of my Ph.D., as I was coming towards writing up my dissertation, lo and behold, I saw the same job being advertised. By this time, I had met Chris Wright at parasitology meetings and was very impressed with him. I put in the application again and was luckily invited to interview. Things went well. As it turns out, at the time, I was also getting together an expedition to retrace some of [Alfred Russel] Wallace's work in the Amazon basin. I think this helped too. I managed to get the job, and ended up here at the Natural History Museum working with Chris Wright. So, I managed to get back to the schistosome area."

I was curious and inquired, "Why the Natural History Museum?" I was thinking that he had not even applied for an academic position. He answered, "Well, I was very keen on working with Chris Wright. He was an amazing draw for me. I was also aware, and this was not scientifically orientated, that the government system at the time offered the kind of career that was good. This idea was also based on the interactions I had had at the Houghton Poultry Research facility, the government facility I mentioned earlier. Finally, I knew the Museum, having been a frequent visitor around the galleries and so forth." I then asked if he had any trepidation about coming to London to live. He replied, "Not really, because although I had been based out in Ascot, I also had connections with the main London college while doing my Ph.D. I had not had any scientific contact with the Museum other than some behind the scenes interaction a few times. I have been very lucky actually, because there were opportunities to do postdocs and so forth, and perhaps get lectureships at the time. It just seemed like the correct course to start my career here at the Museum." Knowing the Museum as I do and having known some of the parasitologists there, both present and past, my very strong opinion is that David made a very wise decision. His work is internationally recognized for its quality and the respect in which his peers hold him is reflected in his reputation. The Museum certainly has not stifled his development as a scientist in any way. He has been nurtured by the experience.

The Museum job is full-time research. Did he miss the opportunity of teaching? For anyone who has met David, they know he is obviously an amiable person, with a good sense of humor, and, seemingly, would be a good lecturer. When I asked about not pursuing this kind of a job, he responded, "That is a very important question. What the Museum lacked back at the beginning for me was student contact, but not so much today. At the beginning for me, I missed the interaction with students and their minds, youngsters asking questions, that sort of thing. The way out for me has always been to collaborate externally. I do teach some at Imperial, at the London School [of Hygiene and Tropical Medicine], and at Liverpool, but just a handful of lectures per year. It is a different working environment than in a university. This struck me on the day that I arrived. It seems that coming here from Imperial it was more of a nine to five research kind of environment. At Silwood, you could work every hour of every day if you liked. But, again, it's the external contact that's important, and I now supervise Ph.D. students at Leeds, York, Aberdeeen, Imperial College, and Oxford. This immediately brings you out into the university environment."

I returned to his research interests by saying, "You have always been interested in molecular approaches for the resolution of evolutionary questions." He responded, "Yes, that's right. When I began working on *Eimeria* with Elizabeth Canning at Imperial, I started with starch gel electrophoresis at a time when it was a brand new technique." He chuckled when he said, "I remember my first electrophoresis gel, the one that actually worked." Then he laughed hard. "We were using what was called a thin-layer starch gel, which had in fact been successful with trypanosomes. I remember running in to Elizabeth's office, shouting, look at this! She was terribly excited about it and so was I, and we went off from there. At some point, we started doing some buoyant density work with *Eimeria* DNA. I had gone to the Liverpool School of Tropical Medicine to see a man named Michael Chance, who had been working with *Leishmania*. However, my introduction to real molecular techniques came down to a colleague named Andy Simpson, who had been over in the States, I think at NIH. So, that's where it all began for me, and I've been taking a hack at it ever since."

In addition to his many other duties, including service as Head of the Biomedical Sciences at the Museum, David had just finished a tour as President of the British Society for Parasitology. As a card-carrying member of the BSP, I think he did a very good job at it. All in all, David has had a very successful career. His work on the schistosomes has carried him into Africa, the Caribbean islands, and South America on many occasions. His enthusiasm for problem solving persists. He continues with the schistosomes, and I see no real end in sight for him.

JOHN HAWDON

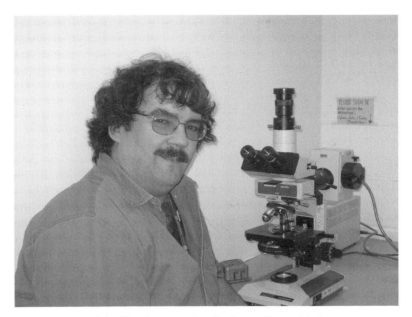

Figure 11. John Hawdon, Associate Professor, George Washington University School of Medicine, Washington, D.C.

Without question, John is the youngest of those interviewed for this book. However, I wanted to include him because he played such a crucial role with Peter Hotez in the development of the hookworm vaccine. I had missed him in early May when I flew up to D.C. to interview Peter and Don Bundy, but caught up with him at the 'hurricane-shortened' annual meeting of the American Society of Parasitologists (ASP) in Mobile, Alabama, in early July 2005. I had met John the first time in Philadelphia at the 2004 ASP annual meeting and knew that I would come away with a lot of good information.

John is a behemoth of a man. He stands at about 6-foot-5 and must weigh in at around 300 pounds or so. I asked if he had played any football while in high school, especially being from a very small, "dying, rust belt sort of town just outside of Pittsburgh, Pennsylvania."

He sort of smiled and said, "No, injuries kept me out of football, but I did wrestle in high school and made the state regionals as a senior." I asked at what weight, and he responded, "Unlimited" – he is a big man! "Our town was out in the country you might say, and I grew up in the woods."

He said that he became interested in biology at a young age and that he had always wanted to go to vet school. I asked what provoked his interest in biology and he replied, "I guess I was always just interested. My Dad was interested too. He liked birds and I guess I kind of picked it up from him. He was an elementary school teacher and just had a natural thing about plants and animals. He taught a lot of biology to his kids. I remember doing my early school projects on lizards, or snakes, . . . you know, 'stuff' like that.

"I guess my real interest in veterinary medicine actually began when I went off to college at Penn State. My first couple of years there, I worked for a veterinarian, but after a while I decided I didn't like it. I discovered that the bottom line for them was profit and it was a business. If I had wanted to become a businessman, I would have majored in it. After going to the main campus of Penn State for my last two years, I became more interested in biology. That was about the time recombinant DNA work was starting, around 1980, and I looked around for a lab in which I could get some experience doing this sort of research, but there was only one lab like this on campus at the time. But I went to the prof anyway and asked him about it. He said I was not qualified, and that I should first take a course in microbiology. So, I took it, but when I went back after that semester, he said I was still not qualified. I got kind of upset when he made that comment. I said, 'Look, buddy, you told me if I went and took the microbiology course that you would let me in.' The prof relented [so would I considering John's size!], and I got to do some piddling sort of work with plasmids. I took the GREs [Graduate Record Examinations] and applied at several of the high-powered graduate schools, but still didn't know exactly what I wanted to do. When I graduated, I kicked around for a year in Philadelphia looking for work in science, but couldn't get anything. So, I ended up going home."

After about six weeks at home, John found a job at what he called a "proprietary technical school", which he explained was something like a trade school, teaching animal health technology. He taught such things as physiology and remedial math. One of the courses assigned to him though was parasitology, even though he had never taken the course himself. He told me that he taught five one-hour lectures each day. For the parasitology course, he had a veterinary parasitology book that he would use for three hours every evening to prepare a lecture for the next day and, accordingly, was dangerously always just one lecture in front of his students.

After a two-year stint at teaching, he obtained a technician position in the lab of Lew Jacobson in the Biology Department at the University of Pittsburgh. Jacobson apparently had a well-funded operation because John said the lab was full of other technicians, postdocs, and students. The study organism in the lab was *Caenorhabditis elegans*, but he said this was before this spectacular nematode found its real niche as the model for molecular biology and genetics that it is today. Not only did he work there for two years, he was able to take courses in biochemistry and genetics, tremendous opportunities to develop what were to become his central skills in biology. He said that by far, "it was the best scientific environment I have encountered to this day." At the end of two years he applied to graduate schools at Georgia, Michigan State, and Penn. I asked, "To do what?" He replied, "To do parasitology. I had taught parasitology and had enjoyed teaching it. And I had worked on worms. So, I thought, maybe I can put these two things together. I liked teaching, but I wanted to work on a worm that might be more important, i.e., medically, than *C. elegans*." Jacobson wanted him to stay at Pittsburgh, but John had made up his mind that he wanted his worm work to be more than just theoretical, as it would have been had he stayed at Pittsburgh.

Penn made the best offer. They said, "Come on down and do research. I was told that I wouldn't even have to teach." Before going though, he checked over some of the people in parasitology at Penn. "I wish I could say I was smart enough to have known I should work with Gerry Schad, but I wasn't." Eventually, however, he and Schad hooked up and it was to become a terrific match. "There was something about his work that I liked. When I got there, we talked. He was working on *Strongyloides* at the time, but that was a difficult one for me to choose. Besides, he already had Linda Mansfield doing work on internal migration of *Strongyloides*. And, I really wanted to something with the molecular biology of infection caused by a worm parasite. One of the problems I had initially was that there was virtually nothing known. There was no in vitro system at all. So, before I could get to the molecular biology, I had to develop some sort of assay to look at what was happening in vivo, during infection. Gerry and I sat and we talked about what could be done. What would happen to the worms in a host? Would they start to eat? Of course they eat when they are L_1s and L_2s, but as free-living L_3s they don't. So, we figured the first thing they wanted to do when they started to develop beyond the L_3 was to start to eat again. We thought this could possibly be a good place to start. While at Pittsburgh, we had done some things with feeding *C. elegans*, like looking at lysosomal packing. So, I said, we could just throw some dye in there and watch to see if they begin to eat, and, lo and behold, I threw serum at them, along with the dye, and they started to eat."

I inquired if this experiment was with *Strongyloides*? He responded, "No, it was with hookworms." They had tried *Strongyloides* first, but it would not work. When I followed with, "Why not?" He replied, "We didn't know at the time. Eventually though, Gerry got that to work. But, serum would stimulate the L_3 of hookworm to undergo development. So, I spent the five years of graduate school characterizing that response."

I then asked John point blank if there were any sort of serendipitous events in any of the research he had done for his dissertation, or in conjunction with Peter. He sort of chuckled, and answered, "Yes. When I was searching for the stimulus in graduate school, I tried everything. I was looking for something that would stimulate feeding by the L_3 larvae, but I had no a priori idea of what it would be. Our whole objective was to get enough larvae to be able to do their biochemistry and molecular biology, so you needed good recovery. We ground up skin and exposed the worms, but this didn't work. We attempted to stimulate them using bile salts of different kinds, but this didn't work either. All sorts of things were tried, but nothing worked until we tried just plain serum. About 50% of the larvae exposed began to feed. I guess the really lucky thing we did was to use glutathione together with serum. It wasn't until we added glutathione that we would see 90–100% response. There was another guy in the Department, Rich Pollock, who was working with *Dirofilaria* and *Brugia*. When he exposed larvae of these two parasites to glutathione, they would molt. He said, 'Why don't you give that a try?'

"So, we tried it alone and 50% responded, but when you do it with serum, the response jumps to 90–100%, a definite synergistic effect. We don't know how it works, but we do know it is not because glutathione is a reducing agent. We know this because we can block its reducing effect by adding a methyl group at a certain position on the molecule and glutathione still stimulates transformation of the L_3 into a feeding stage. So, it has something to do with its molecular shape and not its reducing effect." In reality, John's discovery was not serendipity. It was very similar to Bill Campbell's 'unearthing' ivermectin. Both were looking for something. They each had an objective in mind, but neither knew what the end point was going to be. In each case, there had to be some luck involved. I think more so with Bill and his colleagues though than with John and his.

Of course, the next question that would follow is, what is the receptor? Glutathione is just a tripeptide, with an odd peptide bond in it. As John indicated, "If we could discover the receptor site for the glutathione in the worm and shut it down some way, we could prevent a critical step in the life cycle and stop the parasite from developing properly. Of course, that's another project sitting there in the back of my mind waiting to get started." I asked him if he thought the glutathione might affect the nervous system in some way, but he said, "No, I don't think it gets in. I think it stays on the outside." I then reminded myself of the incredibly impervious character of the nematode's cuticle and his response made sense. John continued, "You know that everything from *Hydra* on up has its own glutathione, probably including hookworm larvae."

By the time Peter Hotez was doing his medical residence up in Boston, John had begun to work on his dissertation with Gerry Schad. Then, toward the end of John's time in Philadelphia, he began providing the larvae for Peter, and they eventually met during one of Peter's trips down to Philadelphia. John said that Peter "knew if I was going to do anything more with hookworm, I would have to go with him, because he was the only one working on hookworm at the time." Peter made John an offer when his dissertation was completed and off he went to Yale. John said that even before he was finished at Penn that he went up to Yale to do some experiments with Peter. They even tried to run some of the incubation media through gels in an effort to isolate secretory/excretory products from larvae, but nothing had worked because John had not yet perfected the technique for stimulating feeding by the L3 larvae.

I was curious about John's status while he was at the Yale Medical Center. It was fairly complicated. Technically, John was in Internal Medicine, and supposedly working with Frank Richards, while Peter was in Pediatrics. However, he was actually was working with Peter. The postdoc lasted three years, then Peter secured for John an appointment as Associate Research Scientist, which meant that as long as you could come up with your salary, you could stay at Yale. While at Yale, he said that he became involved with Peter in writing the proposal for the Tropical Medicine Research Center (TMRC) in Shanghai, the grant from NIH that opened the door for their field studies on hookworm disease in China. John said, "This was the grant that was used by NIH to build infrastructure in developing countries. Peter came to me and asked what I knew about genetic variation. I recall that I told him that I knew a little bit about it and that I was interested in it. I had been thinking about a little side project regarding how hookworms got to the New World. You know, there is a whole body of literature from South America back in the 1920s. There was speculation that *N. americanus* came here with the slave trade, but they found a localized human population that was infected with *A. duodenale* in almost the exact opposite ratio of *N. americanus* for what it should have been compared with the rest of South America. In Paraguay, on one side of the Chaco River, you have fifteen to one *N. americanus* versus *A. duodenale* and on the other side of the river back into the isolated brush country you have just the opposite. It was postulated that hookworm growing there was the ancestral species for the hemisphere and that it came from the trans-Pacific migration of either Japanese or natives of the southwest Pacific", not unlike the speculation regarding pottery shards and *P. vivax* suggested by Bob Desowitz in his book, *Who Gave Pinta to the Santa Maria*. Even though Bob was simply suggesting the possibility of the pre-Columbian introduction of malaria to the western hemisphere, John is adamantly opposed to the idea that hookworms in the New World were pre-Columbian. "My wife is an archeologist, and she turned me on to this trans-Pacific controversy. But it turns out, it's not controversy in the field because there is no archeological evidence to support trans-Pacific contact. It's been pretty much discredited. But some people still use the presence of pre-Columbian hookworm as evidence for trans-Pacific contact. So, my wife and I wrote an article for *Parasitology Today* postulating on how *A. duodenale* could have arrived in the New World as hypobiotic (arrested) larvae in people migrating over the land bridge from Asia. The paper generated some controversy. I always wanted to look at the genetics of the Old World versus the New World worms to see if I could find evidence to support the hypothesis." He has not done it yet, but I think this would be fascinating as well.

John continued, "As I said, Peter came to me and asked me about genetic variation. He told me the grant was about genetic variation in parasites. I indicated that I would love to have some hookworms from China. He said, he was just going to look for vaccine antigens, but would I like to write something about genetic variation? So, I agreed to do it and he added several pages to the proposal. We received the funding. So Peter had the vaccine project and I had the genetic variation project in the TMRC, and we went over there. I don't want to say that it ended up being a disaster, but the Chinese do what they want to do. The data were not very clean. But we did manage to generate enough information for a paper in which we concluded there was genetic structure in the hookworm population in China." He then added a significant punch line, "In the long term, you have to be careful about a vaccine with a sequence in one country, or one part of a country, and having it work in both places because of the potential for differences in sequences in molecules."

I responded by saying that, "This is the dangerous part of the vaccine approach for control." He immediately injected, "It's potentially brutal!" He then said, "There is also a behavioral component in all of this, but it looks like about 30% of the variability in this is genetic. Look, you've got a hookworm in China and one in Africa and they're both supposed to be *Necator*, but are they? It seems to me highly unlikely that these worms are necessarily similar. They could easily be cryptic species."

John continued, "While we were in China doing this study, we did not spend a lot of time out in the field actually doing the work. But we would go out and see the people doing the work. The Chinese have a huge infrastructure of people who actually do the deworming, and they like to keep them busy. So, we would make arrangements with provincial officials to go out and pick up the samples. And they would want to see us, so I was able to travel extensively in southern China. As it turns out, I've been over there twice a year for the past ten years. Wherever we went, they would want to feed us well. We ended up referring to it as the 'banquet circuit'.

"It was funny how all this eventually worked out. The first grant ran out and we were invited to submit a renewal. However, all of the resubmittals were late being turned in, so everyone was given bridge grants until the new proposals could be sent. Then, Peter got the Gates Foundation money. Since he received $15 million, he was essentially poison at NIH, who told him that they weren't going to give him any more money to do the work. So, he got out of the TMRC program and left me to continue with the hookworm project. However, with my constant travels in China, I began to realize what was going on with respect to transmission and that it's not the same as everywhere else. I started thinking about it and it struck me that 'night soil' [human feces] is like a commodity there. It's valuable and it's not given away. It's kept for your own fields. These people would defecate in their own latrines and take their own 'crap' and put it on to their own crops.

Then, they would go out into their own fields and get reinfected when they tended their own crops. So, in effect, they were getting infected with the descendants of their own hookworms. The family's infection from there on down are all descendants of those from 'time x'. This struck me as potentially presenting some weird genetic structure in the hookworm populations at the level of the household. These worms should be genetically different from those of the next household 300 yards away. So, I talked with some people and hooked up with Mike Blouin out at Oregon State who has done some interesting work with nematode genetics. We decided that the best way to look at this genetic business was to use microsatellites. Accordingly, the new proposal was created in two parts. One was to collect worms and look at the vaccine antigens, looking for any variation. As it turns out, the ASP-1 antigen was not very variable. So, we are now looking at ASP-2. The second part was to look at the relatedness of hookworms based on transmission pattern. For comparison, we went to Hainan Island, where they don't use night soil, and people are infected in the usual way. We needed a control for the microsatellite data and we used the worms from people on Hainan Island where the night soil is not used. We are at the point now of beginning the lab work on the microsatellite project, so I don't know what we are going to find."

Of course, Peter left Yale six years ago and John went with him to George Washington University. As John said, "With $16 million of Gates Foundation money in hand, I figured it would be a good ride. So, he got me a good position down there and I'm glad that I went with him. I came as an Associate Professor without tenure, which I am up for this year. He has taken good care of me. I was originally Head of the Antigen Discovery Unit, but it struck me pretty soon that I was not going to get a positive tenure decision doing vaccine trials, because you get these multiauthorship papers and it's very hard to develop a good independent research program in just creating a vaccine. And, it was also clear to me that they were going to get beyond the antigen discovery part of the research as soon as they could, and get to the vaccine, because that's what Gates wanted to do. Gates did not want to fund basic research. They wanted to fund vaccine development. I began to drop out of the vaccine program over a period of time. The vaccine is now into the production phase and I want to do basic research."

So, John is now working on his own, separate from Peter who is aggressively pursuing the application of his vaccine program in Brazil. John is still working on molecular biology of hookworms, using *C.*

elegans as a model, and concentrating on the dauer larva, which is, effectively, an infective L_3. He is in his tenure year now at George Washington. I hope he gets his tenure appointment because he is a damned good parasitologist. (He got it!)

MARK HONIGSBAUM

Figure 12. Mark Honigsbaum, newspaper reporter, *The Guardian*, London, England

This biographic sketch will be shorter than the others that appear in the Prologue because Mark Honigsbaum is not a parasitologist, or even a scientist. He's a newspaper reporter, and a very good one. After Ann and I talked with him on a Saturday afternoon in May 2005, at his home in Shepherd's Bush, London, we began buying the *Guardian* at the local newsstand. His byline was on the front page of virtually every edition we saw.

Now then, why would a nonparasitologist and a nonscientist write a book about malaria, the cinchona tree, and quinine? The answer to this rhetorical question should be easy. Very simply, it is because he is a newspaper reporter. He saw a story and he wanted to write about it.

This seems natural enough for a newspaper reporter, but, first, how and why did he choose this line of work? He said that he was born in west London and grew up there as well. Then, it was off to New College, Oxford, to study "politics, philosophy, and economics", not that he wanted to be a politician, a philosopher, or an economist. He figured this degree would be used as his vehicle to a career in journalism. I asked, "You mean you had no formal training in journalism?" He replied, "Oh, no, in those days, you could go straight from university to a job on a newspaper. Besides, journalism is a job that either you are good at it, or you are not, and you learn to do it as you go along. There are certain things you can be taught, such as shorthand and the law. But the good reporters have an instinct and a certain talent for the work." I immediately inserted, "Yes, but you have to know how to write." His response, "Yes, up to a point, but it is more important to be able to marshal facts concisely and elegantly."

I then inquired, "But, how did you really get started as a reporter?" He said, "Well, when I graduated in 1982, there was a lot going on, and I was always of the opinion that a journalist could make a difference in the world. This was some time after Woodward and Bernstein and Watergate, but journalism still was an exciting profession. You just had to dig around and investigate." I was surprised to hear him say next, "Journalists have a license to go wherever they want, a license not many people have. And, it's been an interesting career ever since. I've traveled the world and met lots of fascinating people." He laughed when he added, "Yourself included."

He then said, "I am interested in all sorts of things under the umbrella of journalism, but I particularly became interested in diseases, especially malaria, and lately in the avian flu disease. I mean disease is fascinating because it crosses over into so many different areas. First of all, there is always a stimulating narrative. This narrative is always an exciting sort of 'who dunnit'? Then, there is a sort of retrogressive analysis. For example, how did these parasites such as those that cause malaria, or how did these viruses from avian sources, emerge and then enter the human population? And then there is sort of the scientific investigation, i.e., how do we develop vaccines or drugs that combat these viruses or these parasites that have this amazing ability to mutate and get around the human immune system? And then, another thing that is fascinating is that they involve these huge social and economic impacts, which bring you right into the field of health policy, and also the dichotomies between rich and poor countries, north and south, east and west. So these subjects can go anywhere. They can lead you into very fertile and stimulating fields for research."

How did he get to the cinchona tree and malaria, and South America? First, he indicated that he is not a specialist in parasitology. He chuckled, saying, "And, I am not a botanist. I mean I'm not even particularly interested in plants." He told us that he was sent to Zurich, Switzerland, by the *Observer*, a Sunday newspaper for which he was working at the time. He was there to do a story about a robbery that had actually taken place in the United States in New Jersey, several years before. It seemed that "the Mafia had purloined all these corporate bonds that had been issued by Citibank and other big American institutions. On the way to the garbage dump and destruction, the Mafia had got hold of them. These bonds were then being distributed all over Europe by persons unknown and were being used by white-collar criminals as collateral to leverage huge loans. There was a large investigation being undertaken by London's Scotland Yard, the FBI, and Interpol. Basically, I had been dispatched to Zurich to see this lawyer who had masterminded the whole scheme." We asked if he had managed to see the guy and Mark said, dourly, yes he had spoken to him, "but not for very long."

He continued, "The more interesting thing about it, at least for me, the day I finished interviewing the Zurich police about their investigation, I went out and looked for a place to eat. All the restaurants were very full, so I finally found a pizza place, which was also full, but the waitress asked if I would mind sharing a table with another gentleman who had just come in and I said that I would be quite happy to. So, as the evening passed, simply in order to make conversation, I asked him, 'What do you do?' And, he said, 'Well, I am a botanist, a naturalist. I work in Zurich.' I don't remember what his name was or where he worked, but I asked him what was the most interesting plant in the history of botany? And he told me the fascinating story of the cinchona tree, which I kind of filed away. Many months later, I thought maybe I should read up on this story a bit more and see who has written about this cinchona tree. So, I went to the Wellcome Library and Kew Gardens here in London, both of which have excellent history of medicine collections. I found many academic studies. The more I read about it, the more interested I became in the stories of the botanists who had been sent to the Andes to harvest the quinine.

"But, in order to know about quinine, I had to gain some understanding of malaria, which meant consulting with textbooks. Very early on, I got in touch with the people at the London School of Hygiene and Tropical Medicine and they were very helpful. Actually, on a trip to Washington, D.C., I also got some help from a military historian who told me about the use of quinine during the American Civil War." It is of interest to note here that there were more hospital admissions due to malaria than there were to bullets or canon shells during this internecine conflict. Malaria continued to be a huge problem during twentieth-century warfare for the United States. Malaria produced more casualties for Americans than in any war (with the exception of WWI) up to the 1990 Gulf conflict, and the one we are in at the moment in Iraq. "The Walter Reed people were also of great value since folks there had developed all of the major new synthetic antimalarials in use since WWII, and they passed on their research to the pharmaceuticals that did the further development and marketing."

My wife Ann then popped in with an interesting question. She asked, "While you were doing all this research, were you still an active journalist?" He responded, "Initially, I was trying to do both, but eventually I had to give up my job on the *Observer* because I was traveling all over South America. When I wasn't traveling for several weeks and months, I was immersed in quite detailed research in London. I was trying to produce the book as quickly as possible."

I asked, "While you were doing all this traveling in the back country of Columbia, Peru, Equador, etc., did you ever have any trepidation? I think I might get sort of worried about running into some rather unsavory characters." His response was that of a reporter. "No, I have been to lots of supposedly dangerous places in the world. Whenever I get there, I nearly always find they are not as dangerous as they are presented in the newspapers [an editorial comment by a real newspaper reporter?]. On the whole, I found people to be very friendly and inviting. The greater worry is protecting yourself against disease. I think of all the places that I went, the most dangerous was the Orinoco River. I retraced the route of one of the heroes of the book, Richard Spruce. This is where he first contracted malaria. Earlier, a colleague of his, Alfred Russel Wallace, had developed malaria on the Orinoco and almost died from it. Since the very early days, a great many white people, Europeans, had visited the Orinoco and had introduced a great many infectious diseases, including malaria, smallpox, measles, etc. I followed Spruce's route along the Orinoco. All along the River, we were picking up people who needed help because they had contracted malaria. There was a great deal of drug resistance there. The other risk was mainly physical because to go to the places in the Andes where these trees are located, you have to climb and walk along quite precipitous rocks and steep mud trails. So, it's quite physically demanding."

I asked Mark if he was writing any new books along the lines of 2001's *The Fever Trail*. He responded that he was not, but at the end of our conversation, he left the kitchen and came back brandishing his second book (published in 2004), this one entitled *Valverde's Gold*.

He signed it for us and said it was, in part, related to his travels in South America. The book is about Pizarro and Inca gold. I have read it and found it to be quite interesting. It does not deal with parasites in any way, except and unless you could consider the early Spanish explorers in this category. Frankly, after reading his new book, I do not see how folks like Pizarro, Cortez, and some of the other early intruders into Mexico, and Central and South America of that era could escape being considered in any way other than as human parasites. Forgive the editorializing. I couldn't help myself.

In closing, I would recommend *The Fever Trail* for your personal library. Mark is not a scientist. He is a reporter and that is the approach he used in telling this story. It is well worth reading.

ROY ANDERSON

Figure 13. Sir Roy Anderson, FRS, Professor and Administrator, Imperial College, University of London, London, England

Without any question, the interview with Roy was the most unusual of any that I did for the book. Roy is employed 5% of the time with Imperial College of Science and Technology's (University of London) School of Medicine. The other 95% of his time is spent as a public servant in Her Majesty's Government. I was not sure of his precise title, but I knew he was serving as the senior science advisor to the Ministry of Defence in the United Kingdom. So, contacting Roy to arrange an interview was difficult, to say the least. Several times, it was through one of his secretaries, but we finally got it fixed for the May trip to London in 2005. It was scheduled for a Friday afternoon at 2:00 P.M. Ann and I arrived early, bought some sandwiches and beverages, and had a nice lunch in a beautiful little park just around the corner from his office. About fifteen minutes before the meeting, I checked in with the security personnel and asked that Roy be told we were waiting. A few minutes later, a very nice young lady met me in the reception area and said that Roy was tied up in a meeting at the Ministry of Defence and could I return around 5:00 P.M.? I responded that I would, but asked for a telephone number so that I could confirm the meeting before I traveled back to his office. Ann and I went back to Kensington and spent the afternoon roaming through some of the local shops before returning to the hotel. About 4:00 P.M., I received a telephone call saying that Roy would be ready for the interview. So, about a half hour later, I hailed a taxi and returned to the medical school. I was standing outside the building, waiting for him to arrive when I felt a tap on my shoulder. I turned, and there he was, with hand extended and a warm smile on his face. We went immediately through a maze of locked doors and hallways, arriving quickly in his office, where he immediately shed his suit coat, removed his tie, and rolled up his sleeves. He asked me to sit down and we began what was scheduled for a three-hour interview. At the outset, though, he told me that he had to catch a train at 6:30 at Paddington Station just around the corner from his office. Apparently, the discouraged look on my face flagged his attention. He immediately apologized and quickly offered to finish the interview in two weeks by long distance telephone, at his expense. I agreed to this arrangement and two weeks later, as promised and on time, he called me from London and we completed the interview.

But, that's not the full story of my interview with Roy. After returning from London, Ann and I flew to Colorado for some R&R at our cabin in the mountains. I took the tapes of our interview, along with several others, so that I could do some transcribing and writing. On my return home, I had to wait in the Denver airport for a while since our plane was about 45 minutes late. So, while biding my time, I got out the first hour of his tape and listened until we boarded. I thoughtlessly placed the tape recorder in the seat beside me thinking I would listen to it some more during the flight. When we arrived in Charlotte after a three-hour trip, the tape recorder was the last thing on my mind because I had about ten minutes to make a fifteen-minute sprint across the airport to catch another plane. You guessed it. I left the tape recorder and the tape in my seat. I realized the error when I finally arrived at home and urgently called the airline to report my loss. Unfortunately, after two weeks, no message came and I considered the tape recorder and tape with the interview lost permanently. So, I meekly sent an email message to Roy and explained the situation. He immediately, and very graciously, agreed to my calling him at his office in London the next Friday afternoon and we completed the first hour of the interview, a second time. I now have the fully taped interview and am ready to start writing. I must say, all of this was a great adventure, but in the end I succeeded in getting what I needed.

I should also note from the standpoint of British national security that nothing was lost with the tape, because this is not what we discussed during any of the interviews. It was strictly about his career and his science, both of which I found to be exceedingly fascinating. Roy is a Scotsman, although most of his adult life and his entire professional career have been spent in England. He was quick to note his allegiance to Scotland, saying, "Whenever England and Scotland play each other in rugby, I always root for Scotland."

He told me that both of his parents had a keen interest in natural history and that he felt their influence was great in his own development as a biologist. He said that he also was strongly oriented toward mathematics, as will be seen later in this short biography and in the essay. In his sixth form (as a senior in high school for us Yanks), he did a project on aquatic ecosystems in a stream near the school in Hertford that he attended. His focus was on the interaction between different species of water beetles. At about this time, he was looking around for a university to continue his studies and he knew by then that he wanted to do biology. One of the places he was interested in attending was the Imperial College of the University of London. The person who interviewed him was Professor O. W. Richards, "a very distinguished entomologist who had actually written the core text on water beetles, so that sealed it really. We had a very good conversation and, as a consequence, he offered me a place. So, I did biology, or zoology, as it was in those days." I asked, "Wasn't it true back in those days that if you did any mathematics or chemistry, that it would be taught by folks in the Zoology Department?" He responded, "Yes, that is correct, although we did have a few lectures from people in biochemistry and a few from folks in statistics, but it was mostly from within the department of your major.

"The first and second year field courses were taught by people at Imperial, but the most influential one for me was in my third year when I did parasitology, taught by Neil Croll. We went down to a field site on the Devonshire coast, a place called Slapton Ley," and the primary research location for nearly 35 years of our mutual friend, Clive Kennedy. It was there that Roy met June Mahon and Liz Canning, both parasitologists at Imperial College at the time. I asked if he had taken his first parasitology course from June Mahon, but he replied, "No, it was from Neil Croll actually, but June Mahon participated in it." He said, "I recall the field course as quite a social experience as well, lots of beer and girls present. We learned a lot, but it was fun as well." I then asked who would have been the primary influence in his becoming a parasitologist and he responded, "I think Neil Croll had a very strong effect, because he was such a good teacher. It was his enthusiasm for the subject. Then, also, the teaching of ecology at Imperial, independently of parasitology, was very good. I just enjoyed it immensely and really did not know what to do after that. I applied to a variety of industries and to the civil service, but ended up getting first class honors with an exceedingly high mark for the field project, which was in parasite ecology. So, June Mahon offered me a Ph.D. studentship and since I didn't have a better idea in mind at the time, I accepted and never really went away. It was a very good decision." I questioned him about this decision because I thought it was intriguing, "So you really hadn't considered graduate school when you finished at Imperial with your undergraduate degree?" He agreed by saying, "I hadn't really thought it through. I had an offer from a very large industry, ICI in the U.K., to go on their management training course and I was contemplating that, but the academic side, especially parasite ecology, had grown considerably and it seemed like a nice idea." He added, "I also had a girl friend in London, a silly reason that one doesn't normally consider. But it was a good decision. Imperial College is in the top three universities in Britain in terms of quality, very close on the heels of Cambridge and Oxford. In fact, it's much less stuffy than Cambridge or Oxford. It's much more international in flavor and horizontal in its structure.

"I had three fabulous years as a postgraduate, helped by June Mahon on the parasitology side, and then by someone named Gareth

Davies, who was interested in the application of computational methods in biology. Another very powerful influence on me was George Murdy, who was a statistical ecologist and a young lecturer in the department at the time." I asked, if when he went to Imperial as a postgraduate, did he have any kind of idea about the research system he wanted to use? He replied that he did because in his honors year at Imperial he had worked on both fish parasites and parasites of small mammals. He thought fish were more interesting because there was a greater diversity of parasitic helminths. He said that he thought that sampling would be easier for parasite ecological and dynamic studies. "You could take a population in a closed lake and obtain appropriate sample sizes. You could also get age structure in the fish samples because of the scale rings. So, it was simply more amenable for study. And, if necessary, you could bring them into the lab for experimental study."

He had chosen a pond in East London in which to do his research. It had a very high density of bream, and was a very easy pond to sample. It was not public domain, so it was much easier to work. Rob Wootten, who was also a graduate student at the time, was working in the Serpentine Pond in Hyde Park, so they would help each other when collecting. A private fishing club owned the pond. He chuckled and said, "I became well acquainted with several of the members who liked to down a few beers on occasion.

"During the time I was working on my Ph.D., I was interested in ecology. And population ecology was going through this growth in mathematical and statistical methodologies in understanding the dynamics of two species interactions." I asked, "How did this turn into mathematical modeling?" He responded, "Well, I realized that to understand the dynamics of population change with birth and death rates and all the rest of it in two species interactions, you really needed to have more mathematical framework to coalesce thoughts there. This was happening in mainstream ecology and I was quite convinced that was the way things would go in parasite population biology. So, when I was going through my Ph.D., I realized that I needed a lot more knowledge in this area. Gareth Davies suggested that I apply for an IBM Fellowship at Oxford in the Department of Biomathematics with a person named Morris Bartlett, who was the world's leading authority on stochastic processes. He had written a really elegant little book called *Stochastic Processes in Ecology and Epidemiology* [1960]. So I wrote to Bartlett, who kindly agreed to support me in my application. I went for an interview and received an appointment as an IBM Research Fellow in Biomathematics at Oxford."

Roy was at Oxford when Harry Crofton published his two seminal papers in *Parasitology* in 1971. In between the two 'biographic' interviews I had with Roy, I went back and looked at some of his dissertation papers and noted a strong similarity between what he had done and what Crofton had formalized in his two papers. I asked, "Am I basically correct in this assessment?" He responded, "Well, Harry didn't formalize it in strict mathematical terms. He was using a computer program, but the ideas were there. You know what I mean, i.e., the conceptual ideas about ecology and population biology, and distributions not being random, etc. Harry didn't specify the mathematics because that wasn't his strength. He had written a computer program. Now, what I did, and then subsequently in a lot more detail with Bob May, was put in formal mathematics and then use formal mathematical tools to investigate properties in a much more generic way. And then we extended it out of helminths into a whole spectrum of infectious disease agents."

I then asked Roy if he had ever talked with Harry Crofton about the two papers. Roy said, "Yes, I did, at the BSP [British Society for Parasitology] spring meeting in 1971. It was very sad that he died because he was a great 'encourager', as it were, of work in that area. I think, irrespective of whether he died or had continued, my part would have diverged because I was interested in a broader spectrum of infectious agents and also very much more into the mathematical end of specifying the problem."

I remarked that during a conversation with Clive Kennedy several years ago, Clive noted that Crofton was really not an ecologist, but that he was a very good parasitologist. Roy said, "That's right. He wrote computer programs and was very good at statistics." I asked Roy if he had ever talked with Harry about this sort of thing? He replied, "Well, if you look at the ecological literature, the negative binomial distribution was already well established. In fact, my supervisor at Oxford, Morris Bartlett, with one of his students, had actually specified the importance of this distribution in biology and epidemiology as a descriptor of observed pattern. Bartlett had also written about the biological processes that could generate the negative binomial probability distribution. So, all this was in the ecological literature and what Harry did was pick it up and apply it to parasitology."

The fellowship at Oxford afforded Roy a wide range of opportunities, including the chance to spend his summer months in the U.S.A. visiting field stations and laboratories from one coast to the other. He was at Oxford for about thirty months. Then, as he explained, obtaining his first academic position was an accidental thing. It seems that Kay Lyons, who had been at Cambridge and the Molteno Institute, was a young lecturer at King's College London when she met "the love of her life, married him, and went off to become a school teacher down in the Cornwall/Devon area of England." This left the lectureship vacant, so he applied for it and was offered the job, beginning what would become a rapid rise in the academic world. At this point, I mentioned, "There was quite a crew of parasitologists down there at the time, wasn't there?" He responded, "Actually, no. Frank Cox and Phil Whitfield were the only two, but they were superb colleagues, stimulating and always encouraging about trying new approaches to parasitological problems. It was a fairly small department, and very friendly, with something like twelve academic staff. The Chairman at the time was a tick authority by the name of Don Arthur. He was one of the world's leading tick taxonomists, and very interested in parasitology. I got on well with him. He was a very keen rugby player and I had put rugby as one of my outside interests on my application. You never know about these sorts of things during an interview. As the youngest faculty member that first year I was forced to teach a course in the philosophy of science, but I still managed to get in some parasite ecology as well." It was during Roy's second or third year at King's College that he met Don Bundy for the first time and began a lasting relationship.

I then said to him, "It sounds to me like you are in transition by this time into 'full blown' epidemiology." He said, "By then, I was equipped with the technical tools, with an ecological and parasitological background. I was a member of a small club, a dining club in Britain, which included a set of young mathematical ecologists. Virtually all of them are in the Royal Society now. We used to meet regularly to talk about mathematical ecology, and share techniques from people in plant and animal behavior through to parasitoids and competition in plant–herbivore interactions, and myself with parasitic organisms. It was then that I began to write down some of these models and realized that you could formulate them for any infectious agent. Sometimes the data were invariably much better for some of the well-studied human systems, and so that was a great influence for me.

"I met Bob May about the same time. He was on sabbatical at Imperial out at Silwood [Imperial College's field station] with the ecology group. Thereafter, we developed a very close working relationship. He and I got along well and we used to do a lot of work together. We would exchange visits between London and Princeton [University in

New Jersey, U.S.A.] on a regular basis." I asked Roy if May was into epidemiology at that point in his career? He replied, "No, he wasn't, but he had an interest via contact with David Bradley at the London School of Hygiene and Tropical Medicine. I was an influence on him to move in this direction, but certainly not the only one. I learned a great deal technically from him. He was a very, very good problem formulator, a good mathematician, with theoretical physics as his background. He was very quick to grasp the essence of a problem and translate it into a mathematical framework that permitted analysis. It was a mutually symbiotic relationship."

After three and a half years at King's, Roy moved to Imperial College. He left to go to Imperial because of his interest in mathematical ecology, as applied to infectious disease, and there was a very strong theoretical ecology group there, with Michael Hassell as the leading young member of staff. He said that he thought he would get more stimulus there than at King's College. Once he arrived at Imperial, he had been publishing a great deal and he rose rapidly through the academic ranks, ending the climb as Head of the Department of Zoology, "at a depressingly young age. But, fortunately, I had a very large research group and I was pretty tough with people. That summer was my time. And, so, I managed to get quite a lot done. My office was in London, but I managed to get out to Silwood quite a lot, and then I would spend the whole summer there."

Being a Chairman myself for twelve years, I had to ask him, "How on earth did you manage to run a department of that size and do the kind of research you were doing at the same time?" He laughed and quickly responded, "I'm good at delegating! I remember having a telephone conversation with Bob May after I had been offered the position and he said I would have to be mad to do it. He said it would really cut down on my productivity. I ignored his advice." Roy then added, "I'm a bit of a workaholic to be honest and also I appointed a very good departmental administrator. Then I took the less research productive, but very able, faculty and encouraged them to take up senior administrative roles. I restructured it in such a way that I only had to take the major decisions. A university is very good at absorbing your time by committees. I'm not a committee person, and I used to delegate most of that to other people to represent the Department. I was very fortunate to be surrounded by very good administrators and a very understanding university environment at Imperial, with rectors [the heads of Imperial College, first Lord Powers and then Sir Eric Ash] who encouraged heads of departments to continue major research careers.

"Another thing about it is that some people are reticent about writing, but I've always enjoyed it. So, once an idea has been worked through and the research done, I'm very quick to write it up. I can sit down in a day and write a paper easily, and just get it out of the way. I'm not a very good third, fourth, or fifth draft sort of person." Having read many of his papers, I will say that he is a good writer. He is clear and concise. He is fortunate in this regard, because I have always believed in, and followed, the old adage, 'the secret of good writing is rewriting'. I guess all of us have our own style.

At this point in his career, Roy was offered the top position at a major university in the U.K. In Britain, the supposed head, or Chancellor, is essentially titular. The person really in charge is the Vice-Chancellor and this is the job Roy was offered. Simultaneously, though, he also was offered the chairmanship of Zoology at Oxford. He said that his decision was a more than a little influenced by personal circumstances. His wife, Claire, had just been diagnosed with breast cancer. "We thought about it, and my own view was that we did not know how long Claire had at that time. We decided to go to Oxford and live in the countryside because we thought it would be a less stressful and more peaceful environment. As Claire would tell you, I also didn't want to become a chief administrator at that point. Fortunately, Claire got through her critical stage and she is still doing exceedingly well ten years out.

"However, I think I knew within two weeks of arriving at Oxford that I had made a mistake in going there." For me, this was a startling and revealing admission. He continued, "It was such a fussy place with sadly a high number of people who had not fully satisfied their career ambitions. It was very different from Imperial, which is very much a 'can do' place. When you have an idea, they help you to do it. Oxford can be rather different. I learned a very important lesson. In some environments at certain times, it is pointless to waste your time trying to change things, because there are so many highly intelligent people with less than fully active research careers and too much time to play politics."

At this point, I inserted my opinion about these sorts of situations by saying, "I don't think there is anything more ugly than academic politics." He agreed, and responded, "Yes, well, Oxford is probably one of the more difficult university environments for this in Britain, and I didn't realize it. I walked into it rather naïvely. So I knew I had made a mistake, but I made the best go of it possible. I left the Department Head's post as soon as I could after five years. It was like five years of penance. It was one of the most difficult periods of my life in terms of administrative achievements, but at least the Department always came in on budget. The most frustrating aspect was the failure to build up a new cadre of outstanding young academics and to shift the Department's interest a little bit more towards molecular and cellular biology, from its base in ecology and evolution, with the aim of exploiting the fertile interface between these different levels of biological study. If somebody very good emerged or wished to move to Oxford, it was virtually impossible to get a new faculty appointment. It was only possible at that time to fill posts arising from resignations or retirements. So, it was really frustrating for someone who wished to build and redirect the Department towards the rapidly expanding areas of biology." I asked if Cambridge was like this as well? He replied, "No, I believe Cambridge is somewhat different. It has a long tradition of outstanding excellence in science. Cambridge is clearly the number one university in the U.K. when it comes to science and technology, certainly in the top five internationally."

After stepping down as Head of the Zoology Department, Roy said that he directed his energies "into developing an application to the Wellcome Trust for a large grant for a new research center to study infectious disease epidemiology. This was successful. I was the first Director and Don Bundy was Deputy Director. I think it was a casualty of its own success because it became dominant over the Department. It attracted more money. It attracted the best students. We had a very successful first year. We had some outstanding people come in. The concept was interdisciplinary. I felt we needed to meld ecology and evolution with mathematics and statistics of dynamic processes, not static ones, and with carefully designed field and experimental study. We managed to put together a superb group of people." I asked, "You pulled a strong group of people from Imperial, didn't you?" He said, "Yes, I did, Don Bundy included, along with quite a few administrative staff, a group of about thirty altogether. A whole range of other people added to it considerably. Then, Don resigned to go to the World Bank, and this left a vacancy. I had a huge difficulty with the University and the Department at this point. They wanted to impose an appointment on the Center, even though the new Deputy Director, Brian Spratt, and I didn't agree on scientific and other grounds. This led to a rather public disagreement, not by my choice, but because of the willingness of others to play out the difficulties in the science popular press and in other media outlets. The upshot was that I experienced a rather unpleasant side of Oxford University.

"I was falsely accused of all sorts of things. It was a great shame, because we spent a lot of money in building that Center, and had recruited an outstanding group of young staff from all over the world to focus their efforts on the epidemiology of infectious diseases from a very broad interdisciplinary perspective, ranging from field and laboratory study to mathematical and statistical approaches. At that point, I decided that Oxford was not the right place for me to develop this new research center, and I resigned my Chair. I then had a very good set of discussions with Lord Oxburgh, who was the new Head of Imperial at the time. With Professor Chris Edwards, the Principal of Medicine at Imperial, they very kindly suggested that I establish a new department entitled Infectious Disease and Epidemiology [at Imperial]. Chris Edwards was very supportive in all sorts of practical ways. They allocated four professorial posts and quite a number of lectureships. It was a very generous offer since these are tenured positions. Then, much to my immense surprise, virtually everyone in the Center said they wanted to move with me, and we brought the whole operation down here to London. We have the largest group of people in the U.K. in helminthology. This group was hugely augmented by the addition of Alan Fenwick, a schistosome person. He had applied to the Gates Foundation for a grant to support a program for the control of schistosomiasis in five major countries in Africa and was awarded $30 million to do that. So we had a major Gates grant from a very early stage running in the Department. Our focus here is helminthology in the parasitic disease area. We also have large chunks of the Department working on bacteria and viruses, with world leading groups in mathematical, statistical, and molecular epidemiology. Our main interests are in helminthology. Soon after I was appointed, Lord Oxburgh retired and Sir Richard Sykes was appointed the new Rector. Richard was Head of GlaxoSmithKline. He has an infectious disease/microbiology background, and has been very supportive of our work." I asked Roy if there was still a functional Center at Oxford and he replied, "No, it died then. Everyone moved with me except the small group who lay at the heart of my disagreement with the University."

I then asked him how he became hooked up with the government job that he holds now? "How did it come about? I know you had to be excited about coming back and getting a new department started at Imperial." He sort of paused at this point before saying, "You know, the Department doesn't really need me now. If I'm going to do something else, and there are a few other things I want to do, it was a good time to switch. The Department is so strong. It is in such a supportive environment at Imperial. It was a good time to accept the government responsibility. I come into the Department every Friday to continue with a part of my research activity with the help of a small group of postdoctoral staff, graduate students, and colleagues. The job with the Ministry of Defence will go on for just three years and after that I wish to return to academic life to continue my research on the epidemiology and control of infectious diseases. There are two senior posts in government at what's called the Permanent Secretary level, which is the highest grade in the Civil Service, and the Chief Scientific Adviser's post in the Ministry of Defence [which I hold] is one of them. It is the oldest chief scientist post in the U.K., established by Winston Churchill, following on from the considerable contributions made by science and technology in helping Britain and its allies to win the Second World War. The first holder of the post was Sir Solly Zuckerman, a medical physiologist by background. So, having a biologist hold the post today is not so unusual! It has a very diverse and large research budget, and a considerable number of very able staff. I negotiated a deal where I would spend one afternoon a week at Imperial and the rest at the Ministry of Defence in Whitehall. That way, I was able to keep some grants and a small research group. At the Ministry, I have responsibility for managing a large number of scientists, engineers, and technologists." I then asked if he made decisions regarding the kinds of research done in there and was it largely directed toward weapons and that sort of thing? "Yes," he replied, "I am very directly involved in research, but the aim of our program is in support of defense. I have a budget. The post is advisory, but I have a very significant research and development budget and am able to direct that in a very practical, 'hands on', way. We need to be technically very sophisticated. The basic philosophy for the defense of the nation is to stay ahead technologically in the fields that are critical to security and sovereignty. So, my research portfolio ranges from the molecular biology of dangerous pathogens through to the engineering and construction of aircraft."

This was about as far as I could go with Roy. His work with the Ministry of Defence is classified and he obviously could not talk about it further. So, our discussion regarding his biographic background ended rather abruptly. As we ended our conversation, however, he emphasized his scientific roots by saying, "I am still a parasite ecologist, turned into an infectious disease epidemiologist!" In fact he was knighted in 2006 "for epidemiological research . . . and providing the government with advice on how to tackle the transmission of infectious disease" and is now Professor Sir Roy Anderson.

STEVE NADLER

Figure 14. Steve Nadler, Professor, University of California-Davis, Davis, California

When I became Editor of the *Journal of Parasitology*, one of the first things I had to do was appoint an Editorial Board. In the immediate past, Editorial Boards of the *Journal* had consisted of 15–20 persons, but none of them was identified on the cover with a specific area of interest. When I was being considered for the job, I was asked how I would go about appointing the new Board. I responded by saying that I wanted to create a Board with each new member being identified as responsible for a specific area of parasitology. One of the areas I wanted to have represented was genetics/evolution. For a long time, I had the feeling that these two latter disciplines were going to really explode in the near future as far as parasitology was concerned (and I was correct). Even though I am not a geneticist, I had an extended and strong interest in the topic and whenever a genetics paper was given at one of our meetings I would make a point of going to hear it. I recall such a meeting in 1989 in Vancouver, British Columbia, and a paper dealing with ascaridoid phylogeny using ribosomal-RNA sequence analysis by a young postdoc parasitologist at the Museum of Natural Science, Louisiana State University-Baton Rouge. I was fascinated by the results and his style of presentation. I thought at the time, 'this guy is going places'. His name was Steven A. Nadler, and my prediction was accurate. As a matter of fact, he will be President of the American Society of Parasitologists by the time this book is published. So, when it came time to appoint an Associate Editor for genetics/evolution, I knew whom I wanted. I asked him if he would serve and he agreed. (Anecdotally, I have been very lucky with all of my appointments to the Editorial Board. In the case of genetics/evolution, Steve was followed by Dennis Minchella, who is our immediate ASP Past-President, and then by Dante Zarlenga, a well-regarded researcher at the USDA in Beltsville, Maryland.)

When I was putting together the names of those I wanted to interview for the book, I knew that genetics and phylogenetics would be included as topics, either together or separately. One of the people I wanted to use was David Rollinson at the Natural History Museum in London. David agreed and we talked during May of 2005. The other was Steve Nadler, now a faculty member in the Department of Nematology at the University of California-Davis. Steve was serving on the graduate committee of one of my students, Joel Fellis, and was due for a trip back to Winston-Salem for Joel's Ph.D. defense, so I decided that would be a good time for the interview. Steve accepted the invitation and we sat down one morning in April of 2005.

Steve was born and raised in the great mid-west of the U.S.A., on the Mississippi River, in St. Louis, Missouri. Naturally, while growing up, he was a St. Louis Cardinal baseball fan. His father worked in one of the major industries in the city, Anheuser-Busch, the makers of Budweiser beer. Steve said that on his way home from a ballgame at the old Busch Stadium, he would sometimes stop at the brewery and have a 'Bud' with his Dad. I asked if he played baseball in high school. He said, "No, I played water polo and swam," making him the first water polo player I had ever met. "This is a tough sport," I replied, and he agreed, saying, "I didn't even know what water polo was when I was a freshman in high school, but a friend wanted me to join him, and I did. It's really a tough sport because there is a lot that goes on under the water that no one can see." A high school teacher kindled an early interest by Steve in biology, especially botany. The teacher was young, enthusiastic, and challenging. Steve said he was lucky to have attended a high school that was teaching a variety of biology courses, ranging from ecology to human anatomy and physiology, and he took them all. He was not sure of his career path at the time, but it was definitely not medicine. He said that he had read all sorts of books as he went through high school, and I asked if one of them had been Paul de Kruif's *The Microbe Hunters*. He said no, but that he did manage to read that one while he was in college, another of the parasitologists I interviewed who, to one extent or another, had been stimulated by the same book.

I asked if he always wanted to go to college. He said that he did. "My Dad had always pressed me and my two sisters to do it because he thought there was something better than working in the brewery all one's life. I started off at Southwest Missouri State, in Springfield. The first college biology course I took was in botany, but the professor really turned me off. It just wasn't any fun. When I was a junior, I took a year off from Southwest Missouri and went to the University of Hawaii." I asked him why he went all the way out there and he explained that his sister was a graduate student studying for her Master's degree in Spanish literature. She invited him out and it was a good year. He took a lot of biology out there, including his first parasitology course from an invertebrate biologist named Sidney Townsley. It was at this point that he became interested in parasites and parasitism. Steve told me, however, that the fascination for parasitology was not driven by Townsley, but by the subject matter. Apparently, Townsley was of the old school and taught the course in a rather dry classical style. Steve said, "It was very much applied. I don't think there was a single thing alive in the lab. After the course was over, I went to Townsley and informed him I wanted to go to graduate school. I told him I was wondering if he thought it was feasible for someone to actually make a living as a parasitologist because I hadn't really heard of parasitology before. He said, 'Sure, you just have to decide what area you want to go into.' At the time, I was thinking about medical parasitology. He suggested a few schools that I should check out, including Johns Hopkins, Tulane, and LSU, as possibilities. After my year in Hawaii, I went back to Southwest Missouri State."

After finishing his undergraduate work, Steve ended up in the Department of Tropical Medicine and Medical Parasitology at the Louisiana State University Medical Center in New Orleans as a graduate student. I asked why he chose LSU. "At the time, it was kind of a typical thing to do. I had received some promotional material describing their programs and they sounded good." I interjected, "So, you picked the school by its reputation, and not a person with whom to work?" He responded, "Yes, that's it. At the time, I don't think I had read a single *Journal of Parasitology*. Before finishing at Southwest Missouri, I remember taking a course in immunology and thinking maybe I would couple that with parasitology."

The change from the conservative, Bible-belt town of Springfield to the 'fast-track' city of New Orleans was something of a culture shock. He told me, "I went there to visit before moving down permanently. Naturally, I went to the French Quarter. I guess maybe what I had heard about the place was just a lot of hype, you know, propaganda to get people to visit the city. But I was kind of shocked, actually, walking down Bourbon Street and seeing all the hookers. I guess I was a somewhat naïve mid-western boy. But then, over the years while I was there, the place was toned down quite a bit. I also remember that I was down there in the summertime and a thunderstorm came up, followed immediately by the sunshine, and there was steam rising from the pavement. Sweat was pouring off me, kind of going nowhere. I was, by that time, wondering what I had got myself into."

I asked if he came to know any of the parasitologists over at Tulane. He said that there were just a few students he met. "There was a real animosity between the two schools. In fact, there was Tulane, sitting next to Charity Hospital, a large teaching facility, and then LSU sitting next to it, and there was no absolutely interaction. I think that way before I got there, something happened to create a long-standing personal feud. The students would sometimes take classes in the other's departments, so there was at least some kind of arrangement to do that. But the faculty, I don't think they ever went to seminars in the other places, even though you could walk from one to another in five minutes. The students would cross over, but not the faculty. It was weird."

I then said, "You must have come under someone's wing when you arrived. Did you do a Master's degree first?" He responded in the affirmative, "Yes, I did a Master's with Joe Miller. I worked on a *Hepatozoon* in cottonmouth snakes. I did that because whenever a new cohort of students came in, a particular professor would try and pick up one for his lab. He would also teach a course in parasite biochemistry when the new students came in. His lab was out at the LSU Dental School, which was off site. He would run his biochemistry labs out there. So, he worked hard trying to convince me that his lab was the place to go, but I really wasn't persuaded. The next semester, I took an electron microscopy course from Joe Miller and you had to do a project. There was another graduate student there at the time, named Carter

Atkinson. Carter was really a hard worker and was doing a lot with different herps. He had found this *Hepatozoon* in a snake and it was very common throughout Louisiana. The parasitemias in the snakes were at about 90%. So, I decided to do my project looking at fine structure of the *Hepatozoon* gamonts inside [red blood cells] of the snakes. By the time I was finishing that research, I was being pushed pretty hard to decide on my Master's research. I couldn't decide what to do, so I went to Miller. I told him that I thought I might be able to develop a thesis project on this *Hepatozoon* because its life cycle wasn't known for example. I think Miller knew what I was up to. It was pretty transparent to him who I didn't want to work with. He said to me, 'Well, you know I'm going to get a lot of heat if I take you on as a student, but I'll do it.' That's how I got with Joe Miller. He had been a student of Horace Stunkard. Miller, incidentally, did end up taking a lot of heat for his decision to help me out."

Steve completed his Master's degree in 1982. He had gone to LSU thinking about medical parasitology, but Carter Atkinson, the student who introduced him to *Hepatozoon*, was still around. He continued, "Harold Trapido was Chairman of the Department, and had obtained his degree from Cornell in herpetology. This was kind of surprising to me because this was supposed to be a medically oriented department, yet here was a herpetologist not only in the Department, but Head as well." Steve explained the circumstances. He said that, "During WWII, Trapido had finished writing his dissertation in latrines, because of the black-out conditions at the time. He was then 'hired' by the Army as an officer and actually did some of the first work on the residual effects of DDT at the Gorgas Memorial Laboratory in Panama. Then he went to work for the Rockefeller Foundation in India doing all sorts of research on arboviruses. So, his career was as a virologist. He was trained as a herp systematist, but abandoned this pursuit as time passed. He ended up at the LSU Medical Center. Carter was working with him, and was all over in the swamps. He was originally a graduate student in ornithology up in Maryland some place and decided he was really interested in the impact of avian malaria on birds. Why he came to the LSU Medical Center to do a Ph.D. I have no idea because he was not going to work on avian malaria down there. He started working on a whole bunch of stuff and was always traipsing around in the field, so I started hanging out with Carter and going along with him on his field excursions. It was then I realized that I didn't want to work in medical parasitology. I was having too much fun working out in the field with Carter."

When he explained his newly developed interest in fieldwork, I would have diagnosed it as a kind of epiphany. I commented, "So you were becoming a real field-type parasitologist at a very young age?" Steve said, "Not really, I guess. I don't know whether I would call myself a parasite ecologist. I wasn't out there shooting and collecting everything in sight. However, the field experience made me realize that I didn't want to do biochemistry or immunology. I had become much more interested in parasite life cycles and systematics instead."

With the transition of interests came that decision that all of us had to make at some point in our graduate careers. What are we going to do for a dissertation? Some of us are lucky in coming to a decision on our own. Others of us have followed in the footsteps of a mentor and taken a piece of his/her research, or done something in parallel with it. In addition to this question, he also "wondered if I should remain there. As it turned out, at the Medical Center, there was another 'displaced person'. Joe Miller was interested in systematics, but most of his research focused on fine structure of parasites. In fact, Dick Lumsden was one of his graduate students. Miller knew of my new interests and indicated his willingness to advise me to some degree. However, as it turns out, a biochemist, Herb Dessauer, had moved into the same building as us. Herb was an interesting guy. During WWII, he was a meteorologist, and then came back to LSU and enrolled in medical school. He told me that in his second year he was taking a course in clinical parasitology. One day he was standing in line for a fecal sample and he suddenly came to the conclusion that he had had enough, so he quit medical school, and enrolled in graduate school. He ultimately became the first student to graduate with a Ph.D. in biochemistry from the LSU Medical Center. He was also a frustrated systematist and had grown up collecting all kinds of herps. However, on finishing his Ph.D., he saw an opportunity of combining his biochemistry and systematics. Some of the very first work using allozymes, or isozymes, in starch gel electrophoresis for systematics and population genetics was done in his lab in the mid 1960s. He was interested in the systematics of snakes and lizards. In the1960s, Dessauer published a paper on isozymes in herps showing that they could use them to look at population polymorphisms and analyzing for Hardy–Weinburg equilibrium and that kind of thing. It was one of the first papers to demonstrate that it was practical to use them in population genetics and evolutionary biology. Herb had spent his entire career at LSU doing systematics and population genetics of herps. Joe suggested I go see Herb about whether I could do some molecular work in systematics. This was 1982 and it was still pretty new.

He agreed that if I wanted to do some molecular parasite systematics that I could work in his lab. He had a couple of other students working at the same time. One of them was Mike Braun who is now one of the key people at the Smithsonian molecular facility in D.C. In the end, Herb and Joe Miller became coadvisors for my dissertation work."

At that point, I asked Steve which organisms had he picked to work on. He said, "Well, that's kind of odd situation because I was interested in protozoans, and *Giardia* was pretty hot. So, at first, I thought I was going to do something on *Giardia*. I was really interested in the fact that there were all these forms of *Giardia* that had not received much attention. So, there was a guy up in Washington State who was doing a fair amount of work culturing human *Giardia*, and I wrote him. I told him of my interest in protein systematics and sent him a summary of my proposal. In order to do this kind of thing, you needed to be able to grow them in in vitro culture. He wrote back and said that he would not advise any of his students to attempt something like this for a Ph.D. because the in vitro culture of *Giardia* is too risky an undertaking. It's too much of an unknown. But, then, I was a naïve student. I was going to get these things and grow them and do my electrophoretic techniques and everything was going to be just fine. When his letter came back and I showed it to Joe Miller, his reaction was the right one. 'I guess you would be pretty smart and avoid that'. At about that time, I had to have a meeting of my advisory committee. They too agreed that it would be too difficult to do, so, *Giardia* was out the window. At the meeting, Herb spoke up and said, 'Look, I've been doing this battle for years and years, in terms of doing systematics and herps. You might have some great project in mind, but it might take a long time to get enough tissue.' I then spoke up and asked rhetorically, but isn't there something big and common that is not too hard to get that would make for an interesting project? I answered my own question. Sure, there are the ascaridoids. Most of them are big and should be easy to get. I said I would look into it and find out. So, that's how and why I began working with *Ascaris*, a matter of tissue."

I said to Steve, "It's kind of unusual to choose an organism and then ask a question for your dissertation research, isn't it?" He agreed. "But initially, I had a pretty good idea of what I wanted to do and it was pretty easy to substitute *Ascaris* for *Giardia*. Actually, a lot of the work I did at first was on *Toxocara canis* and *T. cati*. I also collected some *Baylisascaris*, *Ascaris suum*, and a few North American samples of *A. lumbricoides*. Most of this work was on some real basic geographic sampling and then some on infrapopulations to see if they were in Hardy–Weinburg equilibrium." When his work was starting in the early 1980s, parasite population genetics was in its infancy. There had been some research done using molecular procedures to look at some parasite phylogenies and systematics, but really not that much by then. One could safely assume from this that Steve Nadler was somewhat of a pioneer in this new field. He finished his Ph.D. in 1985.

He then corresponded with two people about a postdoc. One was George Cain at the University of Iowa. There was apparently some sort of internal competition for the Iowa postdoc. He and George proposed to do some work with population genetics of ascaridoids, a natural follow-up to his dissertation research, but it did not go through. The only other place he applied was the University of Massachusetts to work with Bronislaw Honigberg, who called and offered him a postdoc. At that time, he said he was thinking about going back to protozoans, so in that regard it would have been a good fit between Nadler and Honigberg since the latter was one of the premier protozoologists of his era. Steve said that he told Honigberg of his interest in systematics, which was fine with Honigberg, who was then working on the African trypanosomes and on *Trichomonas gallinae* with my first mentor, Robert Stabler at Colorado College. Honigberg actually allowed him a choice of which group on which to work. Steve selected *T. gallinae* because he was just recently married, and the African tryps would have required him to spend extended time in the field in Africa. For a variety of reasons, he stayed with Honigberg for just one year. The primary one though was that, at that time, the NIH had a payback provision with their training grants whereby if you took money for more than a year, you were obligated to provide in-kind service in the form of teaching or research for an equal amount of time or you had to pay back all the salary you received during your tenure of the grant. He said, "I would like to think I would have been able to obtain a tenure-track teaching position somewhere, but there was uncertainty in that proposition because of the job market at the time. I was reluctant to stay beyond the twelve-month grace period, and Honigberg would have run out of guaranteed funds in about eight months anyway. I didn't learn about this situation until I arrived, so I was already looking for another position when I got there.

"As it turned out, just before I left for Massachusetts to start my postdoc, I had met Mark Hafner through Herb Dessauer who had been working with some people at LSU-Baton Rouge for several years. Herb had established an excellent working relationship with the people at the Natural Science Museum in Baton Rouge and was convinced they needed a parasitologist on their staff. He also knew that Ken Corkum, a parasitologist then in the Zoology Department at LSU, was on the verge of retiring. So, he was trying to finagle me into a Museum slot, or into the Zoology Department. Mark was working on rodents, mainly pocket gophers, at the Museum. He managed the collections and simultaneously did electrophoretic work on gophers. Roger Price, an expert on mallophagan parasites at the University of Minnesota, had written Mark and asked him if he could come to the Museum and brush the study skins in the collections so he could look for lice. Mark was looking at the phylogenetics of the gophers and Roger was describing new species of lice from the gophers. Mark began to wonder if there was some way to compare the evolutionary histories of the gophers and the lice. I happened to walk in the door at the same time he was developing this interest" – clearly a serendipitous event if there ever was one!

It was at this point that Steve and Mark Hafner began talking about the possibility of studying the evolutionary histories of pocket gophers and their louse parasites. When Steve left for Massachusetts and a postdoc, Mark went to Washington, D.C. to spend a year working as a Program Officer at the National Science Foundation. By this time, Mark had submitted a proposal for the gopher/louse study and it was funded. Mark called Steve and offered him a postdoctoral position as collaborator on the grant. That is how the Hafner/Nadler connection came about.

Steve began work in 1986 at the Museum at LSU-Baton Rouge and remained until September of 1989, when he left to take a position at the Northern Illinois University. He stayed there for six years when he was offered a job in the Department of Nematology at the University of California-Davis, where he is presently employed. Looking at Steve's career reveals a broad-spectrum phylogeneticist, systematist, and population geneticist. He has worked with a wide variety of protozoans, helminths, and ectoparasites throughout his career and has made his mark with each of these groups. He has had exceptional experience with an equally wide spectrum of parasitologists, almost all of whom approached their science in a different, but interesting, way. Some folks I know think of systematics as being rather boring and somewhat dry as a discipline. By default, many of those with an interest in the subject are frequently considered in the same way. However, I would have to say that Steve Nadler is one of the most broadly trained parasitologists of all those whom I have had the opportunity of interviewing for this book. I said earlier that I had a feeling about him when I heard his first paper at a national meeting in Vancouver, British Columbia, and that I thought 'he was going places'. I was not mistaken.

JIM OLIVER

Figure 15. Jim Oliver, Professor, Georgia Southern University, Statesboro, Georgia

One of the best tick people in the world is James H. Oliver, Jr., at Georgia Southern University in Statesboro, Georgia. I met Jim for the first time in 1975 in Bowling Green, Kentucky, at Western Kentucky University, and under some very unusual circumstances. It was at an annual meeting of the Southeastern Society of Parasitologists. In those days, elections of officers took place at the annual business meetings rather than by mail. Jim Oliver and I were the nominees for President that year. Tick biology was not a big thing for me at the time, so I really did not know Jim from the proverbial Adam. After the votes were cast and the ballots counted by the appointed tellers, we were all informed that the election had ended in a tie. So, what to do? Well, the then current officers got together and announced that one of us would serve that year and the other would be President the following year. Jim recalled that there was a coin toss. He won and became President and I was to do it in 1976–77. That was my first encounter with Jim. After that, we crossed paths many times and became good friends as the years passed. When I became Editor of the *Journal of Parasitology*, I decided I wanted to have two Associates handle ectoparasites, one for aquatic critters and the other for terrestrial ones. I asked Jim if he would help with the latter group and he said yes. He's been with me throughout my tenure as Editor, and has done a superlative job, I think in part because he knows just about everyone in the world who does anything with ticks.

When I signed the contract with Cambridge to write the book, I knew I wanted to do an essay on Lyme disease. Jim Oliver is an expert on the problem here in the southeastern United States. I asked if he would be willing to have an extended dialogue with me regarding Lyme disease and he agreed. We were supposed to consummate our interview at the 2005 annual meeting of the American Society of Parasitologists in Mobile, Alabama, but, as I noted elsewhere, Hurricane Dennis blew us out of town before we could meet.

Jim and his lovely wife, Sue, have a cabin in the small community of Black Mountain, North Carolina, just east of Asheville. I knew he and Sue were heading up that way in the late summer, so I made arrangements to talk with him up there. It was a beautiful drive from Winston-Salem up into those 'hills'. When I arrived in Black Mountain, he drove down to meet me and I followed him back. It's a good thing we did it this way, because I am certain I would have been quickly lost driving all of those 'winding' roads. Their cabin is a beautiful place, in a magnificent setting. It is tucked way back in the woods, not only giving them lots of privacy, but a fantastic and panoramic view of the surrounding mountains. Sue made us some coffee, and then Jim and I retired to his study where we were to spend the rest of the morning talking about him and some of his work, but primarily that dealing with Lyme disease.

Jim Oliver is a life-long southerner, having been born in 1931 in Augusta, Georgia (home of the Master's golf tournament), although his family actually lived in Waynesboro about thirty miles away and that is where he grew up. He told me that he was active in sports, serving as captain of both his football and basketball teams. He said that he graduated in the eleventh grade, but that he could have (and should have, he lamented) stayed in high school and played another year. He chose to leave, but said it was a mistake because he was too young and immature to venture into the world at his young age. He also said that, at the time, he was certainly not academically inclined. His world was sports and that is all he really wanted to do, "get out and run."

When he left high school, he headed for Georgia Teachers College, now Georgia Southern University, where he has been a faculty member for a good long time. He said he was naïve enough at the time to believe he could play 'big time' college football and basketball. However, the year was 1948 and there were a lot of ex-GIs returning home and to school about then and, he said, "They were big," too big for a scrawny kid of 145 pounds from Waynesboro, Georgia. He had hoped to stay a couple of years in Statesboro and move on up to the University of

Georgia. He told me that after the first two or three weeks at a football summer training camp he discovered he was not cut out to play any more ball. The heart was still in it, but he recognized that he was simply too small.

His initial experience at College was not a good one. After the first quarter, he was placed on academic probation. He cut a lot of classes and didn't do very well at all. He said it is an ironic feeling because he now holds a distinguished professorship chair at Georgia Southern. As he put it, "One never knows where one is going to end up."

I then remarked, "You must have been turned around at something at this point or by someone. How did that happen?" Jim replied, "Well, I just kind of bounced around for the first year and a half in college. I even went up to the University of Georgia [UGA] for one quarter, thinking that I might become a veterinarian. I always liked animals." He told me that when he matriculated at UGA, everyone was required to take two years of physical education and that the only class available that particular quarter was in weight lifting. He said he was in the crowded gym one day waiting for a spot to do his work and noticed the boxing team working out in the next room. The boxing coach saw him watching and invited him over to spar for a while with some of the kids on the team. He had finally found his niche in collegiate sports, because he eventually went on to win the lightweight division of the Georgia's Golden Gloves championship in his first year of boxing. He had thought about competing regionally, but his family was against it. So, at the end of one quarter at UGA, he headed back to Statesville where he re-entered Georgia (Southern) College.

This time, though, he was ready. He began to study, came off probation, and the next quarter he made the Dean's list (for those who may not know, the Dean's list is a way of recognizing students for excellence in academics). He had become very interested in biology by then. There were three biology professors in the Department and two of them were highly encouraging. They gave him a job as an assistant to help in the labs and collect specimens to use in teaching. He said they were really good about interacting with their honors students, even to the point of taking them to professional meetings. He soon learned that one could even go on to graduate school in biology, and he made up his mind to do it. So, that's what he did. He applied at Florida State University where he was admitted into the Zoology Department with a graduate assistantship.

I had read his CV and noticed he had coauthored a paper with one of my favorite people down there, Robert (Bob) Short. I said, "You apparently began working on your Master's degree with Bob Short, didn't you?" He responded, "That's right." I then asked, "How did you get with Bob and parasitology?" He said, "Well, I thought when I went there that I would emphasize ornithology, because that is what I had done at Georgia (Southern) Teachers College. In fact, while I was still at Statesboro, one of the professors and I had actually published my first scientific paper on the nesting behavior of brown-headed nuthatches and bluebirds. However, the chemistry between the ornithologist at Florida State and me was just not right. We didn't mesh. I had also always been interested in animal relationships, in particular with symbioses, parasitism, and so forth. I was fascinated in how animals and plants use each other, and adapt to each other, and develop. So, I took a course in parasitology that Bob Short was teaching and I was hooked. He was such a wonderful teacher. He was always so well organized and enthusiastic. I talked to Bob about doing a Master's degree with him and he was encouraging, so that's what I ended up doing."

After finishing his degree work at Florida State, he left Tallahassee. He told me that he had been receiving student deferments from the Army up to that time and the draft was hanging over his head. In addition, by that time, the Korean War had started. He decided to void his deferment, allow himself to be drafted, serve his two years, and then use the GI Bill to support him while he pursued his Ph.D. So, he checked with his draft board and they told him he would be called in about four months. While at Florida State, he had met Don Menzies who had been to Wood's Hole the previous summer. Menzies suggested that the two of them go up to see if they couldn't get a job for the coming summer. The two of them took off for Cape Cod and Wood's Hole, where they spent the summer of 1954. It was a great experience.

He hitchhiked back home and was told to report to Fort Jackson, South Carolina, for induction into the Army. From there, they sent him over to Fort Gordon, Georgia, for basic training. After basic, everyone was shipped out to different bases and forts around the country. Jim said, "My orders were to head for Fort Dietrich in Maryland. Well, I had no idea what this place was, so I asked my commanding officer to check on it. He came back and informed me that I had been assigned something very special. This place is a top secret, biological warfare facility."

Jim continued, "When I got there, I found it was a perfect situation for me. In the barracks where I slept, there were about twenty of us draftees, fifteen of whom already had their Ph.D.s. It was primarily a civilian laboratory, with 500 troops there and about 5000 non-military personnel. We played Army a little bit and had to wear our uniforms while on duty. But we were really given a lot of slack, and our performance was based on our science contribution. This is where I became interested in arthropod vectors of parasitic diseases. When they were assigning us to the different branches at Fort Dietrich, they gave me a choice between virology, bacteriology, pathology, and entomology. I chose the latter and was introduced to ticks, mosquitoes, fleas, and lice, things that were really important as vectors. That's where I first met Willy Burgdorfer, a Swiss scientist who had just come to the United States. He was at the U.S. Public Health Laboratory in Hamilton, Montana. Willy, of course, is the person who found the etiologic agent of Lyme disease [and for whom *Borrelia burgdorferi*, the causative agent of Lyme disease, was later to be named]."

About a mile away from the base was a women's school, rather small in terms of the student number, named Hood College. That is where he was to meet Sue Shuster, who was to become his wife. She was an honors student at Hood, and an honorary marshal. Girls in the latter group would be assigned to special duties around the school during the year. One evening, she was given the task of setting up the projector and screen for a special lecture, which Jim attended. Jim said, "That's when I first saw and met her. Because the soldiers at Fort Dietrich were all educated, the President of the College more or less opened up the College to us. We had library privileges, access to their swimming pool, use of their tennis courts, and so forth."

I asked, "When you finished your two years in the Army, how did you decide where you wanted to go?" Jim said, "Well, there was some serendipity here. Lloyd Roseboom offered me an assistantship at the Johns Hopkins School of Hygiene and Public Health, but I was a little worried this would be too medically oriented. I applied to go to the summer Institute of Acarology at the University of Maryland, which had been formed not many years previous by George Horton, who had been at Duke University. While still at Fort Dietrich, I talked to my military and civilian bosses about going there. I told them that no one in my group knew anything about acarology and that it would be to their benefit, and mine, if I could attend the Institute. They thought that was a good idea, so orders were cut that allowed me to participate. When I came back, I became sort of the resident tick expert. That's when, for example, I first read about acquired immunity to ticks. I did a lot of research on ticks and tick-vectored infectious agents. A problem here though was that later, when doing my dissertation, I tried to get a hold of some of my old notes, but they wouldn't give them to me because they were all classified. They claimed my secrecy clearance had expired by then, and I could not be given access to even my own notes. The secrecy 'stuff' was overwhelming at Fort Dietrich. I was told one year when I went home for Christmas leave that, if I got sick, I was not to check into a local hospital, but was to report back to Maryland where they would take care of me. Out loud I wondered, why? They said most likely that blood would be taken. If they screen your antibody profile, they could tell what diseases we were working on at Dietrich."

His initial decision was to go to Hopkins to work on his Ph.D., which was perfect because Sue had one more year at Hood. If Sue had left Hood without finishing, she was under an obligation to pay back three years of her scholarship money. So that's what they decided to do, him to Hopkins and Sue to finish at Hood before marrying.

However, during the year Jim discovered that Hopkins was too medically oriented, even though there was a good group of parasitologists there, including Clark Read, Fred Bang, and Lloyd Roseboom. But Jim wanted a more liberal arts sort of atmosphere. They also wanted him to take some more chemistry, and that wasn't for him either. In fact, they said to him, "If you want to stay here, you *will* [his emphasis] go out and get these chemistry courses." He continued, "In fact, they lined up *a lot* [his emphasis] of chemistry courses for me to take." I can personally recall this time because it was the beginning of the era in parasitology with a huge emphasis on biochemistry and physiology. Jim was frustrated by some other requirements they placed on him as well. They wanted him to repeat histology and general parasitology, both of which he had taken at Florida State, and both from outstanding professors. He did repeat the histology course, but he thought it to be far poorer in quality than the one he had taken at Florida State. Chuckling, he said, "And I let them know what I thought about the level of instruction in histology at Hopkins, which put me at 'big-time' odds with the person that taught histology."

Jim then remarked, "It just wasn't a good fit at Hopkins for me. So, when Sue graduated from Hood, we got married and headed west to the University of Kansas [KU], which had an excellent entomology program. In addition, one of Lloyd Roseboom's former students, Ralph Barr, was a medical entomology professor at KU, so I knew I would be in good hands if I went there. Lloyd helped me make contact with Ralph and arranged for me to work with him on mosquitoes. The strange thing was that when I got to KU, I had had no formal training in entomology, except the little bit I received in the Army in the short acarology thing I had done at the Institute that summer I was still in the Army. As a result, I had to take two or three undergraduate entomology courses when I got there. Then a new problem developed. Ralph left at the end of my first year at KU and went to California at Davis. He wanted me to go with him, but I decided to stay at KU. I ended up working with

Joe Camin, an acarologist. Joe was an expert on parasitic mites, so that is where my research took me for the Ph.D. degree."

Jim said he didn't do much with ticks until after his Ph.D. In his last year at KU, he had become interested in sex determination in mites, ticks, and other animals as well. Some of the mites with which he was working were parthenogenetic, but produced haploid males and diploid females. Since he wanted to learn some more about cytogenetics, the best person with whom to work was M. J. D. White at Melbourne University in Australia, so that's where he applied. He had also applied for a job as Assistant Professor at the University of California-Berkeley and at Illinois State University at Normal. Jim said, "It was a good time in history because I was offered all three options. I ended up turning down the job at Illinois even though they offered me $4000 more there than at Berkeley. I figured Berkeley would give me better professional leverage than at Illinois even though there was a substantial difference in salary. Accordingly, I approached the Chairman of Parasitology and Entomology, Ray Smith, at Berkeley and described for him the opportunity to go to Melbourne and study for a year. I told him that if I had no choice, I would begin the job at Berkeley if necessary, but I really wanted to exploit the Australian situation. I told him that I would come back to Berkeley the next year at the same salary that he was offering to come now. Would he let me do it?" Jim said that Ray laughed and replied, "You are the damnedest guy. Here you are, you haven't even shown up to work on campus and you are already trying to make deals with me!" Jim continued, "He said, 'Go to Australia and learn all you can, and have a good time, then we'll see you next year', and that's what I did."

Australia was great. He said, "With the Ph.D., I had status. I was not treated like a graduate student. For a biologist to go to Australia, it's just absolutely fascinating because everything is interesting, plants and animals, and the Great Barrier Reef with its marine life. I kept looking at myself like I was a student and, in fact, I still do. Michael White was an eccentric old Englishman who had migrated out to Australia. He knew all the famous geneticists. He was a friend of all of them, from Dobzhansky to Stern, and had interesting stories about all of them. This is when I started working on ticks. It was so easy to get them. The herpetologists and ornithologists at the University would bring me their study animals and allow me to collect their ticks, so I had a ready source of material with which to work. And the ticks had extra large chromosomes and were much easier than mites to work on. Moreover, it was a more 'fundable' group of animals than mites. The year there was fantastic."

On his way home, he and Sue decided on an excursion to the Orient before heading for California. They made a side stop in Singapore, where they stayed with his former mentor, Ralph Barr, who was on sabbatical at the Medical School. They also managed more time with a trip into Thailand, and to Hong Kong and Japan, then home to San Francisco.

Berkeley, in 1963, was in the 'heyday' of political bustle and was bubbling over with all sorts of people and activity. He said that he really loved the place during those years. It was during the "Stop the war demonstrations and movement. It was unbelievably diverse and exciting. It was a 'heady' environment."

He said they were there from 1963 to 1968. I asked him why he left. He responded, "It was one of the hardest decisions I ever had to make. When I left Berkeley, I did not leave to come to Georgia Southern University. I left to go to the University of Georgia at Athens [UGA]. The University of Georgia had received a Center of Excellence Grant from the National Science Foundation [NSF], a grant that ran into the millions of dollars. This program was aimed at taking good biological science programs at universities and trying to move them up to the next level, to make them outstanding. The University System of Georgia and the state politicians agreed that if NSF would give this five-year grant to the University that they would continue that level of funding after the NSF pulled out. In essence, they also could go and recruit new faculty at any level they wanted. They came to Berkeley and recruited me and Mel Fuller in the Botany Department. The offer was especially attractive, for a number of reasons. They had several acarologists in the Entomology Department at Georgia, which really had joint funding from both Arts and Sciences and the College of Agriculture. I was hired as a cytogeneticist. Another reason for leaving Berkeley at the time was that all of the things that had been so charming to us the first few years began to become too excessive. It was beginning to border on anarchy in some ways. The drug culture was really getting bad. These kinds of things were particularly bad for young boys, like we had at the time. Drugs were all the way down to grammar schools in the Bay area and we didn't want our children to be imprinted by their peers with this sort of thing. Moreover, Sue's family and mine were on the East Coast and it was tough to see them as often as we would like." So, basically, everything came to a head at the same time and they left for Georgia.

Interestingly, while he was in Athens interviewing for the job, the faculty at Berkeley met and voted to promote him to Associate Professor, with tenure. The Chairman at Berkeley made every effort to dissuade him from the decision to go to UGA, but their decision had already been made and they left to head east and south.

Jim had been interviewed at Athens by the Dean of Arts and Sciences, John Eibson. However, by the time he arrived, Eibson had moved to Georgia Southern as its President. Jim and Sue were both very happy in Athens. They felt at home, as Jim put it to me. They had even purchased a house. That was a very good thing and something that was out of the question in Berkeley. Real estate in California was way too expensive, and still is, only much worse. And, finally, they were close to his relatives and to Sue's. Very soon after arriving though, Jim received a phone call from John Eibson at Georgia Southern. He said to Jim, "You know, I had a really good interview with you when I was at Athens and we have a new program opening up in the state system. The Calloway Foundation, from Lagrange, Georgia, is giving us several million dollars to establish a Distinguished Professorial Chair at most of the state campuses. We would like for you to come here as the first Fuller E. Calloway Professor." Jim said he interrupted and asked, "Are they going to have any at Athens?" Eibson responded, "There will be one or two at the University of Georgia, but don't hold out for one there, because I can tell you the first one will go to someone who has been there for a long time. Eugene Odum will more than likely get the first one at UGA. When he said that, I thought, that's right. That's what should be done." I agreed with Jim, saying, "Gene Odum was the show down there as far as the biological sciences are concerned. He got the Savannah River Ecology Laboratory going and did the same thing with the field station at Sapelo Island." Eibson continued with Jim, "I know you received your undergraduate degree here at Georgia Southern. Your coming here makes a good fit. Your record, your origin, you were a student here. It would be a natural if you were to come back." Jim said he began thinking about the cons of such a move immediately and began spouting them to Eibson. The latter responded, "Why don't you write down your needs on a piece of paper? We'll see what we can get for you." So, Jim said that he did and sent the information immediately to Eibson, who called him back a few days later. He said that he had looked at Jim's list and said that he could do everything, including almost doubling his salary. With that, Jim said, "I'm coming!"

That was in 1969, and Jim has been at Georgia Southern ever since. During these years, Jim has been exceedingly successful in terms of funding. In fact, he has not been without national research grants for the past 36 years. He also managed to establish his Institute for Arthopodology and Parasitology, and acquire the National Tick Collection from the Smithsonian Museum in Washington, D.C. in 1990.

The latter collection was greatly enhanced by Harry Hoogstral, another parasitology icon, when he donated his personal tick collection. The acquisition of this collection by Jim and Georgia Southern took some finagling. He had to obtain commitments from NIH, which actually owned the collection, and the Smithsonian, where the collection was being stored. It also required Georgia Southern to provide space to house the collection and phase in the salary of a curator, and an assistant, over a period of five years. It was a perfect match and a marvelous coup for Jim and the tick collection. In fact, Jim told me that Georgia Southern had just signed another, ten-year, Memorandum of Agreement with the Smithsonian to keep and maintain the National Tick Collection.

I did not know anything about his early years as a student or while he was at Berkeley. I have, however, watched Jim's career develop and prosper over the past thirty years, and I do know another thing about Jim. I know that John Eibson made one hell of a decision when he offered him the highly coveted Calloway Professorship at Georgia Southern back in 1969. Jim was to become a real prize in the sense that he has clearly reached the status of the greatly esteemed Gene Odum and, moreover, he continues to make his mark. He is a credit to our profession and I am honored to be able to call him friend, and colleague!

PAT LORD

Figure 16. Pat Lord, Lecturer, Wake Forest University, Winston-Salem, North Carolina

Pat Lord came to Wake Forest as a Visiting Professor in January 2000. She was hired initially as a replacement for faculty members in our core sequence who might be going on a sabbatical during a given semester. Gradually, her job has evolved to the point that she can now be considered as a 'permanent' faculty member. Indeed, her title now is that of Lecturer in Biology. I have no doubt that she will soon be carried as Senior Lecturer. Moreover, she is now also teaching a very popular course in virology, her specialty. She is a quality teacher, but no longer does bench work as a researcher. Nonetheless, I have been extremely impressed with her effort to keep up with the literature in her field. Sitting with her and listening to her in our interviews made a huge impression on me. She knows about what she speaks. Since I am not very knowledgeable about viruses, I felt it would be important to have someone like her to flesh out what I had written and to confirm what I had learned for the AIDS–yellow fever essay. It was well worth the time.

Pat's father was a U.S. Marine and that meant she was constantly on the move as she grew up. Anecdotally, Marines are always on the move, and always in the right direction; for example, when they were forced into evacuation of the Chosin Reservoir area during the Korean War, General O. P. Smith, Commander of the 1st Marine Division, was asked about the 'retreat'. His response was, "Hell, we were not retreating, we were advancing in a different direction!" Eventually, her father left the Corps and settled the family in Boone, North Carolina.

When college time came for Pat, it was to North Carolina State University in Raleigh. She said that she was already interested in the sciences when she matriculated at N.C. State and I asked when did she develop the interest? She emphatically responded, "I cannot remember not being interested in the sciences. My mom was a nurse and maybe that had something to do with it, but I was always interested in finding out how things work. I had a lot of great teachers in high school and that probably helped too. We did a lot of 'hands on' stuff." She told me that when she first arrived in Raleigh, it was with a focus on medical technology. When she began at N.C. State, there was a biology department, but she said, "there were also a lot of microbiologists in it. I also was able to do research while I was there, even at the Research Triangle Park in the National Institutes of Environmental Health Sciences facility. I worked with a veterinarian there by the name of Ethard Van Stee. We were interested in the sorts of things that might produce cancer when inhaled. He was great to work with because he let me do my own projects. It was about this time that I began shifting away from the idea of being a medical technologist to wanting to go to graduate school."

When she resolved to pursue a graduate degree, she decided on Wake Forest University's Bowman Gray School of Medicine. I asked why she chose the medical school rather than arts and science side of graduate school. She said, "In hindsight, it was probably because it was a new entity. Of course, growing up in Boone, we knew about Wake Forest as an institution. At that time, I was interested in cancer research, and in viruses, but more in the cancer end of it. I was interested in research that would help people. It seemed like medical school would be a good place for that sort of thing." I then asked with whom she worked. She responded, "I started off with Louis (Lou) Kucera, but transferred to Bill Kilpatrick's lab soon after I got there. In hindsight though, I was not mature enough to go to graduate school. I just don't think I was ready for the hard work. It did not take me very long to understand what I was doing, but being a graduate student in a medical school was hard. You had to be totally self-motivated and self-directed." Pat then made an interesting observation, saying, "Down there, at the medical school, they look at teaching as a chore rather than a privilege. The feeling among the medical students that we taught was that most of us were in graduate school because we couldn't make it into medical school."

We then got back on her biographic track instead of trying to figure out why so many people in the basic medical sciences of medical schools cannot manage their graduate students very well. I again asked, why virology? She said, "I wanted to learn more about the connection between viruses and cancer. When I came to Bowman Gray, Lou Kucera was the only one working with viruses. He was collaborating with Bill Kilpatrick because Bill was using molecular techniques for which Lou didn't have the expertise. So, I went to work with Bill whose teaching techniques were different than most in that he left you on your own to figure things out for yourself. It's interesting that he had the expertise in this area, but he was much more, 'You go figure it out'. This was in tune with Cold Spring Harbor philosophy in that you had to go and figure out how it works." I asked her with what sort of research did she become involved. "When I went to Bill's lab, I worked on human cytomegalovirus. Initially, I worked on the transcription and translation of the virus. What I eventually did was to isolate the host DNA and allow mRNA from the virus to anneal to it and then look at it with an electron microscope and map where the mRNAs are attaching. Then I came back and constructed a transcription map. In hindsight, I feel my training would have been a lot different in a liberal arts setting than in a medical school because, for example, I would have done a Master's degree before the Ph.D. In most medical schools, the Master's degree is skipped."

Having known Pat for several years and then interviewing her as I did, I personally feel she would have enjoyed an arts and science setting more because she would have had the opportunity of interaction with undergraduate students. But, then, she would not have encountered her future husband, Richard, if she hadn't taken the med school and Bowman Gray route. They met toward the end of her degree work, just when he was beginning as a first-year medical student. She said, "We were friends for a long, long time before we married and before he finished with his M.D." This raised an important question in my mind and that had to do with career choices for both of them and I asked her about this. She responded, "You are right, but it was a while later before that occurred. I'll get to that part in just a bit. When we married, Richard was still a couple of years behind me, so I had to find something to do. It turns out that Charles (Cash) McCall, a physician/researcher on the faculty at Bowman Gray, was setting up a laboratory to do molecular biology on neutrophils. I have always been amazed by him. He had always thought that neutrophils were transcriptionally active, when everyone else thought that they were just a 'bag' of enzymes that would kill a bacteria or a virus and then consume the trash. Well, as it turned out, Cash was right and the naysayers were absolutely wrong. Not only are they transcriptionally active, they are translationally active. This became his major focus of research and I found it to be really fulfilling to be a contributor."

Then, when Richard finished, he knew that he wanted to train to be a family physician. However, he also wanted to do OB [obstetrics] work, so when it came time to do his residency, they both knew that he was going to have to be in a large urban area. Pat continued, "After considering several choices, he ended up at Rush Presbyterian St. Luke's Hospital in Chicago, a good move because there were just lots of opportunities for some sort of a postdoc for me in the area as well. I ended up working on *Drosophila* genetics with Robert Storti at the University of Illinois at Chicago Medical School, doing research on muscle gene expression, but, again, not much opportunity for teaching. My experience there was good. I had lots of opportunities to be independent and train others in the lab, but did not receive much mentoring about how to set up my own lab." She then remarked, "I always tell students that I started with a virus, switched to DNA, then went to work with whole cells, and that wasn't enough, so started working with a whole organism. While I was in Storti's lab, I identified a muscle segment homeobox gene that is important in defining body wall muscles in *Drosophila*."

When Richard finished his residency, he and Pat set out on another pathway, this one so that he could practice medicine on his own, while giving Pat the simultaneous opportunity of teaching and doing research at the college/university level. They wound up in Harrisonburg, Virginia. He joined a family practice and she found a job at James Madison University, a somewhat small, but quality, state school. It was an ideal arrangement. Not long after they became settled, they realized that they could not have biological children, so they adopted their first son. It was at this point they realized that their lives had been somewhat overwhelmed with Richard being on call because of his OB work, she working full time in a demanding teaching situation, and they having to care for a new child. They began looking for other options. So, they returned to Chicago. Richard became a staff physician at Rush Hospital where he had done his residency. They loved the new setup. Pat said, "If our families had been close by, we probably would have stayed there."

It was at this point she made the career decision. She knew she had the primary care responsibility for the new baby, and that Richard's practice required him to be on duty for night calls. She also recognized that her life had become chaotic at the wrong time. At this point, she made the decision to give up the academic life and become a 'stay-at-home mom'. They then adopted a second son while in Chicago.

As luck would have it, the residency program at Rush began to collapse at about that time and Richard began making contacts in search of a new position. One of the places he interviewed was at Bowman Gray, Wake Forest University's medical school. Since both boys were about ready by then to start school, Richard suggested, in jest, that it was "time" for Pat to start thinking about getting back into the real world. So, when Richard came down to Winston-Salem, Pat lined up an interview with Herman Eure, then the Chairman of our Department of Biology here at Wake Forest. Herman made an offer that Pat could not refuse and that is her story. She's been with us ever since.

As I expressed early on, I really have a great deal of respect for Pat. I know she is not now a researcher, but she was once, and quite successful. When the time came, she had to make a choice. As a parent myself, I think it was the right one for her to make. I realize that there are other women, several in our Department, who did not make the same decision, but that is okay too, because it was the right way for them to go. Now that Pat is a full-time Lecturer in our Department, she has had the opportunity to take up the gauntlet of teaching that she willingly laid aside a few years ago, and she is making the most of it. It is interesting that the other woman I interviewed for this book, Margaret Ewing, did about the same thing. She taught when she could, but she gave it up to raise her children. Then, when the opportunity presented itself, she, like Pat, came back into academia at Oklahoma State University. I like their stories and admire them for their tenacity. They have each managed to live on both sides of the trail, successfully!

J. P. DUBEY

Figure 17. J. P. Dubey, Senior Scientist, Agricultural Research Service, U.S. Department of Agriculture, Beltsville, Maryland

Jitender Prakash Dubey is his given name, but everyone knows him as J. P. It seems that his dissertation supervisor in England could not pronounce Jitender, so his moniker became J. P. I had the pleasure of interviewing my old friend in College Park, Maryland, not far from Beltsville and the Agricultural Research Service [ARS] laboratory where he has hung his research hat for the last several years. The day before the interview, I had driven up to College Park where he had arranged for me to stay the night at a local motel. We sat for more than three hours, while I 'quizzed' him the same way I had done the other seventeen folks whose biographic sketches appear here. Frankly, I was pleased that this was the final interview for the book because I at last felt completely confident I could meet my deadline with the publisher. However, I also experienced a touch of melancholy because this marked an end to the 'fun' part of writing the book, i.e., the interview phase. I had begun the project almost exactly two years before, when I sat with Dick Seed at his home in Chapel Hill and talked all morning about his research regarding variable surface glycoproteins. As I suspected, my interview with J. P. went exceedingly well, as had all of the others. And, as in the other interviews, I learned an awful lot!

He told me at the beginning of the interview that his uncle had named him Jitener Prakash, which is Hindu for "the one who has won over all of the five senses", reflecting the priesthood heritage of his family. Unlike the religions of most other countries, a priestly identity in India is inherited, meaning that for many prior generations, J. P.'s ancestors were also priests. He told me that he was a member of the Brahmin sect, whose "job it was to teach and to preach. Both my father and grandfather were priests. Even my grandmother was a priest. So, while growing up, everyone wanted me to follow in that line." I asked if it was a disappointment to his family that he had not become a priest. He explained, "Well, to my parents, it was a major disappointment." He went on, however, to say that by taking the route that he did, he felt that it had helped his brothers and sister to move in a different direction than they might have otherwise, i.e., "out of the poverty into which we all were born. My brothers, for example, all became successful professionals, i.e., in engineering, education, etc."

J. P. was born in 1938 in a small rural village near Agra and, ironically, also near the exquisite Taj Mahal, in the northern part of India. In his acceptance speech for the WAAVP-Pfizer Award for Outstanding Contributions to Research in 1995 in Yokohama, Japan, he remarked, "In order to give you an idea of the poverty in which I grew up, the monetary value of the WAAVP-Pfizer Award I received is more than the value of my entire family holdings in India." I asked if he ever goes back to India and he replied, "I go back quite often, every three to five years, to visit my three brothers and one sister who still live there."

He continued his story, "At the age of 11, my father's older brother, my uncle, invited me to live with him and my aunt in Delhi, the capital of India. In fact, they took in all my brothers. They had no children of their own and his uncle saw this as a way of helping his brother and his children. He also brought other children from the village and gave them the same opportunity. The house was run like an orphanage. My uncle was a very progressive person and wanted all of us to receive a good education. Interestingly, he was not wealthy. He had a minor clerk's job in the railway ministry." J. P. said he was involved simply to help these children get ahead.

All of his early schooling was done in India. He said that, initially, he never had any ambition to go further than high school. He told me that he "became a veterinarian just by chance. My uncle retired when I hit high school. He had wanted me to become a typist and get a clerical job after two years of college. However, I had a friend whose father was a major in the Indian Army, and who wanted to study in a newly started veterinary school in Mhow, India. He wanted me to go there with him, and helped me get in. He also helped me financially during my first year of vet school. I studied very hard and stood first in the whole university after the first year. Then, I received a scholarship and I never had to look back in my life after that. It's the initial break that many of us need to be successful in life."

I asked him if this was when and where he became interested in parasites and parasitology. He responded, "Yes, absolutely. It was because of a teacher named Mr. H. L. Shah. He was very much involved with it. He was fascinated by the names of the parasites and their life cycles. I became very much interested in it because he promoted it so hard. We were all very young people and I think that teachers play a very important role in life. And I remember memorizing these life cycles. I was so good at it that I actually tutored my classmates, then you learn even more when you do that." At this point, I chipped in with an old saying I picked up somewhere along the way, "To teach is to learn twice", and J. P. countered by agreeing completely. He said that Shah was not a researcher, although he did have a Master's degree. Mr. Shah did not know much about the diseases caused by parasites, but he really did know about the biology of many of the organisms. His best quality was, however, that "he was a compassionate teacher."

He completed his veterinary degree in 1960, and found a job at the Indian Veterinary Research Institute [IVRI], a very large institution, which would be considered something like the Agricultural Research Center [ARS] in Beltsville, Maryland, where he is presently employed. As it turns out, the Director of the Institute was also interested in parasitology. J. P. related, "I was hired, but my job was not in the Parasitology Department because there was no vacancy. The Director appointed me to the Department of Veterinary Services, where I was placed in charge of horses."

He had been at the Institute for about a year when a most dreadful event was to occur. According to J. P., "There was a very prominent parasitology researcher who worked in the Director's lab, a very big place. Although I don't want to say anything about anyone else's miseries, this incident was actually what got me interested in the transmission of *Toxoplasma*," and we see his initial encounter with a parasite with which he was to do so much over the next 45 years or so.

It seems that the prominent researcher in the Director's lab had stolen somebody else's work. A photograph had been lifted out of another person's research paper, and then published in the *Journal of Infectious Diseases*, together with a totally baseless assertion that *Toxoplasma* was transmitted via chicken eggs. Although plagiarism was a serious issue, the paper had potentially significant public health implications, as well as a possible negative impact on the poultry industry. I asked, "Who published the photograph originally?" J. P. replied, "This person had stolen one of Jack Frenkel's pictures of *Toxoplasma* in mouse peritoneum. The photograph was very fresh in Jack's mind because he used it in almost every book chapter that he wrote. So, he immediately recognized that something was not right about the Indian researcher's study. As soon as Jack saw it, he was outraged, and contacted the editor of the *Journal of Infectious Diseases* to protest. The invalid paper came out in March 1961 and, in the following month, an apology was published." J. P. continued, "If it hadn't been for this incident, I probably would never have known about *Toxoplasma*." He added, "This episode was very bad for the reputation of IVRI. On the good side, this is when I became seriously interested in *Toxoplasma*."

J. P. applied for and received a scholarship from the Ministry of Agriculture to study protozoans and begin work on a Master's degree in the summer of 1961. He left the IVRI and went to Mathura Veterinary College, where he studied with Professor B. Pande, a famous helminthologist. He told Pande that he had a scholarship to study protozoans, and that he would like to do research on *Toxoplasma*. Pande was, however, not in 'sync' with J. P.'s aspirations and told him that he would have to include both protozoans and helminths. As J. P. remarked, "Work on *Toxoplasma* was not politically feasible at that time." Pande asked him to begin a work on the coccidians in Indian jungle cats, *Felis chaus*. He said that he "contacted some people who trapped these cats and they agreed to provide me with carcasses. I looked for both the protozoans and the helminths. I was able to find coccidians that turned out to be *Eimeria* and *Isospora* and described several species that I thought were new at that time. I also came in contact with Professor Norman Levine who helped me to describe these coccidians. At that time, I never realized that I would be working with the coccidians of cats the rest of my life!"

After he received the Master's degree, he decided that he wanted to do a Ph.D. too. By that time, he had become interested in soil amoebae, primarily *Entamoeba histolytica*, and in working with Dr. Singh at the Central Drug Research Institute [CDRI]. Singh had been J. P.'s External Examiner for his Master's degree; he asked J. P. to join his lab because he was so impressed with his research. As a result, he traveled to Singh's lab at CDRI to work on amoebae and his Ph.D. In the meantime, he had applied for a Commonwealth Scholarship to study in England. This was a highly prestigious prize because only one veterinarian from India was selected annually for this scholarship. J. P. competed and, of all the Indian applications that year, only his was successful.

I asked if he and Niti were married by then (1964) and he said, "I was not married at this time. When I won the Scholarship, my name appeared in a very prominent paper, *The Times of India*. My photo also appeared there with the Ambassador from Great Britain. What turned out to be my future mother-in-law saw the picture. Her family was very much interested in knowing if their daughter and I could be married, so I became engaged to Niti. Their incentive was that she could leave India and go to the U.K. So, I got married within a week. The whole thing happened within seven days." At that point, I exclaimed, "Holy cow!" J. P. smiled and reacted, "Holy cow, yes! I headed for England, but had to leave my new spouse behind for a year."

When he had gone for his Commonwealth interview, he was certain he wanted to work on *Toxoplasma*. He told me that he also wanted to work with someone who was an expert with this parasite so that he could benefit from their experience, a very smart move as it turned out. He had already identified the lab of Colin Beattie and Jack Beverley at the University of Sheffield School of Medicine. These two had done some excellent research with *Toxoplasma* and abortions in sheep, and J. P. was impressed. The transmission of *Toxoplasma*, especially in sheep, was to become his primary focus over the next several years and ultimately his penchant for this parasite was to pay great dividends.

In 1992, I visited the University of Sheffield to talk with Peter Calow about a book that Jackie Fernandez and I were writing for the now defunct publishing house of Chapman and Hall. While walking through the zoology building on my visit, I noticed a memorial portrait dedicated to Hans Krebs, who had escaped Germany in the late 1930s because of Nazi persecution and migrated to England where he settled in as Professor of Physiology at the University of Sheffield. J. P. told me that after Krebs had worked out the TCA cycle (for which he later won a Nobel Prize), he sent the paper on it to *Nature*, who proceeded to reject it! J. P. admonished, "I learned a very good lesson from his experience. Now, I am never disappointed. A rejection is only one person's opinion. You should always believe in your own work."

He stayed at the University of Sheffield, successfully completing his Ph.D. after three years of work on *Toxoplasma*. Niti joined him after the first year. I then asked, "How did you get to the U.S.?" He said that after completing his Ph.D., he was required by the terms of his Commonwealth Scholarship to return to India for a year. However, he said, "While I was at Sheffield, I wrote several letters of inquiry regarding a postdoc. One of them was sent to Jack Frenkel, who was then located at the University of Kansas Medical Center in Kansas City, Kansas. Jack responded that he had a place for me in his lab when I was ready."

I told J. P. that I had grown up in Kansas and, as a native Kansan, I naturally asked him how he liked living there. Diplomatically, he replied, "Well, you know, having been raised in India, everything was different in Kansas [a radical understatement!]. It wasn't too much of a shock though because I had gone to school in England. However, since Niti was not with me for several months, it was difficult at first. I had arrived in early July, so it was hot [another understatement!]. It was the experience of my life up to then in having the opportunity of working in Jack's lab."

I asked how long he was with Jack and he replied it was for five years. This was one of the most productive times of his life according to J. P. He said, "I had searched for a job after a few years working in Jack's lab, but couldn't find anything. Then, I had a lucky break. I had applied for many faculty positions. However, in the vet school at Ohio State University, there was a long tradition of working with *Toxoplasma*. They were very much interested in what I was doing in Jack Frenkel's lab, which helped me to get a job there, and I stayed for five years. I next went to the Department of Veterinary Science at Montana State University in Bozeman, and stayed there for another five years. It was a very unusual faculty arrangement at Montana State because there is so little teaching. It's mostly research. They not only provided me with good lab space, I was even given a technician to help me." The stay at Montana proved very useful since, among other things, he was able to get two books written and published. He told me, "They had a very good library facility and provided me plenty of secretarial help. They gave me funds for my research. Actually, Montana State had a separate fund dedicated to support publication, so that whenever anyone wanted to publish a paper, the University would pay for it." He then laughed and said, "Believe me, I really used that fund. But they don't have one anymore." Considering that J. P. published some 150 papers (not counting the two books) over his five-year stay, it is no wonder Montana State's publication fund went 'bust'.

In 1982, he transferred to the ARS facility in Beltsville, Maryland, and has remained there ever since. Nonetheless, Montana State retained lab space and a technician for J. P. for a whole year in hopes he would change his mind and return. During a visit to Bozeman in 1981, Ron Fayer had told J. P. that Harry Herlich was going to retire, that a slot would soon be open, and encouraged him apply for it. When Herlich retired, Ron would become the Parasitology Institute Director and he wanted someone to take care of the coccidian lab. J. P. then quipped, "I hope to stay at ARS and continue working for several more years. I feel we are making great progress and are still very productive."

J. P.'s personal story is a fascinating one and I'm pleased he was willing to share it with me. I believe his personal and scientific achievements indicate that he is in possession of not only a wonderful intellect, but a huge amount of grit as well!

REFERENCES

Bartlett, M. S. 1960. *Stochastic Population Models in Ecology and Epidemiology*. London: Methuen.
Beale, G. 1954. *The Genetics of* Paramecium aurelia. New York: Cambridge University Press.
Chandler, A. C. 1949. *Introduction to Parasitology*. New York: Wiley.
Craig, C. F. and E. C. Faust. 1949. *Clinical Parasitology*. Philadelphia: Lea and Febiger.
de Kruif, P. 1926. *The Microbe Hunters*. New York: Harcourt Brace.
Esch, G. W. 2004. *Parasites, People and Places: Some Essays on Field Parasitology*. Cambridge: Cambridge University Press.
Ewing, S. 2001. *Wendell Krull: Trematodes and Naturalists*. Stillwater: Oklahoma State University, College of Veterinary Medicine.
Ewing and Todd 1961a from p. 75.
Ewing and Todd 1961b from p. 75.
Ewing, S. and A. C. Todd. 1961a. Metastrongylosis in the field: species and sex ratios of the parasites, preferential locations in respiratory apparatus of the host, and concomitant lesions. *American Journal of Veterinary Research* **22**: 606–609.
Ewing. S. and A. C. Todd. 1961b. Association among members of the genus *Metastrongylus* (Nematoda: Metastrongylidae). *American Journal of Veterinary Research* **22**: 1077–1080.

Honigsbaum, M. 2001. *The Fever Trail: In Search of the Cure for Malaria*. London: Macmillan.

Honigsbaum, M. 2004. *Valverde's Gold*. New York: Macmillan.

Spindler 1934 from p. 75.

Spindler, A. 1934. The incidence of worm parasites in swine in the southern United States. *Proceedings of the Helminthological Society of Washington* 1: 40–42.

Stoll, N. R. 1947. On endemic hookworm: where do we stand today? *Experimental Parasitology*, 12: 241–252.

African trypanosomes and their VSGs

Time has fallen asleep in the afternoon sun.

A Boy's Dream, Alexander Smith (1830–1867)

Lyons (1992) describes sleeping sickness as a "classical disease of the savannah and transitional savannah around river systems where tsetse flies and people are forced into close contact by their shared need for water, especially during dry seasons. It has also been labeled a disease of frontier zones as it more generally occurs on the edges of human settlements where transition from the sylvan, or wild, ecosystem to the domesticated ecosystem of man is in progress." One can thus easily view sleeping sickness as a 'pastoral' problem, which it is. It is certainly not an urban disease.

However, when most of us think of a pastoral setting, we probably conjure up a vision of rolling countrysides, green pastures, and a certain sort of tranquility, if you will. This is as true in sub-Saharan Africa as it is in Kansas, where I grew up as a boy. However, lurking in many of these peaceful African locales are tsetse flies and, as a result, anything but tranquility for humans and their domesticated ungulates. The problem is trypanosomiasis: more specifically, sleeping sickness for humans and nagana for the latter.

When we speak of trypanosomiasis, we are actually talking about a wide range of diseases in mammals caused by several species of *Trypanosoma*. Included among these diseases are surra, which affects horses, mules, and camels in India and North Africa. There is nagana (a Zulu word, meaning 'in low or depressed spirits'), in African cattle, oxen, horses, etc. Sleeping sickness is, of course, a serious disease and potent problem for humans in Africa. Then, there is Chagas disease, a trypanosome infection for a wide range of mammals, including humans, in the western hemisphere.

While there are many variations on the life cycle and morphological characteristics for each of the etiological agents of these diseases, they all have several features in common. First, they are all vectored by blood-sucking insects (although the etiological agent for surra is mechanically transmitted). Second, they are all dangerous and capable of inflicting mortality for at least some of their vertebrate victims. Third, the parasites are extracellular (except for *T. cruzi*) and they produce diseases of various tissues, including blood, lymph, and cerebrospinal fluid. Fourth, they are all problems of long-standing consequence. Thus, they certainly are not emerging diseases, at least in the classical sense, although Dick Seed refers to African trypanosomiasis as a 're-emerging' disease. Fifth, except for surra, they are zoonotic and benign in most reservoir hosts, but potentially lethal to domestic animals and humans.

Whereas several of the disease-producing species of *Trypanosoma* cause important economic problems where they occur, none has created the negative economic and health impact of *T. brucei brucei, T. b. rhodesiense*, and *T. b. gambiense* in Africa, with the former causing nagana in cattle, sheep, oxen, etc., and the latter two causing sleeping sickness in humans. For example, it is estimated that approximately 25% (1×10^7 km^2) of the landscape in sub-Saharan Africa is unsuitable for raising livestock because of nagana. Thus, it has had, and continues to have, a devastating effect on many protein-deprived African natives. In contrast, it has been a boon to the conservationists striving to maintain the large herds of African ungulates.

Sleeping sickness in humans has claimed many hundreds of thousands of lives over the centuries and continues to wreak havoc in many parts of Africa wherever the dread tsetse fly is distributed. The disease was largely controlled by the 1970s and early 1980s through active surveillance, the use of insecticides, and chemotherapy. During this time interval, the estimated total number of annual cases was about 25 000. However, I recently saw one account of the disease that estimated a current annual mortality of 50 000, with some 300 000–500 000 individuals infected each year. Moreover, this number expands periodically when epidemic outbreaks occur, usually coinciding with some sort of social unrest, e.g., war, massive migration, breakdown in the public health infrastructure, etc.

Gabriel Valentin, a Swiss physician/microscopist, described the first trypanosome in 1841. Apparently, Valentin stumbled across the organism while examining blood smears from a trout (although Keith Vickerman says it was from a frog). In 1843, David Gruby, a French physician, observed tryps in the blood of frogs and named them *Trypanosoma sanguinis*. The etymology for *Trypanosoma* is a Greek word, 'trupanon', which means auger, or corkscrew, reflecting the motion of the organism as it moves.

Whereas other tryps were seen and described by other investigators over the following years, Griffith Evans, a veterinary officer working in the Punjab area of India, made the next major breakthrough in 1880. Evans observed what he thought to be trypanosomes in the blood of horses and mules afflicted with a fatal disease called surra. Natives in the area had known of the disease for many generations and believed it to be transmitted via blood-sucking tabanid flies, an idea also subscribed to by Evans. I can find no mention that Manson's discovery of the insect transmission for *Wuchereria bancrofti* in 1878 had any impact on Evans' notion regarding the transmission of the etiological agent of surra. Originally misidentified as *Spirochaeta evansi*, its true character was clarified by Edgar Crookshank of King's College Hospital (London), who carefully examined a series of blood films sent by Evans. We, of course, now know this parasite as *Trypanosoma evansi*.

In 1894, there was an outbreak of nagana in Natal, South Africa. David Bruce, a surgeon-major in the British army, had made a reputation for himself by discovering the bacterium causing a disease known as Malta fever, soon to be called brucellosis in recognition of his contributions to the disease's etiology. Bruce and Sir W. Hely-Hutchison, the Governor of Natal (and Governor of Malta when Bruce worked there) colluded to have the former brought to the South African province to study nagana. Accompanied by his wife, Mary, who was also his always-present assistant, they traveled to South Africa by ship. Then, by mule and ox-wagon for 28 days, they made their way to Ubombo in Northern Zululand.

They immediately began examination of diseased cattle. Bruce quickly ruled out bacteria as the cause of nagana. Paul Erhlich had made popular the study of stained blood films by that time, so Bruce refocused his efforts using this relatively new tool. In the blood of cattle with nagana, he frequently observed what he eventually referred to as an "infusorial parasite"; he was initially uncertain of its identity.

At the time, there was also a serious problem in certain parts of Africa with what was referred to as 'fly-disease', which had a fatal impact on cattle, horses, oxen, and other livestock animals. From the time of the earliest explorers and traders up to the travels of

Livingstone, fly-disease posed a constant threat to domesticated live-stock, and the tsetse was well known for its association with the problem. The insect was also known to occur in a sometimes wide, and other times narrow, swath of sub-Saharan Africa from northwest to southeast, always closely identified with lakes and streams and never higher in altitude than about 3000 feet above sea level. Bruce knew that fly-disease was present on the lower slopes of Ubombo, but not up high where he was then ensconced. So, on finding his 'infusorial parasites', he decided on a simple experiment. He sent several oxen and dogs down into the fly-belt for a few hours, and then brought them back. Several days later, he checked their blood and all were infected with his 'infusorial' parasite, which by then he knew to be a trypanosome, and the same one he had observed earlier in association with nagana. By then, Bruce believed that nagana and 'fly-disease' were the same thing, but he needed more evidence to support his hypothesis.

The discovery of his 'trypanosome' and its transmission characteristics opened the door for a whole series of experiments in which Bruce was eventually to show (1): that the bite of a tsetse fly, by itself, could not cause either nagana or fly-disease; (2) that the fly can carry the trypanosomes from one host to another; and (3) that horses could be infected directly with the bite of an infected tsetse fly. Bruce then wondered how the tsetse flies were infected, i.e., what was the source of trypanosomes in the tsetse fly? The African natives knew that if there are no game animals in a given locale, there also is no nagana in cattle or horses. So, he traveled down to the lower Ubombo where he shot and bled buffaloes, wildebeests, bushbucks, and hyenas. Blood from these animals was inoculated into clean and healthy dogs, all of which became infected with the trypanosome, his 'infusorial' parasite. With this discovery, he had conclusively demonstrated that nagana and 'fly-disease' were one and the same, and that the disease was caused by trypansome parasites transmitted by an insect vector, the first time an insect had been implicated in the transmission of a protozoan parasite.

Even though sleeping sickness was well known in Africa, and that it was deadly, it was not until 1901 that a trypanosome was observed in a human. An experienced tropical disease expert, J. E. Dutton, happened to be in Bathurst, The Gambia, studying local mosquitoes and was called upon by a local physician to identify an organism taken from a European sailor. Dutton sent blood films to Alfonse Laveran (of malaria discovery fame), who confirmed his identification. Since they were unlike any trypanosomes he had seen previously, Dutton named them *Trypanosoma gambiense*.

In 1901, a severe epidemic of sleeping sickness swept Uganda, with thousands of people dying from the disease. The Royal Society decided to send a commission to Uganda to study the problem and determine its cause if possible. The commission included the experienced George Low, a protégé of Patrick Manson, and the person who had played such a significant role in working through the life cycle of *Wuchereria bancrofti*. The second was the crusty and arrogant Cuthbert Christy, who had spent a great deal of time in the field working in Africa, India, and Central and South America. The third participant was the very young Aldo Castellani, another protégé of Patrick Manson, but a novice with tropical diseases.

The three commissioners knew nothing about sleeping sickness, although Manson had speculated it might be caused by *Dipetolonema perstans*, a filarial worm. On arriving in Entebbe, Uganda, in July 1902, Low and Castellani established a laboratory immediately, while Christy headed into the field. The latter spent the next six months wandering through parts of Uganda and the Congo where he examined the distribution of sleeping sickness and *D. perstans*. His findings yielded some interesting information. He found, for example, that the spatial distribution of *D. perstans* and sleeping sickness did not match, and that neither mosquitoes nor the filarial worms were associated with sleeping sickness. He also discovered that enlarged cervical lymph nodes were commonly seen in sleeping sickness patients, a condition soon to become known as Winterbottom's sign. It is interesting to note, as an aside, that native Africans and slave traders of the sixteenth to eighteenth centuries had already made a connection between the enlarged cervical lymph nodes and sleeping sickness. In fact, slaves who developed the enlarged lymph nodes while in transit via ships to North America and elsewhere were routinely thrown overboard because the traders knew many of these people were destined for early death. The slaves were discarded in this cruel manner because (disgustingly) they were considered nothing more than 'chattel'.

Meanwhile, Low and Castellani set up shop in a small, native hut in Entebbe. After about six months, Low left to return home and instructed Christy to do the same on his return to Entebbe, which he did. Castellani stayed and continued working on sleeping sickness. Before the arrival of a new commission in March 1903, Castellani made two discoveries, one of not much import, but the other of great significance. In the first case, he had managed to isolate a culture of streptococcus from the heart of a person suffering from sleeping sickness, and thought this might be the cause of the disease. Within a short period of time, he was proven wrong. In the second case, he managed to locate some trypanosomes in the cerebrospinal fluid of a person with sleeping sickness. However, there is serious doubt he made any connection with the disease, at least not until the second commission arrived in March 1903. The new commission included David Bruce and D. N. Nabarro.

All of us in the business of science have an ego; it is almost a required characteristic. In some, however, it is much larger than in others. The latter situation was the case for Aldo Castellani. When Bruce and Nabarro arrived in Entebbe, Castellani told Bruce that he had made two important discoveries, but that he would reveal them only to Bruce and only on the condition that Nabarro not be told. Finally, Castellani demanded that credit for finding trypanosomes in the cerebrospinal fluid would go only to himself, and not to Bruce. The latter agreed, presumably one would think, for noble reasons, i.e., Bruce was interested in solving the problem and not in who would receive the credit. The two then worked 'feverishly' to confirm Castellani's earlier discovery of tryps in the cerebrospinal fluid. After examining the fluid from 34 patients, they found the parasite in 20. In 12 controls, nothing was found. At last, the human sleeping sickness riddle was resolved.

Castellani left for England about a month after Bruce's arrival and published his findings as a report to the Royal Society. Subsequently, however, in his own report to the Royal Society, Bruce revealed the deal he had made with Castellani, with an important note indicating the nature of the agreement. Bruce thus wrote, "At the time of the arrival of the commission, he (Dr. Castellani) did not consider that this trypanosome had any causal relationship to the disease, but thought it was an accidental concomitant like *Filaria perstans* . . . As Dr. Castellani has not entered into any detail respecting these matters in his reports, it is thought advisable to supplement his account with the above, as the history of the discovery of the cause of any important disease must always be of interest."

Before returning to England and filing the report just cited, Bruce and Nabarro succeeded in confirming the etiology of sleeping sickness. First, they definitively associated trypanosomes in the cerebrospinal fluid with the disease. Then, Bruce's experience with nagana, fly-disease, etc., convinced him that the tsetse fly was the insect vector for the trypanosome. So, they contacted numerous missionaries and government employees throughout Uganda and asked them to collect as many tsetse flies as they could and send them to Entebbe. Their idea was to plot the distribution of the insects and sleeping sickness to see if they overlapped in distribution. Not surprisingly, there was clear sympatry.

Over the next several years, other discoveries were made, including a new trypanosome infecting humans, i.e., *Trypanosoma rhodesiense*, by J. W. W. Stephens (although most would now accept the notion that *T. b. brucei* and *T. b rhodesiense* are the same subspecies). With the identity of the vector known, various suggestions were made with respect to the tsetse's control, even eradication. In some areas there were successes, but these were limited and very localized. One of the things they also discovered was the ecological complexity involved in the distribution of the tsetse fly and sleeping sickness, and how much negative impact colonization by Europeans had on the disease in native Africans. The disrupting influence of racial and cultural exploitation was manifest in any number of ways and continues to the present time.

Each year, I teach my parasitology course. Naturally, I give my students several lectures on trypanosomiasis, including sleeping sickness, and I tell them something about the discoveries made by David Bruce and his colleagues so many years ago. Something else I always tell them is one of the reasons why the African tryps have been, and continue to be, so successful more than a hundred years after they were first identified. First, there are several drugs used in treating trypanosomasis, but they are toxic and, therefore, dangerous. Second, there is what I consider the parasites' hugely 'brazen' behavior in challenging the highly sophisticated immune system of their hosts.

Some parasites suppress the host's immunity (so do the African tryps, but not immediately). Others hide from the immune system by sequestration. There are still other successful methods used for avoiding or evading host immunity. The African trypanosome strategy is to 'show the flag', instead of hiding from the immune system, or suppressing it. It is almost like they pronounce their presence and then invite the host to take its best shot, which is exactly what the host does. The African trypanosomes possess genes necessary for producing a series (or repertoire) of what are called variant (variable) antigen types, or VATs. The VATs are expressed by the tryps as variant surface glycoproteins, or VSGs. The VSG is, quite simply, a surface coat that covers the plasmalemma on the blood form of the trypanosome. In response, the host produces an antibody that attacks immunodominant VSG, or surface coat. Parasites with this particular VAT are knocked out and the tryp population declines. However, by the time the old population is eliminated, a new one, produced by a new VAT, has emerged and grows rapidly until the host recognizes the new VSG and responds by producing a new antibody, and the 'see-saw' battle continues. In effect, "the parasite makes use of the immune system to control its own growth, by exhibiting and then changing this immunodominant surface antigen. The benefit of this strategy is that it leads to a persistent infection due to the presence of a relatively constant and tolerable number of parasites in the blood, rather than a rapid killing of its host as would occur in the case of uncontrolled growth" (Pays and Nolan, 1999).

Evidence for the VAT process emerged very early in the twentieth century, although it is clear that none of the investigators involved was aware of what was happening immunologically. Interestingly, a similar process was also described by Rossle (1905, 1909) for the free-living ciliate *Paramecium aurelia*, which continues to receive attention by today's ciliatologists. In 1954, Geoffrey Beale summarized some of what was known about this phenomenon in his now classic book, *The Genetics of* Paramecium aurelia. It seems that certain of the ciliates, including *P. aurelia*, have variant surface glycoproteins on their cilia and can change them in response to various sorts of environmental stressors, e.g., certain chemicals, temperature change, etc., similar to the way in which trypanosomes respond to the host's immune system. During my interviews with Dick Seed and Keith Vickerman, both of them, without prompting on my part, mentioned Beale's book, saying that it was a significant stimulus for starting their research on the variant surface glycoprotein of the African tryps.

The discovery of trypansomes as a cause of nagana and sleeping sickness by David Bruce is a fascinating tale. As I mentioned earlier, the discovery of what we now know as antigenic variation was made just a few years following the work of Bruce. However, it was not until the mid 1950s that serious efforts were made to decipher this phenomenon. Among the many parasitologists who were involved in this research were J. Richard Seed and Keith R. Vickerman. The former made it part of his doctoral dissertation. The latter would have liked to do the same, but was unable to find an appropriate mentor. I interviewed both men, Dick in Chapel Hill, North Carolina, and Keith, some 3000 miles away in Glasgow, Scotland.

Since I was surprised to learn that the notion of antigenic change in African tryps had been around almost since the turn of the twentieth century, I asked Dick Seed about it early during my visit. He replied, "I would assume the early investigators would isolate trypanosomes and serum from let's say a horse early in the infection and then again trypanosomes and serum later during the same infection. Presumably, they found that the early antiserum did not affect the trypansome strain isolated later in the infection. So, it became apparent to them that the trypanosome population was changing over time and this is where the idea of antigenic variation" was born. A serendipitous discovery, I queried? "Yes," he responded, "because they could not have known what they were looking for."

What about the relationship between antigenic variation in the tryp and that in *Paramecium*? Seed replied, "The correlation between the tryp change and *Paramecium* was really quite simple . . . you can take *Paramecium* and add antiserum to it and it will shed its ciliary, or mobilization, antigens. Up will pop another set of antigens, a totally inducible phenomenon. Another similarity between the two systems is that only one surface antigen, the VSA or immobilization antigen, is expressed on the surface at one point in time, the phenomenon of mutual exclusion."

With an answer to my initial questions, I wanted to know something about Dick's early involvement with antigenic variation. So, I asked him about his dissertation research with David Weinman at Yale. He told me, "There were three areas of significance. First, I was probably one of the first to begin to purify the [surface] antigen using modern techniques, i.e., column chromatography, etc. The next thing was, I cloned several lines of the trypanosome that had different antigenic types. Professor Weinman grew tryps all the time. He had them in culture and in animals. I just pulled out one of them, added antiserum to it, pulled up a second, cloned it (or cleaned it up), and now I had two lines that I independently maintained. I harvested them from mice, broke them apart and injected them into a series of rabbits, using Freund's adjuvant. I got antiserum from them and then ran both the homologous reaction and the heterologous reaction, and could show by protection tests in mice that the antiserum was absolutely specific for each line. There was absolutely no cross reactivity whatsoever. Then, in addition, we ran the old agar gel immunodiffusion test [Ouchterlony] and could pick up absolutely specific bands. There was no hint of cross reactivity. Then, . . . there was a second set of bands, two, or three, or four, that were common to both trypanosome lines. So, I was probably one of the first to attempt to show that there were specific antigens involved in protection and there were also common antigens. The common antigens, no matter what strain they came from, were always there and the specific antigens were specific to each clone. That was, probably, the main contribution of the thesis, to show that there were specific and common antigens."

Finally, he wondered, "Could I take culture forms and get them to be infective again?" Weinman "had worked on a whole series of additives to see if he could get the procyclic [found in the insect gut] form to change into an infective [metacyclic] form. We had procyclics of a number of lines in the lab and I was able to get others from people like Theodor von Brand [one of the leading biochemists of the era]. So, I took them and ran them against the antiserum that I had made for the blood forms. Nothing happened. They didn't agglutinate, they didn't inhibit growth, they didn't lyse, they didn't do anything. Procyclics had no relationship to the surface antigens of the blood form. So, I took the procyclic lines and made antiserum to them. Then, I could get agglutination, inhibition, etc. In other words, procyclics didn't react with the antiserum to the blood form and vice versa, so there was specificity there." I asked Dick if the procyclics had a coat and he responded in the affirmative. "But, it switches back to the blood-stream type coat when it gets to the salivary glands of the fly." In fact, he said, "There are all sorts of differences between the blood forms and those in the fly." The antigen of the procyclic trypanosome coat is called procyclin. Biochemically and antigenically, the coat becomes the same no matter what variant type is ingested with the tsetse's blood meal.

He then offered, "Did I follow up on any of these principal findings from my thesis? The answer is, no." Why not, I asked. "Several reasons," he responded, "Mostly though, I was more interested in gene regulation of antigenic variation. That's why I went to work [after his Ph.D. was completed] with the *Paramecium* geneticist at Haverford College, Irving Finger. I thought we were going to be dealing with the equivalent of ß-galactosidase in *E. coli*, but in a eukaryotic cell. That's where my interest rested. My thinking was that antigenic variation was an inducible phenomenon, but until I got to Tulane, I really didn't have an opportunity to investigate the hypothesis."

At that time, there was general agreement on the idea of antigenic variation. Everyone knew it occurred. Vickerman said in our interview, "The work I had done with Cunningham in Uganda and the much earlier work of Ritz had convinced me that the interpretations of Inoki [switching back and forth between two antigens] and Soltys [development of resistance to host antibody] were mistaken." For Keith, the essential question was, "Does the parasite's metacyclic stage, which alone among the stages in the insect vector can infect a mammal, express the basic antigen in the salivary glands of the tsetse vector before it is injected into the mammalian host?" At about the same time, for Dick Seed, the primary question was, is antigenic variation inducible? The two investigators were approaching different problems with very different perspectives.

Dick believed that if Inoki's hypothesis was correct, i.e., switching between two antigens, then he should be able "to take a single tryp, add antibody to it, and get a new variant, just like with *Paramecium*." However, rather than isolating a single tryp, he and his technician, Al Gam, decided to try 1×10^4 tryps, then to work down from that point. "So, we added 10^4 tryps to a culture dish, then added antibody to the culture dish, and waited. All we got was kill, no variants." They tried a number of different numbers of tryps, and other approaches, but nothing worked.

At this point, he began to doubt the original ideas regarding the inducibility of antigenic variation and began thinking in terms of "genetic changes occurring in the tryp", an idea shared by several others as well. In other words, he began to wonder if "it [antigenic variation] was due to mutations in the VSA gene." However, after some more bench work, and based on the variant antigen(s) that was being purified and characterized by George Cross and others, "it was becoming apparent that you couldn't just use single point mutations" to explain it. "There had to be something else. So, attempting to be ingenious, I came up with the idea that it might be some sort of nonreciprocal recombinant event, where you get whole segments of a gene changed, with deletions, or a rearrangement event." In other words, "you get a major change in a gene that would allow for the different variant genes to be produced from a single or limited number of VSA genes." As it turns out, some of his early thinking was fairly close to being correct. The phenomenon is not due to single point mutations, additions, or deletions, but does involve major genetic changes (or rearrangements). What Dick did not see at the time was that there are a very large number of VSA genes that can be expressed in a limited number of expression sites. In *Paramecium*, there appears to be only a limited number of genes coding for immo-bilization antigens.

Presently, it is believed that there are expression sites located on telomeres of the tryp's chromosomes. The genes responsible for the various antigenic types 'simply' move to these sites and then are expressed. A dominant VAT is expressed in the tryp population, one at a time, but other antigen genes are present and can move to the expression site, bumping the previously dominant type and assuming its role. Thus, when one dominant type is overcome and eliminated by the host's immune response, another variant type ascends to dominance, and the sequence of changing the surface coat continues. Dick's ideas about this process of succession are, however, at odds with many of the conventional notions. For example, he believes that, "Every time, depending on the strain, you get 10^3 to 10^6 tryps, you are going to get another variant. So, eventually you may have every single variant that can be expressed in that particular trypanosome's genome sitting out there [in the blood] waiting to ultimately become the dominant VAT."

It is now known that there are at least a thousand genes, each capable of expressing a specific variant type. Over the course of infection in a single host, all of these genes can be expressed, although it is very unlikely this ever occurs. Further discussion with Dick revealed he had come a long way from his early thinking regarding the inducibility of expression and his notions regarding the mutability of these genes. I asked him about these ideas with respect to the order of expression of these genes. He replied that he felt the appearance of each dominant VAT from a trypanosome's repertoire of variant antigen genes differed in their order of appearance in different infected hosts. He also believes "that different VATs can have different growth rates. I have also suggested that different VATs can take on different growth characteristics (or rates) in different host environments," and that these different growth rates are linked to host environmental changes, e.g., glucose concentration in blood, pH, etc., which occur during infection.

However, it has been shown that the appearance of these VATs in the infected host is not necessarily completely random. Certain VATs appear early in infection while others are expressed later in infection. The order of appearance of a VAT would, therefore, be related to the unique interaction between the host and parasite genomes. Compounding the problem of the order of VAT appearance is the immunosuppression of the host by the tryp, "because the more immunosuppressed the host is, the easier it is for minor VATs to hang around. We referred to this phenomenon as 'sneaking through'."

Dick explained that there is a point about which almost everyone agrees, i.e., "the sequence of VAT expression in the mammal does not go from VAT-A, to VAT-B, to VAT-C, it is much more random than that. But is it totally random?", he asked rhetorically. Then, he continued, "Every text or article that I have read would suggest the answer is no. I have never experimentally found anything other than total randomness in the specific order that VATs appear in mice or rabbits. However, there is a general type of order that appears to exist. Following the bite of the tsetse, the first VATs that are found in the host are metaVATs. This is followed by VATs that appear more often early in the infection. Finally, there are VATs that appear more frequently late in the infection. There would, therefore, appear to be a set of early VATs and then a set of late

VATs. So, there is some general pattern to antigenic variation. You don't see just randomness in the system."

I went on to ask him a question related to the idea of random versus non-random sequencing in VAT appearance. If you isolated a single cell, allow it to reproduce four times, producing 32 progeny, and you injected each of these cells into 32 inbred mice, would you expect to see the same repertoire of 20–30 VATs developing in the mice over a period of time? His answer was, "In part, you probably would. If you inject trypanosomes with VAT-A into each of these 32 different mice, then you would expect VAT-A to be the dominant VAT initially. But the real question is, what would you expect to see in the minor population? I do not think you are going to see VAT-B in every animal. Over sufficient time, you would, however, see most of the same antigen types, but they would appear at different times and in different proportions."

As our conversation wound down late in the morning, the topic turned to money and the possibility of ever developing a full understanding of the nature of antigenic variation. "One of the reasons I stopped doing the sort of research in which I had been deeply involved was that we had something like 100+ clones of different variant types in the freezer and the number of animals that were required to continue this work was too great. Even an entire animal facility would not have held what we needed. It simply became logistically impossible." He continued, "Why didn't I stick with the molecular study of antigenic variation? I actually made a conscious decision just before we left Texas [in 1981, to come to North Carolina] that I would not get into the molecular area. I would not become a gene 'jockey', mainly because I really thought that when you explain all there was to know about the genes, the genome, and gene regulation of the African trypanosomes, you still would not have any idea about the phenomenon of antigenic variation. I will also admit that I did not see the full power of molecular biology at that time. However, I still believe that the only way you are ever going to understand the phenomenon is to learn about the in vivo ecology of the African trypanosomes, or, to use the title to one of Keith Vickerman's publications, the trypanosome's 'sociology' and antigenic variation. It is ironic that we now know an incredible amount about the mechanism of antigenic variation (the genetics of variation), but much less about the phenomenon of antigenic variation as it occurs in vivo." Dick believes that the tools for following this line of research are not available and that the budgets of most labs are insufficient even to begin such a program of investigation. Moreover, the funding is not in the pipeline for the African trypanosomes and "certainly not there for studies on the ecology of antigenic variation". He strongly feels the latter approach is the only one that in the long run will give us a complete solution to the phenomenon of antigenic variation.

Based on the discussion with Dick Seed, I was anxious to see how Keith Vickerman felt about his research on antigenic variation and where research on African tryps, in general, might be going today. I had a perfectly marvelous morning in Glasgow talking with Keith regarding his role in the VSG saga.

At the outset, Keith pointedly observed that it was not until the end of WWII that the really penetrating research on the various blood protozoans actually exploded. The reasons for success in these efforts were, for the most part, related to several new and important technological advances. For example, as Keith wrote in 1997 (Hide *et al.*, 1997), "When I entered university in 1952, DNA was regarded by many as so much padding in the chromosome; when I left three years later, few had any doubts as to its cardinal importance in the whole business of life." Another breakthrough was the advent of electron microscopy, which allowed one to actually see inside a cell and take note of its many intricacies. De Kruif's *The Microbe Hunters*, published in 1926, and Beale's *The Genetics of* Paramecium aurelia, published in 1954, also had a significant impact on the direction of Vickerman's career, and that of Dick Seed as well. By the early 1950s, the "Geimsa era", as Keith called it, was over and a revolution in parasitology, immunology, biochemistry, genetics, and most other biological disciplines was well under way.

When Keith finished his three years of graduate school at Exeter, he was still without a Ph.D., without publications, and had none in press. Nonetheless, Peter Medawar at UCL saw talent and offered him a job. Medawar had secured a Research Fellowship from the Wellcome Trust for him, and an Honorary Lectureship. It was perfect in every way, because he was instructed by Medawar to do research on whatever he liked, as long as it was in protozoology. He wrote in the autobiographical sketch that he had prepared for me, "This liberty disconcerted me. My intention was to work on a really fascinating problem relating to one of the medically important parasitic protozoa, and since first seeing them alive, wriggling between red blood cells, I had been captivated by the deadly beauty of trypanosomes." He continued, "In particular, I had been fascinated in reading Geoffrey Beale's book, *The Genetics of* Paramecium aurelia, to learn that the sleeping sickness trypanosomes could evade the host's immune response by changing their antigens. As Medawar's department was a centre of excellence in immunology, I decided that, for a research topic, antigenic variation in trypanosomes would do nicely."

His first tasks involved learning the basic techniques that would be required to pursue this line of work. So, he began by learning how to bleed mice and harvest parasites in R. A. Neil's lab at the nearby Wellcome Laboratories of Tropical Medicine in the Bloomsbury area of London. He also spent time at the Cambridge Veterinary School where he learned the Ouchterlony technique of antigen-antibody precipitation in gels from Elaine Rose. A particularly useful development was the cryopreservation procedure, which he picked up from the pioneering M. A. Soltys, who also taught him about neutralization and agglutination reactions. Learning the techniques was the easy part. As he said, "I was still looking for a new line of attack. The *Trypanosoma brucei* strains available in laboratories then were old syringe-passed lines, which had lost their ability to be transmitted by tsetse flies and had become monomorphic, i.e., had lost the ability to produce the short stumpy forms believed to be infective for the vector. What I needed to make the problem realistic was a recently isolated strain."

As fate would have it, another break was to occur, one that was to greatly influence his skill as a cell biologist–immunologist, and one that was to ultimately play a key role in his career. By the early 1960s, the electron microscope was being used with huge success as a means of unlocking the real intricacies of the cell at the sub light microscope levels. His interest in electron microscopy (EM) had been stimulated during a term at Edinburgh University while still a Ph.D. student at Exeter. He had been asked to deliver lectures to students at University College London on cytology and, of course, because of the new discoveries in cell ultrastructure being made using electron microscopy, he felt compelled to learn the technique. His instructor was none other than David Robertson, a 'Yank', who was the "inventor of the unit membrane" theory. He was Director of the EM facility in the laboratory of the great anatomist, J. Z. Young, with whom Vickerman also established a close working relationship. In fact, Keith was very complimentary of Young's generosity toward the new investigator. He said, "I knew that electron microscopy was an expensive pursuit, but he [Young] told me to make full use of the facilities and to take as many micrographs as I wanted. His only request was that any published papers that resulted be included in the Anatomy Department's list." Keith exclaimed, "What a difference from the 1990s when every trivial cost had to be paid for. In fact, J. Z. included many of my pictures in his book, *An Introduction to the Study of Man*, published some years later [Young, 1971]." While working in Robertson's lab, he focused on some of the soil amoebae on which he had done some of his dissertation research, i.e., "an intracristal body in the mitochondrion" that swelled up during encystations, resulting in a "clampdown on respiration." Vickerman continued, "This was to be the subject of my first publication – a letter to *Nature*!" I should note here that the first publication by Bob Desowitz, the great tropical parasitologist, was also first published in *Nature*, not a bad way to begin a career for either of the young parasitologists!

The electron microscope was to become one of the main weapons in Vickerman's arsenal over the years, but before he could move to the point of his real research interests, he felt it necessary to head for Africa and obtain some new trypanosomes. He also wanted to go there for some field experience. As he said, "I had to confess that working on a tropical disease-causing parasite in Bloomsbury [London] never ever having seen it in its natural environment, did make me feel somewhat like a sham parasitologist." So, he explained, he was put in contact with W. H. R. Lumsden who was in charge of the East African Trypanosomiasis Research Organization (EATRO) and he was awarded a Colonial Development Research Fellowship for work in Uganda. He added at this point, "I belatedly submitted my Ph.D. thesis to London University. It was work that I was not proud of."

Uganda was a great adventure. Everything was new, the sights, the sounds, and the smells. He traveled first to Entebbe and then to the village of Sukulu, close to the Kenyan border where the EATRO lab was located. Soon after arriving at the lab, he teamed with Mathew Cunningham, "a beguilingly amiable Glasgow veterinarian, . . . to whom I very quickly became attached." Their initial efforts were directed at examining variable antigens in a wide range of human patients in the EATRO hospital, along with wild game and domesticated stock animals. "We generated no meaningful results from immunoprecipitation reactions – not surprisingly, I learned later, as this technique demonstrates common rather than variable antigens. But the agglutination reactions yielded interesting results. At the time we were conducting this work, no one was sure of the extent of antigenic variation, or indeed whether it was a real phenomenon." In 1926, H. Ritz had said that he recognized 24 antigenic types that originated from a single clone. Shozo Inoki, a Japanese investigator, believed that the trypanosome alternated back and forth between two antigen types, whereas M. A. Soltys considered it was possible that the tryps became resistant to "antibody in the same way that they did to drugs. Most previous workers had confined their studies to a single stock or isolate. By comparing several stocks, Cunningham and I found that isolates from man, or cattle, or game, had variable antigens in common and that some antigenic types (VATs) were found in quite a wide range of stocks. We also found that stocks isolated from drug-induced relapses showed unique antigens not found in any other stock."

The year following his return from Uganda in 1960 "was probably the most productive of my life; I . . . never got into press so many single or twin-authored papers on so many subjects." He also received his Ph.D. and attended an important international meeting in Prague, but the real highlight was his marriage.

Vickerman's (1997) contributions to various aspects of the biology of trypanosomiasis continued in several areas. One of the most important was his work on metabolic activities of the tryps in the gut of the tsetse fly versus that of the parasites in vertebrate blood. These activities were notably associated with change in the mitochondria themselves, which he was able to document metabolically, and morphologically, via biochemical analysis and electron microscopy. His findings supported the idea that glucose metabolism in the vertebrate bloodstream was anaerobic, i.e., via glycolysis, with a clearly non-functional mitochondrion. In the gut of the tsetse fly, the metabolism was aerobic, i.e., oxidative phosphorylation, with a clearly functional mitochondrion. It was this work that led him eventually to write "one of the most important papers I ever published."

I asked him during our interview, "What led you into the surface coat research?" He responded, "Well, I had been interested in this mitochondrial cycle and I wanted to get some stages in the tsetse fly as opposed to stages in culture. This is what led me to go out to Nigeria," to work with a young vet by the name of Ross Gray who had also been working on antigenic variation in the tsetse-transmitted infections. Apparently, however, finding infected tsetse flies was not going to be an easy proposition. Gray told him that finding an infected tsetse fly would be a "red-letter day", but Keith nonetheless managed to "get some experimentally infected flies with *Trypanosoma brucei*."

He had earlier observed by electron microscopy that some forms of the parasite had 'coats' on them. At that time though, "the general consensus was that the coat was some form of adsorbed host serum. This [idea] was largely inspired by Bob Desowitz", the American [mentioned earlier] who worked in the employ of the British Colonial Service at their research laboratory in the village of Jos in northern Nigeria. According to Keith, Desowitz believed that "*Trypanosoma vivax* adapted to the host by adsorbing host serum proteins, which form some sort of protective layer around it." Continuing, Keith admitted, "So, I hadn't thought much about the surface, but I presumed the coat was adsorbed human protein. If you put the blood stream form into culture, they'll lose this coat. The procyclic forms in the gut of the fly do not have this coat. When they migrate to the salivary glands, the epimastigote forms, which multiply in the glands, initially do not have the coat." But later, they transform into metacyclic trypomastigotes, and "they actually have the coat. So, there beside each other, you've got trypanosomes that have a coat and trypanosomes that don't have a coat. It was then I thought, these metacyclic forms have put on a coat in preparation for going into the mammalian hosts. So, it cannot be adsorbed human serum. It must be, very likely, the antigen coat." And, voilà, the problem was solved. He and A. G. Luckins, with the help of Bob Sinden (now at Imperial College in London), experimentally confirmed his hypothesis and published their results in Nature in 1969. They were able to localize variable antigens in the surface coat of Trypanosoma brucei using ferritin-conjugated antibody.

"Having found, while actually looking at the mitochondrial cycle, these changes in the surface, I then produced this paper [in The Journal of Cell Science] on the surface coat and hypothesized how it was related to antigenic variation." During our conversation that morning, I vividly recall watching him pull this paper out of his files. I asked if he had any copies left. He laughed and said, "No, this is my last copy." I did not ask how many reprint requests he received, but I suspect it was in the hundreds, if not thousands.

This discovery was a clear example of serendipity. He was not looking for the antigenic coat when he cut the ultrathin sections of the tsetse's salivary glands. He was looking for morphological changes in the trypanosomes' mitochondria. However, he was keenly aware of the existence of an antigen coat on the slender and stumpy blood forms in the vertebrate host. The coat was not present on epimastigotes in the salivary glands, but it was present on the surface of the metacyclic trypomastigotes. It then occurred to him that these parasites had not been exposed to vertebrate blood, so no serum proteins from the vertebrate could be covering them. No, they were preparing for entry into the immunologically hostile environment of the vertebrate animal by acquiring an antigen coat, a VSG, a variant surface glycoprotein! Identifying the surface coat for what it was is a perfect application of Louis Pasteur's axiom regarding chance favoring the prepared mind. The observation then led to a hypothesis regarding the antigenic coat. A simple experiment was designed by Vickerman and Luckins, tested, and their hypothesis was corroborated, a classic exhibit of 'Popper's view of the scientific method'. Folks, this is the way it is supposed to happen!

Their original hypothesis was that there was a single antigen type, the same one for all metacyclic trypomastigotes. If this had been the case, then a vaccine would have been easily developed and perhaps we would not still be dealing with sleeping sickness, nagana, etc. Regrettably, this was not the situation. He said, "Later on, we did some work on this question of does the evasive [basic] antigen really exist, largely through the skepticism of a Belgian postdoc who came here [to Glasgow University] to work, named Dominique Le Ray. He was not entirely convinced by this story of reversion to a basic antigen. And so, while he was here, we started the painful process of trying to infect tsetse flies in a little hut across the yard and were successful. We looked at metacyclic trypomastigotes using immunofluorescence and found that there was more than one antigenic type. We found that, in fact, the metacyclic trypomastigotes were a mixture of antigenic types and that there isn't a reversion to a basic type, that there are several." At this point, I asked if this was the reason a vaccine against the parasite would not work. He replied, "Yes," adding, however, "Not only that, but the metacyclic forms are changing. They evolve differently in the different strains of trypanosome as you go on transmitting it."

I did a lot of reading about the African tryps in preparation for my interviews and then spent nearly six hours in conversation with Dick and Keith. This certainly does not make me an expert on antigenic variation, or on the African tryps. However, I can honestly say that I have a much better understanding of this phenomenon and these organisms now than before. My reading and interactions with Dick and Keith also led me to conclude that their research findings were, indeed, formidable.

On a personal note, one of the things I have learned about doing research over the years is that, in answering any sort of scientific question, two or more new questions should emerge as a result of answering the first. Reading their papers and listening to the tapes generated in our interviews would easily lead to the conclusion that this canon is applicable to Dick Seed and Keith Vickerman, and their work over the years. A huge amount of information was generated and many, many new questions were raised.

However, in thinking about their careers, I felt I detected a kind of frustration in both of them, perhaps more so in Dick than in Keith. They both were trying to resolve a basic problem, i.e., how does antigenic variation work? During the course of their careers, both men developed a great deal of insight with respect to that fundamental question. However, I am reminded of what Dick Seed said to me at the end of our interview. "When you explain all there was [is] to know about genes and the genome of African trypanosomes in reference to antigenic variation, you still would not have any idea of the phenomenon." And, at the end of my interview with Keith, he said, "The trypanosomes are always going to win no matter what you do!" I did not hear either of them use the word intractable in connection with the problem of antigenic variation, but it sure does sound like that was what they wanted to say.

REFERENCES

Beale, J. 1954. *The Genetics of* Paramecium aurelia. Cambridge: Cambridge University Press.
de Kruif
Grove, D. I. 1990. *A History of Human Helminthology*. Wallingford: CAB International.
Hide, G., J. C. Mottram, G. H. Combs, and P. H. Holmes. 1997. *Landmarks in Trypanosome Research*. Wallingford: CAB International.
Lyons, 1992. *The Colonial Disease*. Cambridge: Cambridge University Press.
Pays, E. and D. P. Nolan. 1999. Expression and function of surface proteins in *Trypanosoma brucei*. In *The Trypanosome Surface*, ed. E. Pays, pp. 5–51. Paris: De Boeck & Larcier.
Rossle, R. 1905. Spezifische Sera gegen Infusorien. *Archives Hygiene, Berlin* **54**: 1–31.
Rossle, R. 1909. Zur Immunitat einzelleger Organismen. *Verhandlungen der Deutschen Gesellschaft fur Pathologie* **13**: 158–162.
Vickerman, K. 1997. Trypanosomiasis and leishmaniasis. In *Landmarks in Trypanosome Research*, ed. G. Hide, J. C. Mottram, G. H. Coombs, and P. H. Holmes, pp. 1–38. Wallingford: CAB International.
Vickerman, K. and A. G. Luckins. 1969. Localization of variable antigens in the surface coat of *Trypanosoma brucei* using ferritin-conjugated antibody. *Nature* **224**: 1125–1126.
Young, J. Z. 1971. *An Introduction to the Study of Man*. Oxford: Clarendon Press.

2

Malaria: the real killer

Open your mouth: this will shake your shaking . . .
If all the wine in my bottle will recover him, I will help his ague.

The Tempest, William Shakespeare (1564–1616)

Since I have never had malaria, all I know about rigor, or the high fever that follows, is what I read from those who have had the experience. Not long ago, in preparing to write this essay, I spent some time interviewing Robert Desowitz, the American parasitologist who has so marvelously written about his many experiences working with tropical diseases over the years, e.g., *The Malaria Capers*, *Who Gave Pinta to the Santa Maria*, etc. I asked Bob if, while living so long in so many exotic parts of the world, he had ever come down with malaria, or any of the other tropical diseases about which he had written so much or done so much research. He said that he had not been infected with anything except *Plasmodium vivax* and *Plasmodium falciparum*, that both of them came at the same time, and by accident! Acquiring malaria by accident is ironic in his case because he had spent nine years at the West African Trypanosome Research Lab in Jos, in the northern part of Nigeria, working on trypanosomes in the employ of the British Colonial Service. He then spent five more years as Chairman of the Department of Parasitology at the University of Singapore, plus several more years in southeastern Asia while with the U.S. Army at their SEATO (Southeast Asia Treaty Organization) lab in Bangkok, Thailand. Finally, there were many special trips into the boondocks of southeastern Asia after taking his final position at the medical school of the University of Hawaii in 1968. Throughout all of this time, in highly malarious areas and regions with so many other tropical parasites, he had never picked up any of the many dread diseases. It was only after he was hired at the University of Hawaii and had returned to New Guinea on one of his special field trips that he finally came face-to-face with *Plasmodium*. Even then, he acquired the parasite only because he made an unplanned side excursion without some very crucial protection.

It seems that while in New Guinea working on malaria and pregnancy, Bob was invited by a friend of his, Guy Barnish, to accompany him into a remote area of the southern highlands. At the time, Barnish was working on what appeared to be *Strongyloides ransomi* in humans (an unusual find because this parasitic nematode was supposedly restricted to primates in the Congo of central Africa). His friend was planning some additional field studies on this unusual nematode, and thought it would be good to have someone screen the stools for parasitic protozoans while he was looking for *S. ransomi*. Bob agreed to go and do the protozoan work. The site was so remote they had to be helicoptered in. As they were landing, Bob offhandedly asked Barnish if he had brought the mosquito netting? "Yeah, yeah, yeah, I got the mosquito nets," his partner replied, as they climbed from the helicopter. The pilot reminded them he would return in two days to pick them up. As the copter was flying out of site, Guy suddenly turned to Bob and exclaimed, "I forgot the mosquito nets." Since it was a two-day walk to the nearest road and the helicopter was returning in two days, they decided to take their chances and remain until their transport returned.

Several days after completing their trip, Bob said he suddenly developed rigor, but that he immediately took a double dose of chloroquine, after which he said he felt fine. He related that he returned to his research area in the eastern highlands and about two weeks later the rigor broke through again. This time he was treated with a heavy dose of quinine and that was an apparent curative as well. He remarked in our interview, however, that he still has recurring 'ringing' in his ears, a frequent side effect of quinine treatment. He learned later that his *S. ransomi* partner was up in the western highlands at the very time he was being treated with quinine and that "Barnish almost died of cerebral malaria."

Bob said that he finished up his work in New Guinea and returned to Honolulu, but after about a month back in Hawaii, he developed rigor again. He called one of his students immediately and instructed him to "bring a slide to my home, so we can put a blood smear on it." The student was then told to take it to the lab, stain it, and find out which *Plasmodium* he had this time. The student called about 2:00 A.M. and said that it was vivax malaria. He disgustedly lamented in our interview that he had spent more than twenty years in some of the worst malarial areas of the world, but that after just two days in the field without mosquito netting, he managed to pick up not one, but two, species of *Plasmodium*. He told me that the vivax malaria he suffered was far worse than that caused by falciparum, adding, "benign, hell!!" On the bright side, he said, "For teaching purposes, there is nothing quite like experiencing malaria."

For many years, I have taught general parasitology, including of course several lectures on malaria. I had often wondered about the actual discovery of *Plasmodium* spp. by Charles Laveran in 1880, and about Ronald Ross, the person who worked out enough of the details of the parasite's transmission that he received a Nobel Prize for his effort in 1902. However, I had not done much about this curiosity until I decided to include a malaria essay for this book. It was only then that I became seriously interested in the discovery process and made the decision to delve into it. I wanted to know how they did it. Laveran had identified the causative agent for the quartan form of the disease and named the parasite he saw *Oscillaria* (= *Plasmodium*) *malariae*. In 1897, Ross was able to link female culicine mosquitoes with the parasite's transmission to birds and, as importantly, to show that the transfer of the parasite occurred while the mosquito was taking a blood meal. Considering the complexity of the parasite's life cycle and that he had only one protozoan model on which to base it (*Trypanosoma b. brucei*), this was a rather remarkable accomplishment. Despite Ross's discovery, the full life cycle took another fifty years to resolve. Why so long? That is a story in itself. Finally, more recently, the development of molecular technologies made it possible for the phylogeneticists to get a handle on the evolutionary history of these blood parasites, and a fascinating history it is, so we'll look there too. These are a few of the issues with which we will deal in the present essay.

The phylum Apicomplexa, to which *Plasmodium* spp. belongs, is estimated to be more than a billion years old and possess some 4000 species, all parasitic. More than likely, malaria has been with the human species since we evolved between 100 000 and 150 000 years ago. In *Homo sapiens'* written history of about 4600 years, the disease has been cursed mightily by those who acquired it, the decline of more than one empire can be identified with it, malaria has ravaged mostly the young and produced the deaths of scores of millions, the disease has evoked fear and great consternation at one time or another among those living in almost all parts of the world, it has impacted a number of important battles, invasions, and wars throughout history, and it has sapped the wealth of many civilizations. In terms of its impact on humans through-out our history, surely malaria is on par with plague, smallpox, and now HIV/AIDS.

There are two important events that led to identification of the causative agent for malaria by Laveran in 1880. The first was the develop-ment of the microscope. Of course, the primary responsibility for this huge breakthrough is attributed to Antonie van Leeuwenhoek (1632–1723), a poor Dutch cloth merchant who had a penchant for grinding glass into lenses. These lenses were exceedingly primitive, but exacting enough for him to describe cells in his own blood, sperm in his own semen, and even *Giardia* sp. in his own stool, among other things. In 1673, van Leeuwenhoek began sending letters to the newly established *Transactions of The Royal Society* in which he described his many observa-tions. By the time he died in 1723 at the age of 91, some 375 of these letters had been published, not bad for an uneducated Dutchman. At about the same time that van Leeuwenhoek was doing his thing, the Royal Society appointed Robert Hooke as its 'Curator of Instruments'. He became familiar with the primitive microscope and in 1665 published *Micrographia*. Subsequently, on slicing a piece of cork and examining it with his microscope, he observed internal compartmentalization and referred to the spaces as "cellulae."

In 1838, a German botanist, Mathias Schleiden, suggested that all plants are made of cells and in 1839, the German zoologist Theodore Schwann proposed the same thing for animals, creating the first ver-sion of the Cell Theory. Rudolph Virchow expanded this idea in 1858 when he made a compelling case that all cells come from pre-existing cells, i.e., "*omnis cellula e cellula*." This idea was crucial to dispelling the myth of spontaneous generation, although it remained for Louis Pasteur to experimentally quash this well entrenched, but totally erro-neous, concept. The developing Cell Theory was also critical to the then evolving idea that microorganisms could cause disease, i.e., the so-called Germ Theory. A case for supporting the notion of the Germ Theory had been growing since Thomas Fuller (1657–1734) is said to have remarked "that one disease could not change into another 'any more than a Hen can breed a Duck'" (quoted by the great American malariologist, Paul Russell, in his book, *Man's Mastery of Malaria*, 1955).

The name, malaria, was coined from two Latin words, 'mal' (bad) and 'aria' (air), and was frequently written in the early days as 'mal'aria.' The disease was originally associated with the heavy stench of marsh air. The connection between marshes and malaria was so strong that some even held that breathing the foul marsh air caused the disease.

Other terms for malaria included intermittent fever, cattivara, periodic fever, swamp miasmata, paludism, and ague. The disease as an entity has been recognized and written about for nearly 4600 years by many different peoples in many different places. It was to remain that way until microscopy was refined well enough so the parasite could be seen, and until the Germ Theory could be thoroughly tested. By the middle of the nineteenth century, the 'table' was, in effect, set for these breakthroughs.

The first physical clues as to the identity of an organism that might be causing malaria were the frequent descriptions of pigment in the blood of patients suffering from malarious fevers, and the enlarged and heavily pigmented spleens and livers, or the pigmented brains observed at the autopsies of some of those who had apparently died of the disease. Of course, we presently realize that these pigments, what we now call hemozoin, occur in the phagocytic cells of the hyperplastic spleens and livers. In the blood, the actual parasites also were frequently, but unknowingly, seen. In a most interesting twist before Laveran identified and named the parasite, Paul Russell (1955) described the first case of the forensic use of malarial blood and that occurred in 1876. It seems that Professor Joseph Jones of the University of Louisiana testified in a court case that "certain stains on the coat and shirt of an accused prisoner were not paint, as had been affirmed, but were the blood 'of a human being who had suffered and was probably suffering at the moment when the blood was abstracted, with malarial or paroxysmal fever.' Jones described the characteristic pigment and stated that 'many of the particles of melanic pigment were spherical, others irregular and angular, some entirely free, others encased in a hyaline mass . . .'" According to Russell (1955), Jones "was actually, but unknowingly, describing malarial parasites." Others, including Virchow and Frerichs by the mid-nineteenth century, had seen the hemozoin pigment, and Russell (1955) believed the latter had observed the parasite based on drawings published in 1858.

Charles Laveran was a physician in the French army who had been posted to Constantine, Algeria, in 1878. He had developed a keen interest in the tissues of malarious individuals removed at autopsy and in the fresh blood drawn from those known to have the disease. He had observed pigmented cells on many occasions, but on 6 November 1880, he hit the 'jackpot!' In a report to his superiors, Laveran wrote, "I had suspected for a long time the parasitic nature of these bodies when on November 6th, 1880, while examining one of the spherical pigmented elements in a preparation of fresh blood, I noticed with joy at the periphery motile filaments [probably exflagellation] of the animated nature of which there was no room for doubt" (quoted from Russell, 1955).

The subsequent announcement of Laveran's discovery was greeted mostly with skepticism. The first to be convinced, however, were the Italians. Not long after his claim was published, Laveran took some of his blood films with him to Rome and showed them to several workers who recognized immediately that Laveran had won the race. The great Canadian parasitologist and physician, William Osler, was also skeptical at first. He had heard a presentation made by W. T. Councilman at a professional meeting in the summer of 1886. Councilman suggested that Laveran was correct in his identification, and "Osler's confidence in his own view was shaken" (Russell, 1955). Osler immediately set out to discover the truth by viewing many blood smears from malarious individuals over the next several weeks. It took him a while, but when he saw crescent-shaped bodies in several blood smears, he became convinced regarding the validity of Laveran's conclusions for a parasite as the causative agent of malaria. Quoted in Russell (1955), Osler said, "When I first read Laveran's papers nothing excited my incredulity more than his description of the ciliated bodies. It seemed so improbable and so contrary to all past experience, that flagellate organisms should occur in the blood." He continued, "The work of the past six months has taught me a lesson on the folly of skepticism based on theoretical conceptions, and of pre-conceived notions drawn on limited experience." Osler certainly knew how to 'eat crow' without choking!

The next question was, naturally, how do we acquire the disease? How is the parasite transmitted? Up until the mid to late 1880s the idea that insects, let alone mosquitoes, could transmit diseases was a 'real stretch', except for some of the native Africans who had believed for a long time that malaria was transmitted in just exactly that matter. Russell (1955) also, however, refers to a Sanskrit reference published in 1903 by Sir Henry Blake, Governor of Ceylon. This piece refers to "ancient authorities on Ayurvedic medicine", who believed that "the chief causes of the disease [malaria] are impure air and water and the existence of mosquitoes." Using knowledgeable translators and historians, Blake estimated that these writings were somewhere between 1400 and 3000 years old.

Then, along came Manson. Patrick Manson was born in Aberdeenshire, Scotland, in 1844. He became a physician, and traveled to the Far East to practice medicine. While in Formosa, he saw numerous cases of filariasis and began to surmise about its cause. On leave in London in 1874–75, Manson searched the literature and discovered that micro-filariae had been seen in the blood of filarial patients several times in recent years. Equipped with this information, he wondered how these parasites could move from one infected human to another. The best way in his mind was via a bloodsucking insect. He examined the distributions of various bloodsucking insects in Formosa and elsewhere, and compared their pattern with that of filariasis. He was not the least surprised when he found the best fit was with mosquitoes. So, he fed mosquitoes on patients with active cases of filariasis (including his own gardener), then dissected the mosquitoes at varying intervals afterward. He observed the microfilariae in the mosquitoes' gut, watched them develop, and followed their migration into the thoracic muscles where they continued to grow. It was at this point, however, that Manson erred in the life cycle because he believed the infected mosquito would return to water, die, and then accidentally infect humans when they drank the contaminated water. It was left to Thomas Bancroft in Australia and George Low in England to demonstrate successfully that the mosquito inoculated the parasite into the human host as it took its blood meal. Nonetheless, Manson had shown for the first time that an insect could 'vector' a parasite from one host to another. Although Manson did not receive a Nobel Prize for this huge achievement, he was knighted. In 1894, he proposed that malaria too was transmitted by mosquitoes, but was derided by many of his colleagues for such a preposterous hypothesis. Eventually, however, Manson's insights were recognized for their accuracy and he became justifiably known as the 'father' of tropical medicine.

Between 1889 and 1891, Theobald Smith and F. L. Kilbourne discovered that the so-called 'hard tick', *Boophilus annulatus*, was the transmitting vehicle for *Babesia bigemina*, the causative agent for Texas cattle fever. This protozoan piroplasm was causing up to 75% mortality in some endemic areas and was particularly devastating to the ranching interests in southwestern parts of the U.S.A. An interesting aspect to this finding was that, locally, ranchers had believed for a long time that ticks were responsible for spreading the disease, but most scientists did not, at least until the brilliant experimental proof provided by Smith and Kilbourne. This was a real breakthrough in another way too because it was the first time an arthropod had been 'fingered' in the transmission of a protozoan parasite.

Then, working in Zululand from 1894–96, David Bruce was able to show conclusively that nagana, a disease produced by a trypanosome parasite was transferred from wild ungulates to horses and other domesticated stock, thereby demonstrating protozoan transmission by an insect for the first time. Tsetse flies, *Glossina* spp., were identified as the culprits. However, like Manson for his filarial worm transmission, Bruce was wrong in his thinking about the trypanosome life cycle. He had postulated that mechanical transmission was the mechanism of transfer for the trypanosome. In 1900, the tsetse fly was found to be the true intermediate host for the trypanosome.

The door was open for Ross to take the final step in understanding the transmission of *Plasmodium* to humans. At the very beginning of his pursuit though, Ross had several problems. First, he was not a trained entomologist. In fact, he didn't even know the difference between a culicine and an anopheline mosquito. Second, he was not a well-trained scientist; indeed, he was not even a good microscopist. Third, he was an underling physician in the Indian Medical Service and could be posted almost anywhere, then be instructed to do whatever his superiors might want him to do. In fact, just before his major success, he was to be transferred away from the malarious area in which he had been working, and was told to begin research on leishmaniasis (Manson played a role in bailing him out of that predicament by intervening with Ross's superiors). Fourth, he was in serious competition with highly competent and hard-working French and Italian scientists, including Laveran and Grassi, among others. Finally, he would have preferred to write poetry, though he was not very good at that either! In fact, during my interview with Desowitz, Bob referred to him as "an absolute nutcase!"

And, so it was when Ronald Ross visited Patrick Manson at his home in Cavendish Square, London, in April 1894, and was shown by Manson a malarial parasite in a human blood film, at last, Ross at least knew for what he was looking. Now convinced that *Plasmodium* and malaria were linked, Manson next persuaded Ross that mosquitoes were the vectors for the parasite during a stroll down Oxford Street in London in November 1894. In early 1895, Ross returned to India, equipped with a microscope and fired by Manson's revolutionary idea that mosquitoes were the vectors of *Plasmodium*. However, as noted above, Ross was not an entomologist and knew very little about mosquitoes, at least at that point in time. In fact, when he fed mosquitoes on humans exhibiting the characteristic malarial signs and symptoms, he incorrectly used species of *Aedes* and *Culex*, which he had reared in the lab. The most he was able to observe was exflagellation by microgametocytes, because in the gut of the foreign mosquito the parasites always died. Moreover, he had no real understanding of exflagellation either. He even tested

Manson's long-held mantra that infected mosquitoes died in water, and that the then 'contaminated' water was the vehicle for transmission of the parasite to humans. Oddly enough, the first person on which he tried this approach died of malaria, but he could not get it to work again, so this line of research was eventually dropped.

The process of exflagellation in the mosquito gut had been observed by a number of investigators, including Laveran, Grassi, and Ross. However, no one could explain it until William George MacCallum, a Canadian-borne physician, trained at Johns Hopkins, decided to look at *Haemoproteus* in the blood of a crow in the summer of 1897 while on vacation in Ontario. The late and great Canadian parasitologist, Murray Fallis, in his book on the history of Canadian parasitology, describes MacCallum's observations as follows, "He [MacCallum] noticed as others had for species of *Plasmodium*, the flagella (microgametes) breaking free from the parent cells (microgametocytes), wriggling among blood cells, bumping into other cells and, finally, contacting and penetrating another cell (macrogamete)." This was, of course, part of the sexual cycle of plasmodid parasites. Interestingly, not long after, MacCallum confirmed his findings in crow's blood when he observed the same phenomenon associated with the sexual stages of *Plasmodium falciparum*.

Ross continued his search in India. On two consecutive days in August 1897, he made the discovery that would inexorably lead him to success with the life cycle of *Plasmodium*. He finally obtained and used mosquitoes whose wings were dappled and had four dark spots; in other words, he finally had some anophelines. On these two days, he observed cysts growing on the stomachs of anopheline mosquitoes that had been fed on a patient who had crescent-shaped bodies in their red blood cells. At that moment, he knew he was on the right track. Ross was now certain that the female anopheline mosquito was the definitive host for species of *Plasmodium* infecting humans as soon as Manson transmitted MacCallum's observations regarding exflagellation and the sexual phase of *Haemoproteus* in the blood of a crow.

On the verge of a major discovery, Ross was then confronted with another dilemma. He learned that he was to be transferred to another part of India to take part in work on an epidemic of leishmaniasis that was killing hundreds of locals. However, Manson intervened and persuaded Ross's superiors of the serious nature of his research and he was instead transferred to Calcutta, another highly malarious area. But, he was blocked from continuing his work because the locals refused to provide him blood and, therefore, the parasites that would enable him to continue the research. Ross brilliantly proved his newly acquired research mettle at that point by switching his focus from human to avian malaria. Using *Culex* sp. and infected sparrows, he was able to follow the entire course of the parasites' development in a mosquito host, i.e., exflagellation and fertilization in the insect's gut, oocyst formation and sporogony, and then migration of sporozoites to the salivary glands. The *coup d'état* in his research came when he inoculated sporozoites isolated from experimentally infected mosquitoes into uninfected sparrows and obtained an infection. This finding was not only a 'shocker' to Ross but to Manson as well because the latter still firmly believed humans acquired malaria when they drank water in which infected mosquitoes had died. The final proof was made on 4 July 1898. Manson gave the report to a meeting of the British Medical Association on 24 July 1898, in Edinburgh, Scotland. Substituting for Ross, Manson was said to have received a rousing, and standing, ovation.

In 1902, Ronald Ross received his Nobel Prize, and was subsequently knighted! The Italians were very upset that Giovanni Grassi did not share in the Prize, both because Grassi had done so much ground-breaking work with the malarial parasites, but also because Grassi resolved the life cycle of *Plasmodium falciparum* in 1898, the first species of human *Plasmodium* for which the entire cycle was determined. In fact, Grassi began the experimental work on *P. falciparum* within two months of Manson's announcement in Edinburgh. Apparently, though, there were some real political shenanigans that played a significant role in Grassi being denied by the committee designating the prize. Bob Desowitz, in his book *The Malaria Capers*, provides details regarding the mess created, and who did what to whom. First of all, the Italians strongly believed that Ross had accomplished hardly anything, claiming that "birds don't count!" when it came to human malaria (Desowitz, 1991). According to Desowitz, "when Grassi published his paper in November 1898 [describing the life cycle of *P. falciparum*], he cited Ross's work only at the very end, as a scant, grudging afterthought." Then, when he published a more complete description a little more than a month later, Ross's work wasn't cited at all. The Englishman was hugely annoyed with the slight.

The 'waters' surrounding these discoveries were muddied even further by the entry of Robert Koch (of Koch's Postulates) into the fray. A German, Koch was the transcendent microbiologist of the day, but certainly not in the same league as Grassi or Ross when it came to protozoology and malaria. According to Desowitz (1991), all three of these men not surprisingly also had monstrous egos. Grassi became greatly incensed with Koch when he stopped in Rome (on his way home from an official mission of the German government to Africa) to 'lend a hand' to the Italian malariologists by doing some, so-called, 'final' experiments. It is clear, however, that Koch contributed nothing to completing the *P. falciparum* life cycle. Then, when Grassi published his work, he is reported to have said, "I gave Koch a Christmas present," (by sending him a reprint of his falciparum paper). In 1902, the Nobel committee was deciding the value of Ross's contribution and that of Grassi, and the initial sentiment was that the prize should be shared. According to Desowitz (1991), "Koch threw the full weight of his considerable authority in insisting that Grassi did not deserve the honor." So much for Christmas presents!

Desowitz (1991) tells the story of riding in the back seat of a cab in Lisbon at a professional meeting in 1958, sandwiched between Professor Saul Adler and Count Aldo Castellani, both with significant reputations in tropical medicine, when an argument broke out regarding the merits of Ross versus Grassi in the race to success with malaria. Adler was deriding Grassi as a "liar" and a "thief" of Ross's work, and Castellani was trying to defend Grassi's honor by asserting the importance of his contributions. Finally, Castellani said, "It was those Germans, those damned Germans, who hated us Italians and denied Grassi the honor he deserved." Desowitz (1991) also recalls being with Bernardino Fantini, Professor of Medical History at the University of Rome, some thirty years later. Fantini was giving Desowitz a tour of the Compagna and Pontine marshes where malaria had once ruled the countryside. The Italian professor and Desowitz began discussing Ross and Grassi, and the significance of both their contributions to understanding the malaria problem. However, as they talked, again the Germans entered the conversation. After agreeing on the valuable works of both Ross and Grassi, Fantini turned to Desowitz and remarked, "It wasn't Ross; it was that damned German, Koch, who denied Grassi a share of the Nobel Prize and the recognition he deserved." 'Peace on Earth . . . !'

It is amazing sometimes to see the errors that make it into the peer-reviewed scientific literature, but it does happen. It is even more amazing with respect to errors that are then transferred from the primary literature into scientific textbooks. I recall taking a general zoology course in the first semester of my freshman year at Colorado College from a very good parasitologist, Robert M. (Doc) Stabler. Our textbook for the course was entitled *General Zoology*. Tracy I. Storer of the University of California–Davis and Robert L. Usinger, University of California–Berkeley, were coauthors. Both were very well respected in their fields.

The book was in its third edition [Storer and Usinger, 1957] and was, as I recall, widely used at the time. On p. 52, however, there is a figure (Fig. 3-9) depicting a human karyotype. The figure legend reads, "The 48 human chromosomes."

Yes, I learned that the diploid number of chromosomes for *Homo sapiens* was 48 in my first semester as an undergraduate student. I think, though I cannot recall for certain, that I 'unlearned' this piece of information before I finished college. I still have the book, and just recounted the chromosomes in the figure. There are still 48 of them. Where the extra pair came from is impossible to say because the Literature Cited section at the end of the chapter does not cite any papers that would suggest an original source. Somehow, this bit of misinformation appeared in the primary literature and became 'gospel' when it was published in a textbook. Not long ago, I was interviewing Jim Oliver for the present book and we talked about this issue. Jim told me that he had been a faculty member at University of California-Berkeley and a friend of Robert Usinger. He said that the karyotype that appeared in the Storer–Usinger textbook was real and he recalled that Usinger told him that it had come from someone who had been institutionalized in a mental hospital. As he recalls it, he believed that Usinger had said the karyotype was representative of someone with some sort of non-disjunction problem and this was why there were 48 chromosomes rather than 46.

A similar kind of 'fraud' was perpetrated in 1903. That year, Professor Fritz Schaudinn, a noted German protozoologist and microscopist, published a paper in which he said that after the sporozoite of *Plasmodium* spp. is inoculated into the blood of its human host, it circulates for about thirty minutes and disappears – into red blood cells! In the published paper, he said that he saw it happen. Max Hartmann, Schaudinn's lab assistant, reportedly told a certain Professor Reichenow, another of the great malariologists of that era, that he had seen the same thing that Schaudinn had observed. Within a short period of time, Schaudinn's observation began to appear in parasitology textbooks. In some, there was even a diagram depicting the fictitious event. However, everything reported by Schaudinn regarding sporozoite penetration of red blood cells was 'bogus'! I asked Bob Desowitz how in the world Schaudinn could have made such a mistake. He replied, "I heard an interesting conjecture about this when I was a student at the London School of Hygiene and Tropical Medicine. I recall listening to a lecture by Phillip Manson-Bahr, a son-in-law of Sir Patrick Manson and a well-regarded tropical medicine figure. Manson-Bahr said, almost jocularly, that maybe Schaudinn had too much good German beer when he sat down at the microscope. Bahr was German, so he should know."

This erroneous observation of Schaudinn created huge problems in resolving the final step in the malaria life cycle. Over a period of some forty years that followed, several post-Schaudinn investigators reported that the parasite disappeared from the blood for a time, then reappeared at species-specific intervals afterward. For example, Sir Neil Fairley conclusively determined during WWII that blood removed from individuals infected with *P. vivax* was not infective for nine days when inoculated into uninfected volunteers and that, similarly, it was not infective for six days in the case of *P. falciparum*. When blood from people with patent infections of each of these species of parasite was transferred to uninfected volunteers, the blood recipients then suffered the consequences of the disease. But where were the parasites from inoculation to the time of patent parasitemia? This was a very troublesome, some would say vexing, problem, especially in view of Schaudinn's report and the fact that major textbooks of the day indicated that sporozoites penetrated red blood cells, even to the point of diagramming the phenomenon. Interestingly, it is probable that several early investigators, including such stalwarts as Grassi, Golgi, and MacCallum, had observed tissue forms of the parasite, but did not understand what they had seen. Mostly because of Schaudinn's report, none of them pursued the issue.

In 1930, however, Clay Huff found stages of *Plasmodium elongatum* in the avian red blood cell (rbc) precursors of hemopoietic tissues. Then, five years later, he and William Bloom found the same parasite "in all cells of the lymphoid and myeloid series" (cited from Russell, 1955). In 1938, S. P. James and P. Tate described a non-rbc form of schizogony in *Plasmodium gallinaceum*, referring to it as "exoerythrocytic schizogony." This kind of multiple fission refers to schizogony in any tissue other than that which occurs in blood. Pre-erythrocytic schizogony occurs in tissues before blood is invaded. The latter kind of division was discovered independently in 1940 by L. Mudrow in Germany and H. E. Shortt, K. P. Menon, and P. V. Iyer in India, all working with *P. gallinaceum*.

Henry Shortt had recently retired from the Indian Medical Service (IMS) and joined the faculty at the very prestigious London School of Hygiene and Tropical Medicine where he was to team up with P. C. C. Garnham for the momentous research effort to which both were to contribute equally. Shortt had made his reputation working on both *Plasmodium* spp. and *Leishmania* spp. while in India before retiring as a Colonel in the IMS (Desowitz remarked during my interview with him that "the rank of colonel wasn't that unusual, they were all colonels").

Garnham had just returned from East Africa where he had made a considerable contribution to the apicomplexan literature with his work on *Hepatocystis* sp. Because schizogony is restricted to the liver in *Hepatocystis* sp. and only gametocytes occur in the blood, some folks have speculated that it was Garnham who pushed the liver idea on Shortt as the site for an exoerythrocytic stage of *Plasmodium* spp. However, Bob Desowitz told me that he had written a piece for the WHO Bulletin in a commemorative issue in which he argued that Garnham's *Hepatocystis* studies did not direct examination to the liver in their *P. cynomolgi* experiment. Later, Bob Killick-Kendrick, a student of Garnham (and Desowitz' technician in Africa in his predoctoral days) read the piece and told Bob that he was correct. It seems that Killick-Kendrick had interviewed Garnham with a view to making an oral history (which never materialized) and said to Desowitz that Garnham had told him that he thought the exoerythrocytic stage would be found in a reticulo-endothelial cell as in the avian malarias. Still, there was a friendly dispute between Garnham's students and those of Shortt about the issue. Desowitz emphatically disagreed with those who claimed that Shortt was the follower. He told me, "It wasn't that way at all!" He said that, together, Shortt and Garnham literally "took that monkey apart." So, their first vertebrate model for this groundbreaking research was a monkey and the parasite was *Plasmodium cynomolgi*, a species very closely related to *P. vivax*. Essentially what they did was to collect a large number of infected mosquitoes, grind them up, and inoculate the 'mess' containing the sporozoites into the monkey. They waited six days and killed the animal. They then very carefully searched the various organs and tissues for the parasite. The parasites were found in the liver parenchymal cells. The experiment was repeated using a second monkey and the results were the same. The parasites were definitely in the liver.

Desowitz said that, inevitably, word of the Shortt/Garnham discovery got out before they could publish it. Word also leaked out that Frank Hawking, who worked for the National Institute for Medical Reseach (NIMR) at Mill Hill in London (NIMR is the English equivalent of the NIH in the U.S.A.) went to his collection of old tissues on hearing of the new discovery, and he found it too, thus confirming the site of exoerythrocytic schizogony. "But it was too late by that time. Hawking had wanted to outmaneuver Shortt, and Shortt never forgave him for that," according to Desowitz. Apparently, Hawking had developed a reputation for confirming the discoveries of other researchers over the years. Bob told me that in a somewhat contemptuous way, Shortt referred to Hawking as "the Bishop", for this reason.

Eventually, Shortt and Garnham had to support the exoerythro-cytic hypothesis with a species of human *Plasmodium*. So, they began a collaboration with G. Covell and P. G. Shute of the Ministry of Health's Malaria Laboratory at the Horton hospital for Mental Diseases, Epsom, England. They used *P. vivax* and a human volunteer. They had to find someone who was not only willing to suffer the consequences of hav-ing malaria, but who was also willing to undergo major abdominal surgery so that subsequently a piece of liver could be excised for histo-logical examination. The man who volunteered, with his wife's per-mission as well, had been given malaria nearly two years previous for a medical reason, but had recovered. He was inoculated intra-venously with sporozoites isolated from 200 salivary glands removed from some 2000 infected *Maculipennis atroparvus*. A week later, the vol-unteer was operated on and a piece of tissue was removed from his liver.

Desowitz recalled that when he arrived in London in September 1948, "Shortt was away. He and Garnham had just done the exoerythro-cytic cycle infection and the volunteer had had his liver biopsy com-pleted, but Shortt decided he had to go and 'kill' a couple of salmon in Cornwall first." Apparently, Shortt was an avid sportsman, even to the point of shooting tigers when he was in India. According to Desowitz, "This guy would kill anything. So the tissues just sat there. No one was allowed to look at them. When he came back, they cut the tissues and, *voila*, there were these things [the parasites] in the liver," just as they had predicted. Desowitz said that Shortt then "called all the grad-uate students down to his lab and they all peered into his new Letitz microscope."

A year later, in 1949, they obtained a known strain of *P. falci-parum* from a Hungarian colleague, infected 770 mosquitoes, and then fed them on another volunteer. Approximately six days later, the oper-ation took place and a piece of liver was removed for sectioning and staining. Again, the parasite was found. Incidentally, the infected volun-teer was treated with chloroquine and recovered, though not without subsequent incident. Russell (1955) reported that the volunteer later developed a duodenal ulcer. When operated on to fix the ulcer, the sur-geon reported that adhesions from the first operation had covered the exit of the ulcer into the body cavity, preventing peritonitis from devel-oping and probably saving the man's life, "a sort of reward for services rendered!" The exoerythrocytic stage in the parasite's life cycle has been found to be a consistent phase in the life cycle of all *Plasmodium* species that have since been checked.

Following the discovery of the liver stages for *P. vivax* in 1948, and *P. falciparum* in 1949 by Henry Shortt, P. C. C. Garnham, and their colleagues, a great deal more was to occur in the malaria world. Perhaps the first momentous event came with the World Health Organization's announcement in the 1950s that they were going to eradicate malaria from the earth by eliminating the mosquito vector, not a bad goal. However, the insecticide selected to do the job was DDT. The full ramifications of this choice were not clearly understood until the WHO announced several years later that instead of eradication, they were just going to 'control' the mosquito. They had run into some unanticipated roadblocks. First, WHO did not figure on mosquitoes developing resistance to the insecticide, but they did. Second, they did not figure on *Plasmodium* spp. developing resistance to most of the old 'tried and true' drugs that had been in use for several years (aside from artemisinin, the only drug that has remained consistently effective over the years is quinine and now there are even indications that its effectiveness is diminishing in certain parts of the world). Finally, the mathematical models of George McDonald used to predict the demise of the mosquito were totally inadequate; they were not even close to being accurate. According to Bob Desowitz, projections indicated that "reducing" mosquito numbers via indoor house spraying with DDT could interrupt malaria transmission. Then, according to Bob, "After 5–10 years of the attack–consolidation phase, vector numbers would be allowed to rise again, but then would be only nuisance biters without parasites." Why was it that those in charge of these programs never realized, or were never told, that the mathematical model is just that, a model, a mathematical hypothesis, nothing more or less. Any hypothesis, mathematical or not, is always based on a certain set of assumptions. If the assumptions are incorrect, the models will be as well, in some cases by many orders of magnitude. The old computer cliché definitely applied for the models used for mosquito eradication in the 1950s, i.e., "garbage in, garbage out," because they certainly did not work. Finally, of course, Rachel Carson published *Silent Spring* in 1962, and this did not help the DDT cause very much either, even though the target was vectoring a parasite that was killing several million people each year, and mostly children at that!

But, let's not castigate Rachel, or anyone else, because the idea for mosquito eradication was well intended. However, it was doomed to fail from the beginning because, very simply, it will not work. Killing all the mosquitoes is way too far out. In dealing with a similar, but certainly far less serious, problem, I heard a famous malacologist say once that trying to reduce snail populations to control swimmer's itch in Michigan was like trying to kill all of the poison ivy in the woods. But then, the road to hell is paved with good intentions, isn't it?!

Nonetheless, success has been achieved with reduction of malaria in some parts of the world. In the U.S.A., for example, we had >1 000 000 cases as late as the 1930s, but it is gone now, except for the occasional outbreak adjacent to a military base, especially in the southeastern part of the country. At least two programs helped with the knockout of malaria here, although neither was directly designed for that purpose. One was simply the installation of hardware cloth, i.e., 'screens', to cover the windows of our homes. The second was construction of the TVA (Tennessee Valley Authority) dams in the southeast. This series of dams was originally designed for flood control and power production. However, it was also learned that manipulation of water levels in the reservoirs could also control mosquitoes during their breeding season.

Another major contribution to the study of malaria over the years, mostly since 1949, has been the development of various technologies that have led to major breakthroughs in microscopy, X-ray crystallography, and molecular biology. Electron microscopy has allowed us to peer inside cells with much greater resolution and magnification, and inside plasmodial cells as well. X-ray crystallography permitted Rosalind Franklin to take beautiful photographs of DNA, which in turn permitted James Watson, Francis Crick, and Maurice Wilkins to figure out the helical nature of DNA. This allowed them to postulate the structure of DNA and, with this information, molecular biology was born as a discipline. With it came recombinant DNA technology and genomics. With the latter came the successful Human Genome Project. Then, very recently, the same procedures were used to map the genomes of *P. falciparum*, which is the highly lethal subtertian form of malaria, as well as *Anopheles gambiae*, which is the primary vector of the parasite in sub-Saharan Africa. While too early yet for any breakthroughs using the detailed information of the parasite's or the mosquito's genomes, most predict that it will happen relatively soon.

Interestingly, there are two ideas tied to recombinant DNA technologies that had their origins in the nineteenth century. One is that mitochondria were 'derived' from an endosymbiotic relationship. The second is like the first, and it is that plastids were also endosymbionts, and that the latter, more specifically, evolved from a cyanobacteria-like organism. There are several ways in which knowledge of the endosymbiotic plastid has significance to current research dealing with *Plasmodium* species. The most important has to do with the fact that since plastids are derived organelles, they should have retained at least some enzymatic activity that is unique to their evolutionary ancestors, even though strong evidence suggests that at least some of its genome has been transferred to that of the host cell. For example, Roberts *et al.* (1998) have shown that at least some apicomplexan parasites possess a "shikimate pathway" that is used in the synthesis of certain aromatic amino acids, e.g., tryptophan, phenylalanine, and tyrosine. Chorismate, an intermediate in the pathway, is involved in the production of folate, which is essential to sustain life of the rbc forms of the *Plasmodium* parasites. The significance of this pathway is evident because it is not part of the metabolic circuitry of the mammalian host. Since this is the case, scientists have recognized that enzymes in the pathway are natural targets for development of possible ways for dealing with the malarial parasites. In effect, *Plasmodium* spp. have a 'relic' organelle that offers a number of opportunities for possibly creating new drug therapies.

The second line of research using the new molecular technologies provides for understanding evolutionary relationships among the various species of *Plasmodium*. This approach has been well documented in a recent review by Stephen Rich and Francisco Ayala (2003). Phylogenetic analysis, whether based on perceived morphological or genetic similarities or differences, is used to formulate hypotheses that project an organism's evolutionary history. Most of the time, the basis for phylogenetic analysis will have a solid foundation and should not be misconstrued, even though they frequently are. Why? In part, it is because parasitic protozoans and helminths have no fossil record. However, I also think it is, in part, because some folks get some things too set in their thinking. It is, after all, difficult to give up an old idea, or an old friend. In fact, during my tenure as Editor of the *Journal of Parasitology*, most of the really fractious 'interchanges' that occurred between authors and referees have dealt with phylogenetic issues.

There is general agreement that host switching played a major role in the dissemination of *Plasmodium* species throughout their history, and it is apparently a long one. There is also good evidence to suggest that apoicomplexans have been around for more than a billion years – yes, that's a billion, with a 'B', not a million, with an 'M'. However, it is not my intention to rehash the phylogeny of *Plasmodium* species. I will deal, instead, with two other issues raised by Rich and Ayala (2003) and by Bob Desowitz in our recent interview and in his delightful (1997) book, *Who Gave Pinta to the Santa Maria*? The first is, who gave what to whom? The second is, how did human *Plasmodium* get to the New World? Was it already here when Columbus came, or did the Old World bring it to the New in 1492?

There have been many phylogenetic analyses of the hemosporidian group. As far as I am concerned, and Rich and Ayala (2003) agree, one of the best is that of Susan Perkins and Joe Schall (2002) published in the *Journal of Parasitology*. It is the one on which I will base my position. Several years ago, Waters *et al.* (1991) suggested that *P. falciparum* had an avian origin; then, via host switching, it was acquired by humans. This is an intriguing idea, one supported in part because it is the highly virulent species among the four species traditionally considered as primary for humans. Waters *et al.* (1991) asserted that humans acquired *P. falciparum* when they stopped being 'hunter-gatherers' and turned to agriculture for sustenance. However, Perkins and Schall (2002) disagreed, saying *P. falciparum* has a "deep root within the *Plasmodium* species infecting mammals." They thus concurred with Escalante and Ayala (1994) who had previously indicated that *P. falciparum* was more closely related to *Plasmodium reichenovi*, a species strictly associated with African chimpanzees. This is the same position taken by Bob Desowitz in *Malaria Capers*, and again during our conversation in his Southern Pines home. But, what of the other species? According to Perkins and Schall (2002), *Plasmodium ovale* appears to be alone and more closely related to *Hepatocystis* spp., which are parasites of African rodents.

For many years, there has been a lively discussion as to whether *P. vivax* and *P. malariae* were present in the New World prior to the arrival of Columbus. If they were here, how did they arrive? For *P. vivax*, there are two possible explanations. In the first case, some have assumed that the parasite might have come out of Asia with the first native Americans who crossed into Alaska via the Bering Straits about 15 000 years ago. However, considering the cold temperatures that far north, the amount of time it would have taken to reach more temperate parts of North America, and the biology of the parasite, most agree this is not a viable hypothesis. The second idea suggests the parasite came with immigrant Japanese, a sort of reverse 'Kon Tiki', if you will. Bob Desowitz, in *Who Gave Pinta to the Santa Maria*, describes the work of Betty. J. Meggars and Clifford Evans, a team of archeologists, who claim it is probable that natives from the Japanese Island of Kyushu made their way, by accident, to the shores of Ecuador some 5000 years ago. Apparently, shards from a dig at Valdivia, a seaside town on the Pacific coast of Ecuador, are identical to those recovered on Kyushu Island from about the same time period. Desowitz thinks these folks could have brought *P. vivax* with them, as well as their pottery. In contrast, the quartan species *P. malariae* had to come from Africa. There are some who believe that native Africans were here in the western hemisphere prior to Columbus and they could have brought *P. malariae* with them. However, it is much more likely they would have brought *P. falciparum* since it is this species that ravages sub-Saharan Africa. Tongue-in-cheek, Desowitz even offers the suggestion that perhaps *P. falciparum*, along with *P. malariae*, both came with wandering Africans prior to Columbus and that the former was eliminated because of its virulence for the immune-naïve native Americans, leaving the more benign quartan form of the parasite to fester.

There is yet another interesting species identification problem that could shed some light on resolving this question. It has to do with *P. simium*, which looks exactly like *P. vivax*, and *P. brasilianum*, which is a perfect match for *P. malariae*. Both *P. simium* and *P. brasilianum* are in New World monkeys, and both can be transmitted to humans, or is it the other way around? Were *P. vivax* and *P. malariae* transmitted to monkeys after Columbus and his cohorts arrived in the fifteenth century? Answers to these questions are obviously lacking. The serious phylogeny of Perkins and Schall (2002) place *P. vivax* and *P. simium* together, but they omit any reference to *P. malariae* and *P. brasilianum*. For some reason, they even exclude mention of the latter species. Rich and Ayala (2003) suggest the question will be eventually answered, "by comparing the genetic diversity of the human and primate parasites." Continuing, they state, "Genetic diversity will be greater in the donor host than in the recipient host of the switch. If the transfer has been from humans to monkeys, the amount of genetic diversity, particularly at silent nucleotides sites and other neutral polymorphisms, will be much greater in *P. vivax* than in *P. simium*, and in *P. malariae* than in *P. brasilianum*." Desowitz still believes the transfer went the other way. He said to me, "humans gave it [*P. vivax*] to monkeys. I got this from Garnham, and Garnham always had the last word. *Plasmodium malariae* is a great mystery with regard to its origin." Most, however, feel that *P. malariae* came from chimpanzees and that it is definitely African in origin. There are two main problems in determining which way the transfer occurred, i.e., human to monkey, or monkey to human. First, we have no written documentation of the disease being present in the western hemisphere prior to Columbus. Second, we have very little information on the genetic diversity for any of the four species with which we are dealing, so it is difficult to make a call one way or another.

As I close this essay, I look back in awe at the discoveries that have been made with respect to malaria and *Plasmodium* spp. over the past

125 years, and I ask the question, were these discoveries serendipitous, or otherwise? Laveran's discovery was not serendipity. He was looking for the parasite. Ross's discovery was also not luck; it was highly calculated. He knew what the parasite looked like in human blood. He also strongly suspected the mosquito's involvement in transmission. If there was any luck, it involved his switch from human malaria to that of birds once he was based in Calcutta. If anything, it would seem this switch speeded his resolution of the parasite's life cycle. Contributing significantly to his discovery was the work of McCallum in understanding the sexual phase of the cycle. If there was any serendipity in the overall progress with understanding the *Plasmodium* spp. life cycle, it was Schaudinn's mistake in saying the parasite went immediately into red blood cells on inoculation by mosquitoes into the human host. This delayed completion of the final step in the life cycle for 45 years. I am absolutely convinced that discovery of the liver phase and pre-erythrocytic schizogony would have occurred much sooner than it did otherwise. After all, Clay Huff knew about the exoerythrocytic steps in avian malaria as early as 1930, nearly twenty years before it was known in monkeys, and then in humans. Recall also that the Italians ran the cycles of vivax, quartan, and falciparum malaria through mosquitoes and into humans within a matter of two years following Ross's discovery. If Schaudinn had not 'poisoned the trough', complete success almost certainly would have been achieved much more quickly. So, except for Schaudinn's mistake, serendipity definitely was not a factor in pursuit of the *Plasmodium* spp. life cycle. The research here was more like careful plodding. Each step was based on a foundation created by a previous investigator. In reality, we can follow the trail all the way back to the uneducated Dutchman Antonie van Leeuwenhoek, and his passion for grinding lenses.

REFERENCES

Carson, R. 1962. *Silent Spring*. New York: Houghton Mifflin.
Desowitz, R. S. 1991. *The Malaria Capers*. New York: Norton.
Desowitz, R. S. 1997. *Who Gave Pinta to the Santa Maria?* New York: Norton.
Escalante, A. A. and F. J. Ayala. 1994. Phylogeny of the malarial genus *Plasmodium*, derived from rRNA gene sequences. *Proceedings of the National Academy of Sciences USA* **91**: 11373–11377.
Perkins, S. L. and J. J. Schall. 2002. A molecular phylogeny of malarial parasites recovered from cytochrome b sequences. *Journal of Parasitology* **88**: 972–978.

Rich, S. H. and F. J. Ayala. 2003. Progress in malaria research: the case for phylogenies. *Advances in Parasitology* **54**: 255–280.

Roberts, F., J. J. Johnson, D. E. Kyle, *et al.* 1998. Evidence for a shikimate pathway in apicomplexan parasites. *Nature* **393**: 801–805.

Russell, P. 1955. *Man's Mastery of Malaria*. Oxford: Oxford University Press.

Storer, T. I. and R. L. Usinger. 1957. *General Zoology*. New York: McGraw-Hill.

Waters, A. P., D. G. Higgins, and T. F. McCutchan. 1991. *Plasmodium falciparum* appears to have arisen as the result of lateral transfer between avian and human hosts. *Proceedings of the National Academy of Sciences USA* **88**: 3140–3144.

3

The HIV–AIDS vaccine and the disadvantage of natural selection: the yellow fever vaccine and the advantage of artificial selection

Cruel as death, and hungry as the grave.

The Seasons: Winter, James Thomson (1700–1748)

One of the things I have always been curious about is why vaccines are effective for certain viruses and not others. The major viral scourge of today is, of course, HIV, with influenza probably a close second. Yellow fever was a major problem in the world until the early 1930s. In the case of HIV, a vaccine has not yet been created, despite an investment of hundreds of millions of dollars. An effective yellow fever vaccine was developed some 75 years ago, in the early 1930s. In fact, since then, nearly 300 000 000 doses of the latter vaccine have been administered without adverse effect. The question I am going to ask in this essay is, why has there been a vaccine success for one of these viruses, but not the other? In a curious way, the answer is decidedly ecological.

The primary sources for my information came from several very good virology books, discussions with a virologist colleague, plus some literature searches in our library. The first tome I used was Topley and Wilson's Volume I of *Microbiology and Microbial Infections: Virology*, edited by Brian Mahy and Leslie Collier (1998). A second was a popular general ecology textbook, *Evolutionary Analysis*, written by Scott Freeman and Jon Herron (2004). I also had a series of very productive discussions with Pat Lord, a very solid virologist in our Biology Department here at Wake Forest University. Finally, I tracked down a number of invaluable literature sources that provided some very insightful and useful information regarding some of the earliest work on yellow fever. Pat also shared a couple of papers with me.

Quite simply, viruses are little more than nucleic acids wrapped in a protein coat. The nucleic acid within the coat may be single stranded or double stranded and it may be either DNA or RNA. Every virus, no matter the kind of nucleic acid, is totally dependent on the host's cellular machinery to do its thing.

The yellow fever virus (an arbovirus, one that is arthropod borne) belongs in the Flaviviridae (Gubler and Roehing, 1998). As it turns out, the nucleic acid in the virus causing yellow fever is an average-sized piece of single stranded RNA. It consists of about 10 000 bases, as compared with HIV, which is 9.1 Kb. (The RNA viruses typically range from 7 to 12 Kb in total size.) Yellow fever RNA looks just like a strand of mRNA normally found in the host cell. According to Pat Lord, "All of these viruses have an RNA polymerase that can copy their genome. None of these RNA polymerases has a proof-reading mechanism. It is thought that, evolutionarily, one of the reasons these RNA viruses all have this smaller number of bases is because if they were any larger, there would be more reading mistakes than there are already. Plus, if they were larger, they would take longer to replicate, which would not be very good for getting into a host cell and getting out. Interestingly though, the STARS (DNA) virus has a whopping 32 Kb, which is huge compared with the yellow fever and HIV viruses. So, DNA viruses tend to be much larger. The reason is that all DNA polymerases have proof-reading mechanisms and are not, therefore, nearly as vulnerable to mutation and change as are RNA viruses." However, there is an exception to the 'rule' regarding mutation among RNA viruses and that exception resides with the yellow fever virus, to which I will refer later in this essay.

As soon as the virus enters the cell, the yellow fever mRNA will cause the synthesis of viral protein-using ribosomes of the host cell. As with several other arboviruses, it is zoonotic, with monkeys serving as the primary reservoir. Interestingly, the yellow fever virus can kill South American monkeys in nature and Asian monkeys in the lab, but it is not harmful to African monkeys, one of the reasons yellow fever is believed to have an African origin.

In contrast, the HIV virus, a member of the Retroviridae, includes two copies of its RNA genome. Each RNA strand of the virus is equivalent, in terms of its function, to a strand of host mRNA. The viral genomic RNA is copied by viral reverse transcriptase into double-stranded DNA. It then integrates into the host cell DNA. The cellular RNA polymerase II then synthesizes viral mRNAs (some of which can serve as genomic RNA). Retrovirus RNA is not totally double stranded;

they have regions that will anneal to each other, making them partially double. For this reason, retroviruses are often referred to as being pseudodiploid. As Pat Lord explained, "These viruses carry with them several of their own enzymes, including a reverse transcriptase that will stimulate the synthesis of double-stranded DNA. The mRNA of the virus and the newly synthesized DNA are, however, not exactly alike. There are unique sequences in two places within the new DNA. Thus, at each end of the viral DNA, there are terminal sequences that act as recognition sites for insertion of its DNA into the host's genome. Moreover, the insertion location is not totally random with regard to the host's DNA. Apparently, there are insertion sites in the host DNA that can be considered as 'hot spots', i.e., regions that tend to be transcriptionally active. Insertion is possible because of a second viral enzyme, this one called integrase, which cuts the host DNA and allows for the viral DNA to be inserted. Once the double-stranded DNA is in place, the reverse transcriptase is degraded."

Pat continued, "The advantage of being a retrovirus is that once it is inserted into the host genome, it is there for the life of the cell. This is perhaps the greatest problem with this virus and at least one reason why we have not been able to come up with a vaccine. We have not been able to prevent this step from occurring." I then wondered out loud if she "felt that HIV/AIDS is an 'intractable' problem?" I also said, "I know this is an impossible question to answer because of the speed with which science and technology are moving in these days, but what do you think?" She responded, "Certainly, we have come a long way in treating this disease. I mean people are living a lot longer than they were 15 to 20 years ago, but only because of the drugs now being used to treat the disease. The reason that drug therapy is working is because scientists have recognized that the only way to attack this virus is to give people a regimented drug cocktail, to the extent that virus production is slowed way down. As soon as the virus starts making new copies, however, and because the enzyme [reverse transcriptase] is so lousy and keeps making mistakes, it is going to produce new strains." I then asked, "Is HIV cell specific?" She replied, "HIV will only infect CD4 T-helper cells, the very ones that trigger the immune response. And here is another trick of this virus. The T-helper cells are usually 'sitting around' not doing very much, unless they are exposed to an antigen to which they are sensitive. As soon as they recognize the antigen, they proliferate. If the T-helper cell happens to be infected with HIV, the virus inside the cell will be duplicated when the cell begins dividing. It is then passed on to all of the progeny cells. What an ideal place for this virus to hang out!" she exclaimed. She then said, "The group of viruses to which HIV belongs is also called the lentiviruses. It is a subfamily of the Retroviridae. The lentiviruses have this characteristic long-term association with their host. HIV, unlike a lot of other lentiviruses, has a lot of other accessory proteins that help it to subvert immune detection or that help it to proliferate. It is evolution at its highest! HIV has also evolved another protein called the viral infectivity factor [vif]. When it was first identified, they didn't know what it was doing. However, they noticed that mutants that didn't have this protein were not as infective."

I wanted to know more about vif because I was totally in the dark, so I asked Pat. As she described it, "vif is involved with circumventing a natural defense system associated with a universal cellular enzyme called APOBEC. Remember, lentiviruses stay in association with host cells for a long time, so it behooves host cells to have a mechanism to prevent the spread of the infection. When lentiviruses are being packaged inside the cell, the APOBEC protein is incorporated inside the new virions. When the new virion containing APOBEC spreads to a new cell and has its RNA copied into double-stranded DNA, APOBEC acts on the reverse transcriptase, which causes mutations in the double-stranded DNA. In effect, it causes the reverse transcriptase to 'screw up' even more! So, now then, even though this is now inside the cell, it is so highly mutated it cannot direct synthesis of new virions, i.e., new genetic RNA. In effect, it stops it and, for most lentiviruses, the processes involved with virion production are halted. However, HIV does not play by the same rules as other lentiviruses. It codes for this protein called vif, the sole purpose of which is to bind to APOBEC, preventing the cellular enzyme from causing mutations." On hearing this story, I muttered under my breath, "Well, I'll be damned!" Pat's tale about vif and HIV immediately re-enforced my thinking regarding how devious parasites can be in their relationships with their various hosts and how far up and down the evolutionary scale this Machiavellian behavior extends.

There are some workers who feel that we simply do not know enough about the biology of the AIDS virus to develop effective vaccine designs. Pat reminded me, "Moreover, we do not have a good animal model for vaccine trials with HIV and this is also a serious handicap. There is a similar virus, SIV, in monkeys, but it does not resemble HIV close enough for it to be of any real value."

Evolutionary trees suggest HIV was first transferred from a chimpanzee to a human in 1931. It took 50 more years, i.e., 1981, for the viral disease to reach epidemic, then pandemic, proportions. I was interested in the timing here, because of something that Bob Desowitz said to me during our interview. He remarked that in Africa during the 1930s, some European researchers made a habit of inoculating *Plasmodium*-infected simian blood, e.g., chimpanzees, etc., into unknowing human volunteers, "to see what would happen." This was of course, back in the days when there were no Animal Care and Use Committees to regulate such activities. I was given the strong impression that some of the science done in Africa during that time frame was rather 'sloppy', and even disgusting. During my conversation with Bob, I asked if HIV could have possibly been transferred to humans during some of these 'experiments', and he replied that he did not see why not.

Shankarappa *et al.* (1999) reported the details of a 12-year case history of a person with AIDS. During the first seven years, they discovered that 8% of the nucleotides in the virus underwent a change when compared with the original genome. This is an exceedingly rapid rate of mutation, more so than for most viruses, certainly than for any eukaryotic organism. As noted by Freeman and Heron (2004), between years six and eight of infection, the rate of mutation slowed radically. As I mentioned above, CD4 T-helper cells function in activating other cells in the immune system and are essential for the system's proper function, so these populations were also followed during the course of infection. At year six, the CD4 T-helper cell count in the blood was $1200/ml^3$ and, by year eight, it was $200/ml^3$. Freeman and Heron (2004) indicate that the decline in the number of CD4 T-helper cells reflects the disintegration of the host's immune system. They state, "The collapse of the patient's immune system meant that the patient's body was no longer producing new kinds of antibodies and new kinds of killer T cells. This freed the HIV population from the selective power that was forcing it to evolve. There was no longer any benefit to having novel epitopes. Instead, the strains most capable of rapid replication simply spread and those less capable became rare." In other words, at this point the host was immunodeficient. The result in a person infected with HIV would then be total vulnerability to infection by a wide array of pathogens, or cancers, or both. The overall scenario with respect to viral mutation and host immune response can be considered as natural selection running 'full tilt'.

Vaccines work by taking advantage of the host's anamnestic capability. What does this mean? It simply refers to the immunological capacity of a vertebrate animal to remember. If an antigen is presented to a host, antibodies will be produced under normal conditions. If the same antigen is presented to a host again, antibody production is greatly speeded up and, hopefully, quickly enough to neutralize the antigen's potentially negative impact. On the surface of the HIV virion, the infective agent for the AIDS virus, is an envelope protein known as gp120. The diversity of gp120 epitopes on the surfaces of HIV virions is huge. Therefore, any vaccine developed for HIV must be effective for all possible gp120 epitopes, a colossal, if not impracticable, challenge. Another problem is that continued SIV transmission from chimpanzees to humans is also a real likelihood, and this means the constant acquisition of new and different epitopes by the human population. In many ways, this constant change in epitopes, or transmission of new ones from the chimpanzee, reminds me in some ways of the manner in which the variant surface glycoproteins (VSGs) of the African trypanosomes operate, i.e., they are changing constantly in response to the host's immune challenge. Moreover, immunosuppression is, as in the case of AIDS, also an important feature of African trypanosomiasis.

Another approach in the effort to control HIV/AIDS should be antiviral drugs, but this has proved to be nearly as challenging, and frustrating, as the vaccine route. One of the earliest drugs to be tested, and initially thought to be effective, was called AZT, an abbreviation for azidothymidine. Normally, a reverse transcriptase from the virus uses the viral RNA to synthesize new DNA. The nucleotide building blocks incorporated into the new DNA are 'stolen' from a pool maintained by the host to construct its own nucleic acids. AZT looks enough like the real thing (thymidine) to fool the viral reverse transcriptase into using it, as well as normal thymidine in this synthetic process. The problem is, if AZT is incorporated into the new DNA, additional nucleotides cannot be added and normal DNA synthesis is stalled. However, after several months of effective use, the drug stops working. Freeman and Herron (2004) explained the problem, again in terms of natural selection. Over this period of several months, the HIV population evolved. As indicated by Larder *et al.* (1989), "In most patients, the evolution of AZT-resistant HIV takes just six months."

But what goes wrong and encourages drug resistance to develop? The answer lies with the viral reverse transcriptase itself. As stated earlier, it makes mistakes, many, many mistakes. According to Hubner *et al.* (1992) and Wain-Hobson (1993), in excess of 50% of the DNA transcripts made by reverse transcriptase of HIV have errors, with each possessing at least one 'erroneous' nucleotide in place of the correct one. Freeman and Heron (2004) note that the HIV system is very sloppy in this regard. Since so many HIV generations are produced within an infected individual, it is inevitable that a mutant strain resistant to the AZT drug will eventually emerge. They go on to say, "This process of change over time in the composition of the viral population is called evolution by natural selection" (Freeman and Heron, 2004).

In these ways, and others, AIDS is cruelly successful, with estimates of about 8000 deaths now occurring per day. Freeman and Herron (2004) cite data from the Joint United Nations Programme on HIV/AIDS suggesting that by 2020 the AIDS pandemic will have claimed 90 million lives, and virtually all of these since 1981 when the disease's epidemic status moved to that of a pandemic. As near as I can tell, for as long as HIV mutation and natural selection in the chimpanzee and the human host continues, the virus will be with us.

Whereas success in fashioning a vaccine for HIV has proved to be impossible to date, notable victories have been achieved for other viruses, for example, smallpox, polio, and tuberculosis. The yellow fever vaccine is another excellent example, even though at least 200 000 cases of this disease and some 30 000 deaths are known to occur annually. None of these people, however, mostly in sub-Saharan Africa, received the vaccine. As indicated earlier, the obvious question is why vaccines for some viruses can be developed, but not for HIV? There is a simple answer to this question, one to which we will now proceed.

The first recorded outbreak of yellow fever occurred in Mexico in 1648. It was most probably associated with the slave trade since most evidence indicates the yellow fever virus was first transferred via mosquitoes from monkeys to humans, and most likely in Africa, not in the western hemisphere. Although it can be transferred from monkey to humans via mosquitoes, making it a zoonotic virus, it can also go from human to mosquito to human. It is this latter route that accounts for most of the major epidemic outbreaks recorded over the past 375 years. The mosquito vector in the latter instances is *Aedes aegypti*. If one reads the historical literature describing these outbreaks, they have most frequently occurred in seaport cities involved in inter- or intracontinental trade. For this reason, *A. aegypti* has become known as an urban vector. This mosquito breeds in artificial containers, e.g., flower pots, birdbaths, etc., and, although it prefers human blood, the mosquito will take monkey blood as well.

In addition to the urban cycle and a long list of terrible epidemics, there is also a jungle, or sylvatic, cycle, and another one known simply as 'intermediate'. In the jungle cycle, which is restricted to South America, the virus is transmitted by species of *Haemogogus* that normally live high in the jungle canopy. Loggers who cut these trees will come in contact with infected mosquitoes and, when bitten, acquire the virus. If these infected individuals venture into nearby villages, towns, or cities, the disease will be transmitted via *A. aegypti* to local residents and a localized epidemic can start in this way.

The intermediate cycle occurs in humid or semihumid savannahs and is now the most common method of spread in Africa. In this pattern, several small-scale outbreaks of the disease will occur simultaneously in a localized geographic area, although fewer people die in these endemics. It is said that semidomestic mosquitoes transmit the virus in these cases, biting both monkeys and humans. These localities are referred to as 'zones of emergence' and may spread to more urban areas, producing severe epidemics in the process. It is interesting to note that *Aedes aegypti* also occurs in Asia. However, no epidemic outbreaks have ever been reported in these vast areas with so many people potentially at risk.

In 1898, the battleship Maine blew up in the harbor of Havana, Cuba, resulting in the death of 260 crewmen. Not long after, the U.S.A. declared war on Spain, after which the Spanish-American War ensued. Fortunately, it did not last very long and fewer then a thousand Americans were lost in battle. However, close to 5000 were killed by disease, primarily yellow fever. In the years immediately preceding these events, yellow fever had made its presence felt in a number of coastal cities of the U.S.A., i.e., New Orleans, Savannah, Charleston, Norfolk, etc. For example, in 1878 alone, the disease was reported in 132 cities where it produced nearly 16 000 deaths from among 74 000 sickened Americans. It was, in other words, a real scourge. General Leonard Wood, who was appointed to head the occupying contingent of the U.S. Army in Cuba following the war, had recognized the potentially serious hazard of yellow fever for his troops and persuaded the then Surgeon-General, Michael Sternberg, to appoint a Yellow Fever Commission to investigate and determine the cause of yellow fever on the island of Cuba. The Commission was to be headed by Surgeon Walter Reed and included Assistant Surgeons Aristides Agramonte, James Carroll, and Jesse Lazear, all members of the Medical Corps, U.S. Army.

The Commission considered a number of possibilities for the cause of yellow fever, but all save one were almost immediately dismissed. Among those quickly excluded was direct transmission. A bacterial etiology for the disease was also eliminated. Yet another idea was transmission via mosquitoes, a notion originally promulgated in 1881 by Carlos Finlay, a Cuban physician. During the course of the Commission's investigation in Cuba, Carroll and Lazear allowed themselves to be bitten by mosquitoes known to have recently fed on patients with active cases of the disease. Agramonte did not participate in this trial since he had the disease as a young child and was immune. Reed was away in Washington and also did not participate in the experiment. Both of the volunteers became very ill with yellow fever after a prepatent period of about five days. Carroll survived, but Lazear died. Even though they had not identified a virus in the etiology of the disease, when mosquitoes were identified as the agents of spread, a successful control program was initiated using quarantine of infected patients and mosquito extermination. By the way, the same sort of program was used in Panama during construction of the Canal, and the Americans succeeded where the French failed because they knew mosquitoes were the transmitters of the disease! The explanation for this huge success was simple: control the mosquitoes and quarantine anyone who became ill.

Shortly before the discovery of yellow fever transmission, a group of scientists had discovered a new class of infectious agents, called viruses. They had found, for example, that foot-and-mouth disease was caused by one of these newly discovered 'organisms'. However, no one suspected that any of these new viruses could cause problems for humans. Although transmission of the urban form of yellow fever by mosquitoes had been resolved, it was not until 1911 that another group of investigators demonstrated yellow fever could also be transmitted in nonurban areas, producing a condition known as 'jungle fever' and, moreover, without the involvement of *A. aegypti*. This meant that other mosquito species were definitely involved in transmitting the virus and that elimination of *A. aegypti* alone was not going to prevent outbreaks of yellow fever in nonurban settings. To control the disease, a vaccine would be necessary.

Max Theiler was born in South Africa, where he was first educated. He subsequently studied at St. Thomas' Hospital and the School of Hygiene and Tropical Medicine, both in London. He spent a brief period of time at Harvard before joining the staff of the International Health Division of the Rockefeller Foundation. In 1927, Theiler and his colleagues proved that yellow fever was caused by a virus and successfully passed it to monkeys in the lab. Then, in 1930, Theiler was able to use the yellow fever virus to infect white mice in the laboratory.

Early during the research on yellow fever, two strains of the virus were isolated. One was referred to as the Asibi strain and was used primarily in rhesus monkeys where it produced a disease not unlike that in humans where the main targets were the heart, kidney, and liver. A so-called French strain was isolated in Dakar, West Africa. The latter was used by Theiler to infect ordinary white laboratory mice, a significant advantage because of the reduced costs for experimental work. Even though the French strain was highly virulent in mice and rhesus monkeys, in the former host it had to be introduced via cerebral inoculation, where it caused a serious encephalomyelitis. Based on this preliminary research, Theiler made three discoveries that were to have an impact on his really significant findings that were to follow. First, with continuous passage in the laboratory, the prepatent period for producing disease symptoms in mice was reduced until it became fixed. Second, he observed that with extended passage in mice, the virus became more and more virulent (a clear example of natural selection). Third, in monkeys, where parenteric inoculation was required to establish a visceral disease, the virus became less and less virulent (another example of natural selection). These latter results, especially, immediately raised in Theiler's mind the possibility of a vaccine.

To develop a vaccine against the yellow fever virus, Theiler and his colleagues turned to tissue culture techniques. The objective was to obtain an attenuated virus, one that could be grown with relative ease and that would stimulate production of sufficient antibody levels in humans to prevent the acquisition of a 'wild-type' yellow fever virus. It was the Asibi virus with which he achieved success. This virus had the characteristics he needed to evaluate its effectiveness, i.e., it was highly pathogenic in mice by intracerebral inoculation and for rhesus monkeys via parenteral insertion. The tissue culture media in which he detected the desired change in the virus were chick embryos containing small quantities of nervous tissue. The attenuated virus was designated as the 17D strain and is the one still in use to this day for production of vaccine. Attenuation of the 17D strain of the yellow fever Asibi virus was achieved at some point between 89 and 114 passages in the culture system. The vaccine produced minimal side effects and elicited high antibody production in humans. I asked Pat about this because it almost sounded to me as if there is some sort of molecular clock in the yellow fever virus that is either turned on, or turned off, in such a manner as to say, 'this is where attenuation begins'. She replied, "I do not know of anything like that. However, the yellow fever virus is unbelievably stable. There are five RNA viruses for which there are licensed live attenuated vaccines. These include mumps, measles, polio, rubella, and yellow fever. The first four have much higher rates of reversion to virulence than the yellow fever virus. In fact, we do not use the live attenuated polio virus any more because the risk of reversion is much too high."

Theiler had said, "The reason for the rapid change in the 17D strain . . . still is completely unknown." He also indicated, "Occasionally, however, for some unknown reason, a mutant appears with marked reduction in both neurotropism as well as viscerotropism. This mutant is comparably stable, but it too has been observed to undergo change on two occasions. The first time, . . . the cultured virus was found to have become so attenuated that it failed to produce immunity in a fair proportion of persons vaccinated, and the second time, . . . the virus had regained some neurotropism so that it actually produced encephalitis in a small proportion of persons vaccinated." In other words, the virus was still capable of change through mutation. These quotations are from the Nobel Lecture given by Max Theiler on 11 December 1951, in Stockholm, Sweden!.

I asked Pat about this kind of reversion. She replied, "As we noted previously, this group of viruses has an RNA genome. They have to carry a gene with them to code for their own RNA polymerase so they can copy their own genome. Well, RNA polymerases have no proof-reading mechanisms. So, all of these RNA viruses are much more error prone than the DNA viruses. The fact that the yellow fever virus gave rise to a mutant that could not raise an immune response probably indicates that it had become so attenuated that it could no longer attach to the cell. In other words, it never got inside the cell so that it could make more viruses. Accordingly, the host immune response could not be triggered, or maybe it could still get in, but the viral RNA polymerase was defective and it couldn't make any more viruses. The latter seems to be the type of mutation that the yellow fever virus accumulates. This would account for the greater stability in an attenuated yellow fever virus. These mutations are simply not tolerated because the virus cannot copy its own genome." I then asked if there was any sort of change in the virus in the mosquito because I had not seen any sort of reference to such a phenomenon. Pat said she believed there was no change and that the vector was simply a passive carrier of the virus.

Another question I had, sort of as an aside was, why is the tissue culture method of growing viruses used rather than growing them in eggs? Pat responded, "It's so much easier. It's really no more than inoculating the tissue culture with the virus, allow them to penetrate the cells and exit, then simply harvest them from the growth medium. It is a labor-intensive process, but not nearly so as using chick embryos."

The main question raised at the outset of this essay was why a given virus cannot be made to produce an appropriate vaccine when another virus can be made to work? In the case of HIV, the answer is simple. First, it is much too variable antigenically and, second, it mutates too fast, which accounts for the antigenic variability. The host can keep up immunologically for a while, but at some point, the immune system collapses and the host becomes immunodeficient. For me at least, the most interesting feature of the interaction between the HIV virus and the host is that it operates entirely within the context of natural selection. In the case of the yellow fever virus, the reason for success is likewise simple. During in vitro culture, the yellow fever virus is manipulated in such a way that it eventually becomes attenuated and avirulent. Although Theiler could not provide a molecular explanation for why the virus became attenuated after x-number of passages, the general explanation for the process is also rooted in the process of natural selection, followed by artificial selection of the attenuated virus. In other words, he discovered that with continued passage in tissue culture, eventually there will be a mutation in the virus, one that could then be manipulated for vaccine production. It is interesting that Pat Lord said that one of the ways to develop an attenuated virus is to pass it through a different host. She, however, likened it to an evolutionary chance regarding the way virulence goes, toward either attenuation or greater virulence.

I should note at this point that when I began the research for this essay, I knew that there was not an HIV vaccine, even though a very extensive attempt had been made to make one. However, I also knew that there was a yellow fever vaccine and that it had been in use for a very long time. What I did not know was that both HIV and the yellow fever viruses were of the RNA variety. What I also did not know was that there was a strong tendency for RNA viruses to mutate, which, in turn, made it very difficult to develop a vaccine. With full knowledge of this information now, two natural questions occur. First, why is the yellow fever virus so stable? Second, how can it be that the 17D strain isolated so long ago is still used as a vaccine source today, while HIV remains what appears to be enigmatically unstable and, for this reason, useless for creating a vaccine? I asked Pat about this apparent dichotomy in a second interview. She responded, "It's the same thing that you have already written about. The yellow fever virus is evolutionarily conserved, even though is does have a polymerase that still makes mistakes. When a virus infects a cell, if there are mutations that occur, like in the polymerase gene so that the polymerase would not be functional, you'll never see that. Thus, if that virion goes and infects another cell, there will be no virions made. All you see are expressed virions; you do not see how many virions are not expressed because of an inexpressible gene mutation. Somewhere I read that in the yellow fever virus, mutations tend most often to occur in the polymerase gene, so those would not survive." After thinking about this for just a moment, I quickly responded, "So I just got lucky when I chose these two viruses to write about?" Pat answered, "You picked out two very good viruses to contrast in an evolutionary context."

Before closing, there is one last point I would like to make and this has to do with the delivery of the vaccines to people who need them. I mentioned earlier that an estimated 200 000 people in sub-Saharan Africa are infected annually with yellow fever and that about 20 000 of them die each year. It is probable that many of these same people did not have access to the vaccine. Some may not have even known there is such a thing as a vaccine. However, there is another roadblock in this vaccine 'business', and this one is just really beginning to emerge as a real problem. Moreover, after trying to consider the overall vaccine situation as objectively as I can, I think it is unquestionably the most pitiful aspect of the overall difficulties we have with yellow fever, or HIV, or any of the virally induced diseases for which effective vaccines are presently available. It relates to the fact that some, perhaps many, in Africa and Southeast Asia, are being told that yellow fever, and especially HIV/AIDS, ultimately emanates from the white man of America and other western countries, and that there is a conspiracy being promulgated by the West to spread these diseases into the underbelly of Africa. In other words, many people in Africa honestly believe that HIV began in America and has been introduced into Africa, not the other way around. As I said, I had heard of this idea previously, except I cannot remember the source. During my second conversation with Pat, this phenomenon came up during our interview. She confirmed it was true, based on a personal experience she had while teaching virology just a short time ago. It seems that one of her students was a native of Nigeria and another the daughter of missionaries in Kenya where she was raised as a child. During the semester, she and her students became engaged in a discussion regarding the epidemiology and the spread of disease in the world and Pat was informed by these particular students that the idea of a white man's conspiracy for the introduction of disease into Africa via vaccination, etc., was pervasive in Africa. In other words, there is a widespread belief among native Africans that the white man of the West is responsible for the tragic pandemic caused by HIV. I am not sure about the long-term ramification of such 'gossip', but it seems the potential for serious fallout is genuine, and is something that must be considered by local health officials and epidemiologists, especially if an HIV vaccine is ever developed.

For HIV and AIDS, the lack of success in producing a vaccine is attributable to rapid mutation of the virus, accompanied by natural selection generated from a host response to the changing HIV virus. For the yellow fever virus, mutation is again involved, except, in this case, the outcome of artificial selection is an attenuated virus. Max Theiler resolved the yellow fever vaccine problem roughly 85 years ago, but the HIV situation remains intractable.

REFERENCES

Collier, L. H. and B. W. J. Mahy (eds.). 1998. *Topley and Wilson's Microbiology and Microbial Infections*. Vol. I: *Virology*. London: Hodder Arnold.

Freeman, S., and J. C. Herron. 2004. *Evolutionary Analysis*. Pearson Education, Inc., Upper Saddle River, New Jersey, 802 p.

Gubler, D. J., and J. T. Roehing. 1998. Arboviruses (Togaviridae and Flaviviridae). In *Topley and Wilson's Microbiology and Microbial Infections*, Vol. I: *Virology*, ed. B. W. J. Mahy and L. Collier, pp. 579–600. London: Arnold.

Hubner, A., M. Kruhoffer, F. Grosse, and G. Krauss. 1992. Fidelity of human immunodeficiency vurus type 1 reverse transcriptase in copying natural RNA. *Journal of Molecular Biology* **223**: 595–600.

Larder, B. A., G. Darby, and D. D. Richman. 1989. HIV with reduced sensitivity to Zidovudine (AZT) isolated during prolonged therapy. *Science* **243**: 1731–1734.

Shankarappa, R., J. B. Margolick, S. J. Gange *et al.* 1999. Consistent viral evolutionary changes associated with the progression of human immunodeficiency virus type 1 infection. *Journal of Virology* **73**: 10489–10502.

Wain-Hobson, S. 1993. The fastest genome evolution ever described: HIV variation in situ. *Current Opinion in Genetics and Development* **3**: 878–883.

4

Lyme disease: a classic emerging disease

Nature is a mutable cloud, which is always and never the same.
                    Essays: First Series, Ralph Waldo Emerson (1803–1882)

Jim Oliver began work in 1988 on *Borrelia burgdoferi*, the etiological agent of Lyme disease, and has been a leading figure in its study since that time. According to Jim, "The primary symptom of Lyme disease is a bulls-eye lesion on the skin that continues to expand." This lesion is not just a localized hypersensitivity "reaction like you would see with the bite of a mosquito or a chigger. Clinically, this is the single-most diagnostic feature of infection with the spirochete that causes the disease." I then asked if this is due to inflammation, or an indication of bacteria within the skin? Jim responded that he was not certain, but probably both are involved. "I say that because if I want to isolate spirochetes, a biopsy at the margin of the skin lesion would give me the best chance for success. The spreading of the lesion is referred to as erythema migrans." This characteristic of the disease was initially described in Europe in the late nineteenth century. At the time, it was not associated with any other symptoms of the disease, or with an etiological agent. The disease in North America was first noted in Wisconsin in 1970. Subsequently, there was an outbreak of the disease in Old Lyme (hence Lyme disease) and surrounding counties in Connecticut in the mid 1970s. It seems that a group of children presented juvenile arthritic symptoms. According to Allen Steere, an arthritis specialist at Yale, the epidemiology of the disease in these children suggested the cause as possibly a pathogenic organism. The epidemiologists did not believe a mosquito was vectoring the etiological agent, but something like a tick instead.

Willy Burgdorfer, while a student in Switzerland, had studied a spirochete that causes relapsing fever before he immigrated to the U.S.A. and went to work for the Public Health Service in Hamilton, Montana. In 1982, he discovered similar spirochetes in the midgut of the black-legged tick, *Ixodes scapularis*, on Long Island in New York. At that time, it was incorrectly known as *Ixodes dammini*, the deer tick. Not long after Burgdorfer isolated the spirochete, the new organism was named *Borrelia burgdorferi*, thereby honoring the person who first isolated it (Johnson, 1998). Then, subsequently, a connection was made between Lyme borreliosis and the new spirochete found by Burgdorfer.

Prior to our interview, Jim had prepared a list of species of *Borrelia*, which can cause Lyme disease, or relapsing fever. One of these, *B. burgdorferi sensu lato*, consists of at least eleven genospecies. Three of these can cause Lyme borelliosis (disease) in humans, including *B. burgdorferi sensu stricto* in North America, with *B. burgdorferi s.s.*, *B. garinii*, and *B. afzeli* in Europe, and the latter two also in Asia. In fact, Jim Oliver estimates that Lyme borreliosis accounts for roughly 80% of all the arthropod-borne illnesses in the U.S.A., with roughly 15 000 cases annually. I have seen others place it as high as 90%. Endemic areas for the disease include the northeastern parts of the U.S.A. and the north-central states where *I. scapularis* is the vector, and the Pacific coast where *I. pacificus* transmits the parasite. Both of these species are known as bridge, and enzootic, ticks because they not only transmit the parasite, they maintain it as well. Jim Oliver also believes, strongly and with very good evidence, that Lyme disease occurs in southeastern areas of the U.S.A., but more about this latter contention in a bit. The disease we know as Lyme borreliosis is circumpolar in the north temperate parts of the world. As is the case of so many parasitic diseases, its geographic distribution is determined in large measure by the distribution of the tick vector(s). The major reservoir in the northeastern U.S.A. is the white-footed mouse, *Peromyscus leucopus*. Most claim that deer are not reservoirs and Jim agrees, but he also wonders if they might serve occasionally for a short time in certain situations. In the south, the main reservoirs are cotton mice (*Peromyscus gossypinus*), cotton rats (*Sigmodon hispidus*), and wood rats (*Neotoma floridana*). The Lyme spirochete has been isolated from at least 18 species of wild mammals, 3 domestic animals, and 8 birds.

Once inside the mammalian host, the spirochete moves via the bloodstream to several sites. These primarily include the heart, nervous system, and musculoskeletal system. Complaints of patients with confirmed cases of Lyme borreliosis include fatigue, myalgia, headache, joint pain, fever, nausea, vomiting, etc. Skin at the point of the tick bite may also be affected, producing the classic erythema migrans mentioned earlier. It is of interest to note that the disease is manifested through effects on any of these organs/organ systems singly, or in some combination. For example, I have seen it said in the literature that up to 20% of Lyme disease patients do not exhibit skin lesions, but they may suffer from various sorts of problems, e.g., neurologic, or arthritic, or both, among others. Jim Oliver believes that many of these variations are due to strain differences among the spirochetes.

I asked Jim how the tick becomes infected. Does it acquire the spirochete from infected humans? He quickly replied, "No, the tick almost never gets it from humans. It's at a dead end in us. Instead, it is a classic zoonotic disease." I then asked about the reservoir hosts in nature. He responded, "The usual situation is that the spirochetes are obtained by ticks that feed most commonly on rodents and birds. The life cycle of the hard tick [Ixodidae] goes from larvae, to nymphs, to adults. Each tick stage feeds just one time. The usual situation is that larvae feed on a mouse, then molt to the nymphal stage before they can transmit it to us. The spirochete must be able to stay with the tick when it molts to the nymphal stage and the nymph must then feed on a susceptible animal, including humans, where it transmits the parasite. Incidentally, adult ticks can also transmit the spirochete." I then wanted to know about the transmission process itself. I said, "When the mosquito takes a blood meal, it 'spits' an anticoagulant into the host before it feeds. How does a tick do it?" Jim answered, "The same way." I then asked, "When a larva feeds on a mouse and picks up the spirochete, then what?" He replied, "A blood meal activates the spirochete in the gut." At this point, spirochetes in the midgut are stimulated to migrate to the salivary glands of the tick. After the tick attaches and begins to feed, it is of interest to note that at least one of the antigenic outer surface proteins (Osp) of the spirochete 'down-regulates' and another one ([Osp]C) up-regulates. Most agree that expression of the latter antigen is necessary for the spirochete to become infective to the mammalian host. (Sounds somewhat analogous to the variable surface glycoprotein change by the African trypanosomes on their arrival in the salivary glands of the tsetse fly prior to entry into the vertebrate host.) Jim continued, "It takes a while for transmission to occur after the tick attaches and begins to feed. The stimulation is blood in the midgut. It is also at this point that they begin to multiply. It depends upon the species or strain of the spirochete how quickly they are activated. In the past, it was always assumed that when a tick bites a human, let's say, if you remove that tick within 24 to 36 hours, the tick hasn't had time to activate spirochetes. The old rule of thumb is that if you come in at night and do a body check and you remove any tick that might be on you, that you will not get Lyme disease. Presently, however, there is some controversy developing from ongoing research in Russia that this may not be the case."

"The next question," I inquired, "is a natural one. Do the spirochetes produce adverse conditions in hosts other than humans?" Jim responded, "Some animals show no ill effects, particularly rodents. I have kept infected cotton mice, cotton rats, and wood rats in the lab successfully for seven to eight years. Now, this is not the mean life expectancy, but it is at the long end of their potential life span. At the end of these periods, I can feed uninfected ticks on these animals and obtained infections with spirochetes. In other words, they remain infected for life. The spirochetes can be maintained in cell culture indefinitely too, but they lose their infectivity with passage in culture. In my transmission experiments, I never use an infectious source that has been subcultured three times or more. Some researchers think that twelve transfers is the limit, but I try to work on the safe side of that number. Infectivity is a multigenic phenomenon, and is associated with plasmids inside the spirochete. About 40% of the spirochete's genes are on these plasmids. In culture, you lose unpredictable amounts of this genetic information, and at unpredictable speeds, via plasmid loss. So, you may subculture and you don't lose any plasmids, or you don't lose enough, to have any effect on infectivity. But, if you do it continually, you are going to lose some of those plasmids, which is going to impact on their precision for reproduction. And, to answer your question, both dogs and horses, among others, can be adversely affected by the spirochetes that cause Lyme borreliosis."

I then asked if Lyme disease is deadly or is it just debilitating? Jim ducked the answer to this query by saying, "That's controversial. The prevailing thought of the most experienced clinicians is that it is usually not fatal."

Something else I wanted to know about had to do with the mechanism for the disease pathology in humans. I asked, "Is there anything known about how the spirochete actually impacts an infected host. Do they produce toxins, destroy cells mechanically, or what? How does it work?" Jim said, "I don't know and I don't think it is known." I then said, "If you took a snippet of tissue from the edge of a skin lesion, would you be able to find spirochetes in the lesion?" Jim replied that you could, but you would not see them inside the cells. In extra reading that I did in preparation for the interview, other experts agreed with Jim's assessment. No one knows if there is a potent toxin produced by the spirochete, or if a host's "inflammatory response is a result of amplification by potent host-derived cytokines" (Johnson, 1998). Jim continued, "And the spirochetes get into protected places, like knee joints. There is a big debate about how that works. Some people claim that when you treat the disease, you really are not reaching some of these protected sites, so that some of the spirochetes survive. Other people say, no, you are killing them, but as an antigenic information site it remains and, therefore, whether the parasite is alive or dead, it is going to continue causing immune problems for the host in the form of inflammation and joint difficulties, i.e., arthritis. So, if you get Lyme disease, you may suffer from its effects for several years."

I was curious about the epidemiological character of the disease and asked Jim if the there was a particular group of humans that was more vulnerable to infection than another, but Jim said that there was not. Lyme borreliosis seems to be more of a matter of exposure than anything else. I asked, "If I am exposed to the spirochete, will I get sick?" Jim said, "No, that's not necessarily so. Not everyone who is exposed will come down with Lyme disease. It's that way with almost any disease, certainly the vector-borne ones anyway. Individual host immunity is so variable. That's just from the standpoint of the host too. When you consider the variability in the spirochete virulence, then that must be factored in as well. Everyone is susceptible at some level. Everyone is potentially vulnerable, assuming that the particular genospecies of *Borrelia* and strain is infective in the first place."

At this point in our discussion, I recalled for Jim a joint meeting of the ASP and the American Society of Tropical Medicine in Atlanta, Georgia, in 1993. In a symposium dealing with Lyme disease, the story goes that Jim and Andy Spielman of Harvard University engaged in what can only be described as a 'verbal war'. The argument centered around two things, i.e., the identity of a new species of tick described by Spielman and whether Lyme disease was present in the southern United States, or not. I asked Jim to tell me about what had become a long-standing feud between the two men. I told him it was the 'talk of the meeting'. He roared with laughter when I said that. I inquired, "What was the contentious issue?" He said, "That particular confrontation wasn't about spirochetes at all. Let me tell you the story. Andy and I were good friends. He has visited me in Statesboro, I have taken him out to St. Catherine's Island off the coast of Georgia where I conduct some of my research. He has given lectures at Georgia Southern, and I have gone up to Harvard and given lectures for him at the School of Public Health, and stayed with him in his house. In summary, we were once pretty good friends. I still have a lot of respect for Andy and the many good contributions he has made over the years. However, he had described a population of ticks from New England that he thought represented a new species. They are very similar to a species that was described back in 1812, or something like that. It was similar to *I. scapularis*, the eastern black-legged tick. Andy believes that he saw enough differences in the New England area population to warrant calling them a different species, separate from *I. scapularis*. I went along with it initially because he pointed out some morphological differences that are best seen in the nymphal stage of the tick. If you take one from Boston and one from Savannah and look at two or three characters of the nymphal stage, you would say, yes, this one is from the north and this one is from the south. Plus, he says there is a difference in the ecology of the tick in that there is a greater tendency for the nymphal stage to feed on a human than there is for the southern stage. He named the tick *Ixodes dammini*, and called it the deer tick.

   "Subsequent to the new species being described by Andy, several people talked to me a good deal about ticks that they found in North Carolina, South Carolina, Virginia, and Maryland. If you look at Spielman's description and pictures of these ticks, the differences are not as clear in ticks from the Carolinas as you see between those from Savannah and Boston. Those characters are judgment calls a lot of times. So, I asked the question, 'Is it possible that we are talking about differences in specimens from two ends of a north–south cline?' I mentioned that to Andy and he didn't agree with the idea, at all. I said to myself, 'What we really need to do is to ask the question, is this so?' With that, I set about doing hybridization experiments using populations from Massachusetts and Georgia. As a control, we used *Ixodes pacificus*, the western black-legged tick. We were very careful with the crosses in the lab with the F1 progeny. These ticks then successfully produced F2 progeny. We then quit the experiment." At this point, I interjected, "Then they are not different species." Jim reacted, saying, "That's what I contend. Andy said, 'Well you know that under artificial conditions something like that may happen and they wouldn't do that in nature.' In response, I said, 'Look Andy, we'll do some additional things. Maybe it's a behavioral situation.' So, we set up an assortitive maze and give the ticks a choice. We took males from Savannah and males from Boston and gave them choices of females of their own type versus the other geographic type. In general, they showed no preferences of mating with their population versus another population. So, from actual hybridization they are not different, from assortative mating they are not different, and it's not a behavioral thing. So, next, I said to myself, 'I'm an old chromosome man, let's check their chromosomes', so we did that too. As a control group, we used *Ixodes pacificus*, the western black-legged tick. We could not see any differences in the chromosomes, or C-banding of the chromosomes between Massachusetts and Georgia." They also did some isozyme analyses comparing northern and southern forms, and *I. pacificus*. The results clearly lumped the northern and southern ticks, but separated these two forms from *I. pacificus*. I asked if they had done any sequencing and he responded by saying that they were doing that at the present time. He also said that they had done a thorough multivariate statistical morphological analysis of ticks from a geographical gradient by looking at samples from north to south and east to west. "There is a morphological cline both ways, so in Missouri, for example, they are very different from North Carolina, just like they are different from Georgia to Massachusetts."

Jim continued, "Andy is just unbelievably stubborn and was mad as hell with me. So I called him one day and said, 'I know you are mad, but I'm not wedded to this concept. If you can prove to me that what I have presented is incorrect, then that's fine. I'll switch horses.' Andy responded by saying that the populations in the north and south were different morphologically and behaviorally, and were different species. That was several years ago. In the interval between then and now, I sometimes see him at meetings. We shake hands and exchange small talk, but clearly our friendship is not as warm as it was at one time and as I would like it to be. He has stuck to his guns and continues to refer to the northern populations as *Ixodes dammini*, or deer tick, although most of the scientific community accepts that the northern and southern populations are a single species, *Ixodes scapularis*, the black-legged tick. Unfortunately, the common name, 'deer tick', has not been universally dropped and continues to be used by many. This sometimes results in the incorrect combination of *I. scapularis* and deer tick. The term deer tick has been so widely used in the press, the public, and among uninformed scientists that dropping the name deer tick is proving difficult."

Spielman also says that Lyme disease does not occur in the south, something that some scientists at Centers for Disease Control and Prevention (CDC) in Atlanta also believe. I asked Jim about this issue. He said, "I have no data proving that Lyme disease occurs in the south.

However, I have 250 separate isolates of *Borrelia burgdorferi* from South Carolina, Georgia, Florida, and Missouri in my deep freeze in Statesboro right now." I blurted out, "What is the problem with all of these folks then?" Jim reacted in a frustrated way, "I think it's just that people do not like to be wrong and it is a very complicated situation. From an historical perspective, there were no data indicating the presence of *B. burgdorferi* in the south, and people accepted that the vector tick in the north was a separate species that did not occur in the south. At that time, it did not seem wrong to believe that if that were true, probably *B. burgdorferi* did not occur in the south. Of course, absence of data does not mean negative data. It appears that some people drew their conclusions too early.

"The dogma of the 1990s was that the origin of *B. burgdorferi* was in the coastal areas of New England and that the spirochete and tick vector were gradually spreading out from there. It was quite a shock when we demonstrated that the northeastern vector tick of *B. burgdorferi* was the same species that occurs in the south (Oliver, *et al.*, 1993a) and that we had made isolations of *B. burgdorferi* in Georgia (Oliver, *et al.*, 1993b) and three other southern states. When I told Andy Speilman about the isolates, it was hard for him to believe. He questioned the correct identity of *B. burgdorferi* and suggested that if the spirochetes were truly *B. burgdorferi*, they were probably transported south via birds from the northeast.'

"I responded that the wide geographic distribution of *B. burgdorferi* from many sites in four states, and from several species of mammals, argued against chance transport by birds. He asked if we had been able to successfully transmit some of our isolates via southern ticks in the lab. I confirmed that we had done this, but apparently he did not believe me.'

"What I think has happened is that the ticks have not moved from north to south, but that they have spread up from the south to the north. I think originally that the ticks were down here and that now they are going north. My working hypothesis is that *I. scapularis* ticks and *Borrelia burgdorferi* have been in the south longer than in the north, and what happened was that very infectious strains of *Borrelia* happened to spread north. Clearly, there is greater genetic variation in tick vectors and in *Borrelia* in the south than in the north and, therefore, some of the strains present in the south almost certainly can cause the disease and others probably do not. From an evolutionary perspective, one would expect greater genetic variation in a location that has had this parasite for a longer time. I think what you see up north is more like a 'founder effect', with less genetic variation. These sorts of data support my hypothesis regarding colonization and distribution. Moreover, when birds are coming through the south going north, they are migrating at a time when larvae and nymphs are most active in the field and are more likely to pick up the spirochete as a result. In contrast, when birds are going south from the north in late summer and early fall, the larvae and nymphs are not as active. So, collectively, all of these reasons suggest to me that the parasite has been around longer in the south than in the north."

As I mentioned earlier, CDC in Atlanta had said originally that Lyme disease did not occur in the south. But Jim said that there now appears to be a diversity of opinion (Oliver, *et al.*, 1993a,b). Jim said, "Our work clearly shows that *B. burgdorferi* does occur in the southern U.S. We have further shown that it is geographically widespread, in a lot of different animals, and we have done transmission experiments to show that it is infectious. I believe that most CDC scientists interested in Lyme disease now acknowledge all that. But, they say we have not proven that it occurs in humans. They point out that no one has isolated it from humans in the south. That, of course, is the gold standard and we would like to do all that. However, my lab is not set up to do that kind of thing. We do not see patients. Most physicians in the south who do have access to patients with the symptoms do not have the time, inclination, or money to do it. There are many more cases of Lyme borreliosis in the north. There are several reasons for this situation. One is that the nymphs of *I. scapularis* have many more options regarding the choice of hosts. Many select lizards that are noncompetent reservoir hosts for *B. burgdorferi*. There are more than a hundred species of animals from which these ticks have been reported, and a great many of them are native to the southeast. Wherever you find an ecological situation where there is a restricted range of potential hosts for vector-borne parasites, there is a greater focus of the enzootic cycle and often there is an increase in the likelihood of transmission to humans.'

"There is one more caveat regarding Lyme borreliosis in the south. Thus, the problem in diagnosing *B. burgdorferi* in humans in the south is complicated by a disease called STARI [Southern Tick Associated Rash Illness], also known as Masters' disease. Ed Masters is a physician in southeast Missouri who has made significant contributions to knowledge dealing with tick-borne pathogens. The etiological agent for STARI is *Borrelia lonestari*, and its vector is another hard tick, *Amblyomma americana*. It seems that in many of these cases, there is the typical bullseye lesion in southern patients, which are associated with bites by

A. americana, not *I. scapularis*. However, I have found *B. burgdorferi* in *A. americana* along the coast of Georgia, although rarely. It has been suggested that a lot of the cases of Lyme disease reported in the south are actually STARI, especially those associated with *A. americana*. Alan Barber found *B. lonestari*, which he identified molecularly, but he couldn't culture it in the standard medium which he and others developed [BSK medium]. I also failed to culture it in BSK medium when I tried. I then tried to grow it in tick cell culture, but it would not work either. Recently, however, some people in Athens at the University of Georgia attempted the same thing and were successful. They identified it as *Borrelia lonestrai* and, molecularly, it was the same as that which Barber had described. So, it looks like STARI is caused by *B. lonestari*, not *B. burgdorferi*, and it forms the classic bulls-eye lesion. These people also get flu-like symptoms, and have fewer arthritic problems but, in general, it is a milder form of disease than Lyme borreliosis.

"In the south, there are four closely related species, i.e., *Borrelia andersoni, B. lonestari, B. bissettii*, and *B. burgdorferi s.s.* In the U.S.A., only the latter has been isolated from humans. However, there is a medical group in Slovenia in eastern Europe that has reportedly isolated *B. bessettii* from humans. Thus, although *B. bissetti* has not been identified in humans in the U.S., it must be suspect. Almost all the biomedical community agrees that *B. burgdorferi* is very common in humans with Lyme disease in the U.S., with *B. garinii* and B. *afzelii* in Europe. In Japan and Korea, it is primarily *B. garinii* in humans."

Lyme borreliosis is clearly an emerging disease and, while not generally considered as lethal, *B. burgdorferi* can produce long-lasting and nasty morbidity. Although found mostly in northeastern and north central states, as well as the Pacific coast, Jim Oliver and his colleagues have definitively demonstrated its presence in the southeastern United States as well. Persistence in pursuit of the spirochete and its relationship with tick vectors has resulted in Jim's identification with Lyme disease and has earned for him the recognition he so well deserves.

My drive to Black Mountain and the 'hills' of North Carolina that day in August 2005 was well worth it.

REFERENCES

Johnson, R. C. 1998. Borreliosis. In *Bacterial Infections*, Vol. 3: *Microbiology and Microbial Infections*, ed. W. J. Hausler, Jr. and M. Sussman, pp. 954–967. London: Arnold.

Oliver, J. H., Jr., M. R. Owsley, H. J. Hutcheson, *et al.* 1993a. Conspecificity of the ticks *Ixodes scapularis* and *I. dammini* (Acari: Ixodidae). *Journal of Medical Entomology* **30**: 54–63.

Oliver, J. H., Jr., F. W. Chandler, M. P. Luttrell, *et al.* 1993b. Isolation and transmission of the Lyme disease spirochete from the southeastern United States. *Proceedings of the National Academy of Sciences USA* **90**: 7371–7375.

5

The discovery of ivermectin: a 'crapshoot', or not?

Healing is a matter of time, but it is sometimes also a matter of opportunity.
The Physician's Oath, Hippocrates (460–377 B.C.)

What is ivermectin? Chemically, and technically, ivermectin is aligned with a family of "16-membered macrocyclic lactones with a disaccharide attached at the carbon-13 position" (Campbell, 1989). Scientists at Merck, Sharp & Dohme Research Laboratories (MSDRL) in Rahway, New Jersey, initially isolated it. These molecules most closely resemble milbemycins, discovered in Japan and first thought to have toxic effects just for mites. It is now known that these drugs are effective against certain parasitic nematodes as well. Biologically, ivermectin is a broad-spectrum anthelmintic, acaricide, and insecticide.

For me, the story of this drug's discovery is an interesting tale, for a number of reasons. First, one of the folks deeply involved in it is an old friend of mine, Bill Campbell. I have known Bill for more than thirty years. Second, before I began writing this essay, I knew very little about this sort of applied research. So, I had to sit down and do some serious reading. Quite honestly, I found it to be rather intriguing. Third, I was told by a colleague of mine, who should know, that ivermectin is considered by many in the agricultural industry almost as a miracle drug, primarily because of its toxic breadth for both ecto- and endoparasitic organisms. Finally, part of the reward for venturing into the chemotherapy arena was being able to sit with Bill Campbell one morning in Philadelphia for almost three hours in July of 2004 and listen as he told me about himself and about ivermectin.

The treatment of disease caused by protozoans and worms has a number of approaches. Of course, one is to prevent infection before it can occur. For some parasites, we can go after the source of infection

175

via sanitation measures. For others, we can eliminate the source by killing the vectors. Immunotherapy is a useful method for some etiological agents, mostly viruses and bacteria, however. Then, there is chemotherapy, which can be used as a prophylactic to prevent infection or, if infection has already occurred, as a way of killing the infective agent.

One of the earliest proponents of immuno- and chemotherapy was the great Paul Erhlich (1854–1915). Born in Strehlen, Upper Silesia, in Central Europe, Ehrlich's early education was in Strasbourg; he then received his Ph.D. at the University of Leipzig in 1878. Throughout his career, he was fascinated by chemical dyes and the various ways these agents could be used in the treatment of contagious diseases, or in their diagnosis.

While strongly identified with chemotherapy throughout his career, he also believed in immunotherapy, saying at one point, "The antibodies are to some extent magic bullets which seek out their own target without harming the organism. Consequently, in all circumstances where it is feasible, the immunization method is preferable to any other method." He noted though that with many disease organisms, especially tropical parasites such as those causing malaria and African sleeping sickness, immunization was inadequate (and still is). He thus wrote, "In all these cases, an attempt must be made to kill the parasites within the body by chemical agents. In other words, chemical agents must be used where serum [immuno-] therapy is impossible."

Paul Ehrlich became the most successful of all the early chemotherapists, winning a Nobel Prize in 1908. His pioneering achievement with '606' in 1909 (dihydroxy-diaminoarsenbenzene) for treating syphilis was nothing short of remarkable (the same drug was also effective for trypanosomes, but not to the extent it was for syphilis). This feat was accomplished using the technique of 'empirical screening', the same procedure to be used so effectively by Bill Campbell and his associates in their search for ivermectin. An arsenical compound had been reported as active against the parasite that causes trypanosomiasis, and now Erhlich was brilliantly investigating the potential of arsenicals in the treatment of that and other diseases. He wrote at one point, "What we want to do, therefore, is to strike the parasites in isolation. This means we must learn how to aim, that is, how to aim in a chemical sense." He had a supplier of potential drugs. After a new one was synthesized, he and his colleagues would test it in laboratory animals, i.e., 'empirical screening'; '606' was simply the next in line, after 605 others! The complexity of drug testing in today's pharmaceutical industry has changed in an extraordinary manner; 606 'drugs', and even more, can now be easily screened on a single plate in the span of a few hours.

About sixty years after Paul Ehrlich and several others began blazing the trail into the unknowns of chemotherapy, Bill Campbell left the University of Wisconsin with his Ph.D. degree in hand and headed east for Rahway, New Jersey, home of the Merck Institute for Therapeutic Research. Even though he had no thought about it at the time, Merck was to become his 'place of business' throughout his scientific career. His new boss at Merck was to be the most amiable Aston Cuckler.

As I listened to Bill in Philadelphia, I could tell that he had a feeling early during his career that he was going to be happy at Merck. He had high praise for Ashton Cuckler. He said that one of the things he looked forward to each Friday, for example, was the group luncheon in Cuckler's office where they would gather and discuss knew projects, their current research, or simply exchange new information. He praised Cuckler for having, "a great feel for applied experimental chemotherapy. He also had a broad background. He had worked with strigeid trematodes at Nebraska with the great Harold Manter. He also worked with George LaRue at Michigan and Alicata at Hawaii." Bill remarked that at Merck, at least in his early days, "somehow the atmosphere was one of tremendous learning."

Bill was hired to work on schistosomiasis, even though he had never dealt with these parasites. However, by the time he arrived, a technician had already been hired to set up a schistosome life cycle in his lab. He said, laughing, that he was probably motivated by his new technician, because by the time he arrived, she already knew more about the system than he did. During our conversation he kept referring to how much fun it was in learning about all these new parasites, about the work others were doing, etc.

We talked a while about the research atmosphere at Merck. It was apparently quite liberal, in the sense that he never felt constrained in what he could do. He said that he had a standing order of 300 mice per week with which to work and that he could do just about anything he wanted. There was never anyone telling him what to do, and no one asked for an accounting of his research animals in regard to schistosome drug testing. "I found the ordinary routine of empirical drug testing very exciting in its own right. It was super!" He then quickly added, "I think one of the reasons I enjoyed my new career was that I was at the bottom of the heap. There were no politics of any sort. There was no administration to worry about." He continued, "I worked on schisto for seven years and . . . had an absolutely unblemished record of failure."

Before turning to the actual discovery process for ivermectin, we talked about a word I had used in an exchange of correspondence Bill and I had prior to our Philadelphia discussion. I had given him the tentative title of the new book and told him I was interested in finding out how much serendipity was involved in parasitological discovery over the years. When we got to this point in our discussion, it was like I had pushed a button, because he lit up like a lamp. He said that he was editing a volume for a journal published by the National Academy of Sciences. The focus of the new tome was going to be on the relationship of serendipity and discovery as it involved laboratory animal research, in effect, the same sort of thing I am doing here with serendipity in parasitology. We turned immediately to whether there was any serendipity involved in the discovery of ivermectin. He referred to his own definition of serendipity as a "rational exploitation of chance observation to find something that was unsought." Throughout the course of our conversation, we examined this idea and the role of serendipity in his search for the new drug.

At the outset, I should say that Bill claimed serendipity played no real role in their discovery. The reason is very simple. They were looking for a new drug, and they found it. This would not be serendipity. In this regard, he said, "We began looking for something that was not just incrementally better than an existing molecule. We were looking for something that was radically different, with properties against parasites, and we found what we were looking for . . . something . . . radically different." To accomplish this goal, by design, they also used a radically different approach, several in fact. "New analogues of existing anthelmintics, or even moderately adventurous departures from known structures, might provide better products – but only a truly novel structure, with a truly novel mode of action, would yield a true breakthrough" (Campbell, 1992).

As we talked, and as I read before the interview, I began to see that there were several things they did to break with the traditional lines of research then being employed in applied parasitology. First, they devised a new assay procedure, primarily through the efforts of John Egerton. It was what they referred to as a "tandem assay". They infected their mice with a coccidian, *Eimeria muris*, and with a nematode, *Nematospiroides dubius*. This gave them two targets to hit, a parasitic protozoan and a parasitic helminth. Another break with tradition was the size of the experimental treatment groups. He indicated that the size of each of their experimental groups receiving the test compound was one! (So much for statistics.) He said, "When I was working with schistosomes, I would use 10 mice per group. Looking back on it, I was crazy!" I remarked that if you get a hit in one animal, you are going to have to go back and confirm it anyway. He agreed, saying, "No matter what you get, if you find a mouse that is now parasite free, you can follow up on it. On the other hand, if the mouse turns out to be totally free and there is no evidence of having had parasites, you would have to ask yourself, did it get skipped? But you can find out. This procedure allowed us to do a tremendous number of things." When I said this would save a lot of money too, he just laughed while shaking his head in the affirmative, and saying, "Yes, yes!"

A really good question is, what made them go off in a completely new research direction? In Campbell (1992), he wrote, "By the early 1970s, it was evident that empirical testing of synthetic chemicals was reaching a point of diminishing returns. The history of antibiotics, however, suggested that microorganisms, unencumbered by the restraints of human knowledge, were capable of producing wildly exotic chemical structures. Novelty, after all, would be the key to success." So, they moved into the fermentation business. In essence, what they did was to take an isolated bacterium and grow it in in vitro culture. Then they would take broth from the culture and feed their mice with it. More precisely, they impregnated mouse food with the broth and then fed their infected mice with it. In effect, they used empirical testing via their new "tandem assay".

Bill explained, "Fermentation had been used as the mainstay in bacteriology where the emphasis was on in vitro systems. With helminths we have bioassays, if you believe in bioassays, and I do, but it's very difficult to use fermentation products. This is because microbial cultures in nature, as opposed to the ones that occur in culture collections, tend to produce things in extremely small amounts. So, people don't use them in bioassays where you feed animals, or you inject animals . . . and look for an effect. About the only time you might make an exception would be if you were using refined or purified material from the culture, but to do it with crude culture broths . . . no. But, we decided to go ahead and do it with crude broths and then feed the mice with the material." I asked him why they would do it that way knowing it might not work for the reason he had just given me. He replied, "Well, Cuckler was always big on medicated food, and I am still very big on it as well." The fermentation broth procedure, he explained, "was not a very sophisticated approach and no one was using it at the time, but it worked for us."

In 1973, another large step was undertaken in the Merck drug search program. In March of that year, they entered into a collaborative venture with the microbiologists at the Kitasato Institute in Japan. Shibasaburo Kitasato was one of the early pioneers in chemotherapy and had worked with both Robert Koch and Paul Ehrlich in Germany in the late nineteenth and early twentieth centuries. Because of this close working relationship between the Japanese and German workers, Robert Koch was venerated in Japan. According to Bill, "Koch, when he visited Japan on a tour with his new young bride, was feted and celebrated by the Japanese. In fact, there is even a shrine in Japan dedicated to Koch." Interestingly, Kitasato and Alexandre Yersin were the first to successfully isolate the plague bacillus in 1894. Nearly eighty years later, those in the Institute were still in the business of isolating new kinds of bacteria. The Institute thus had an agreement with Merck to supply soil microorganisms isolated in Japan, and they already had several thousand cultures on hand when the two groups began the collaboration. Bill said, "They [Kitasato] did not send them to us for antiparasite testing. The agreement was, simply put, for Kitasato scientists, under the leadership of Satoshi Omura, to select unusual isolates and send them to Merck. That was it. But here was our chance now, with this new assay, for the first time to tap into a resource like the one we were offered from the Japanese microbiologists."

In March 1974, a batch of 54 isolates was received from Kitasato at the Merck Institute and their laboratory in Rahway, New Jersey. As Bill describes it, about a year later these bacterial lines were inoculated into broth cultures and allowed to ferment for three days. The broths were then mixed into the feed of laboratory mice. The mice were fed impregnated food from the broths for six days. Of the 54 mice used in the tandem assay, one animal was found to be free of *Nematospiroides dubius*. This mouse was, however, not in very good shape. As Bill indicated, "The mouse nearly died." He said, "At the time, we all thought, this isn't very exciting – we have a sick mouse on our hands, so what." Moreover, it had eaten only half of its food, and it had clearly lost weight. But, it was still uninfected. So, they followed up with additional mice to determine if the result was spurious or if, in fact, the isolate had any useful efficacy. This experiment was also successful. They were surprised to find, in fact, it was successful over an eightfold range of dosages. The broth clearly possessed anthelmintic properties. As it turns out, this bacterium, OS3153, had been isolated from dirt dug up on a Japanese golf course! Despite similar efforts, this bacterium (now known as *Streptomyces avermitilis*, i.e., "the streptomycete that helps create an averminous condition" [Campbell, 1992]; in other words, the streptomycete that kills worms) has not been isolated anywhere else in the world. In our Philadelphia conversation, Bill said, "No one has ever found it again. In Australia they did find an actinomycete that produced a similar molecule, but that was it." Moreover, of the many thousand isolates tested from the Kitasato Institute, only one, OS3153, exhibited any sort of anthelmintic effect. This was not serendipity, but it was luck, at least in my judgment. I mean, come on, if they had dug dirt from the ninth tee rather than from the fairway of the thirteenth hole, they could have missed it completely! Of course, I'm not certain what tee, fairway, or green was used as a source of the dirt, but I think my point is well taken nonetheless. In fact, why would anyone look for bacteria capable of killing parasitic worms or protozoans in dirt taken from a Japanese golf course?

As Bill wrote, "The crude broth of that actinomycete culture was a potent toxin. The only thing more striking than its toxicity for mice was its even greater toxicity for worms" (Campbell, 1992). But what made it so potent? This is where the chemists would make their mark. They first determined that the drug's efficacy rested with a component inside the mycelium, and not with anything the organism released to the outside. Using thin-layer chromatography, mass spectrometric analysis, reverse-phase high-pressure liquid chromatography, and nuclear magnetic resonance spectroscopy, the active components of the molecules (now called C-076) were determined to be "glycosidic derivatives of pentacyclic sixteen-membered lactones."

They decided at that point that a new name for this 'stuff' was required. Bill's choice was "avermecticin, the ending '-icin' indicating, by convention, an actinomycete origin, and the rest of the word suggesting antagonism for worms (vermes) and ectoparasites" (Campbell, 1992). Jerry Birnbaum suggested avermectin, which was finally accepted by the U.S. Adopted Names Committee.

He (Campbell, 1992) wrote, "My chemist colleagues appear to take the view that, while nature can make a truly novel compound, only a chemist can make it right!" So, after determining the structure of the active, averminous molecules, Merck chemists set about to create something that would be even more effective than the ones they isolated from the streptomycete. There were problems with the molecule isolated from the bacteria. For example, these compounds were ineffective against at least one major nematode species. Furthermore, they were ineffective when inoculated subcutaneously for control of another nematode, but worked orally in the same case. In other words, they wanted something better than what the 'bug' produced. After considerable effort, the chemists finally developed something called "22, 23-dihydroavermectin B_1." It was effective against all the nematodes they tested and it worked whether given orally or injected subcutaneously. They had their drug, but, again, they needed a name. According to Campbell, (1992), "The obvious choice for this hydrogenated avermectin was *hyvermectin*, but this was overruled by those of our number to protect our global respectability. Apparently, that name would sound slightly naughty to certain Eastern European ears, and so it was emended to *ivermectin*."

By the time they completed their testing trials, they "knew this new entity [even] without any molecular modification or formulation work, was the most potent anthelmintic known; it acted orally or parentally; it had an unusually broad spectrum of activity; it apparently had a wide therapeutic index; and it probably had a novel mode of action" (Campbell, 1992). Thus, based on these trials, they knew it was effective against at least nine species of enteric and tissue-dwelling nematodes, although not against cestodes. One of the most remarkable finds, though, was the drug's ineffectiveness against adult forms of *Dirofilaria immitis*, the dog heartworm. As it turned out, however, the drug is effective against both the preadult stages and microfilariae. This was truly fortuitous since it is risky to kill adult worms in the heart of dogs for fear of having the dead worms be carried to the lungs where they could easily produce an embolism. Being able to kill preadult stages meant 100% prophylaxis. Moreover, killing the microfilariae would reduce the threat of recruitment by mosquitoes and, thereby, potentially impact the parasite's transmission locally. When we talked in Philadelphia, I said to Bill, "That's got to be serendipity. You obviously weren't looking for something like this." Bill agreed completely, saying, "If you wrote a script for something like this, of course you would say that it couldn't happen in real life.

During the interview, Bill related, "Oddly enough, that kind of thing was my one lab contribution to the whole ivermectin story." It seemed that when Bill returned from an Australia sabbatical, he was able to set up a heartworm lab, in the same way he had set up schistosome and liver fluke labs several years previously. Bill followed with, "I had suggested the potential of a heartworm drug as part of our program, not in connection with ivermectin, but just to have a heartworm program." He said that he "happened to be at a cocktail party at one of these meetings [American Society of Parasitologists] and I happened to be at the bar with Lawson [E. J. L.] Soulsby. He was telling me about a veterinarian who had a client who had brought in a ferret that had died and the veterinarian had discovered *D. immitis* in the dead animal." He then thought to himself, "My goodness, that's a real opportunity. So, I went back and ordered some ferrets and we found, in fact, that the ferret is a fabulous model for filarial infections." Continuing, he said, "That was new. And that was serendipity in the sense that someone just happened to bring in a dead ferret with the parasite, and that Lawson Soulsby just happened to have a conversation with the veterinarian, and that I just happened to meet him at the bar. As a result of all this, I was able to show the effects of the drug on the preadult phases of the drug with my assistant, Lynda Slayton Blair." He then said, "I can remember the first experiment in dogs that showed this effect, the one that turned out to become such a big commercial thing, because my assistant was on vacation at the time and I had a vet student summer intern helping me. The two of us were out at our farm doing a necropsy on these dogs and doing them just by the number on the dog. All of them were, of course, coded. We didn't know which dog was which. At the end, I can remember having a sense of a sort of dissatisfaction as we were doing this because the numbers were low on some dogs and some didn't seem to have any heartworms. I thought that maybe this just wasn't a good batch of dogs. Then, when I put everything together, it was just so clear cut! To me, that was our lab's contribution."

One more point: Bill sent me a letter after reading the initial draft of the essay in which he noted a couple of other important features of ivermectin in treating heartworm in dogs. He wrote, "At the time of ivermectin's discovery, heartworm was controlled by *daily* [his emphasis] medication. (The drug used was diethylcarbazine, DEC.) It was thus controlled with much inconvenience (and in some dogs, with great difficulty) and with limited success. The big deal, scientifically, was that whereas DEC killed one-day-old migrating larvae, ivermectin killed one-month-old migrating larvae (larvae that were still precardiac and thus not causing heartworm disease). The big deal, sociologically, was vastly greater convenience. The big deal, commercially, was that people were willing to pay for that convenience. The big deal, epidemiologically, was that fewer dogs developed patent infections, so mosquito acquisition of parasites was reduced by that means (as well as by suppression of microfilaremia in dogs treated for that purpose)." He continued, "As an aside for clarification, it is commonly believed that ivermectin gives once-a-month prevention of heartworm by hanging around in the dog's body for a month and providing continuous plasma level or tissue level of active drug. In fact, however, the drug is rapidly absorbed and rapidly excreted. When the next treatment comes due, the dog has long been free of the drug, but the new treatment kills any larvae that have come in at any time during the preceding month (whether they now be one day old or 31 days old)." For me, these latter comments were hugely useful in seeing the advantages of the new heartworm treatment versus the old approach.

For the applied chemotherapist, and certainly the pharmaceutical company that must consider marketing a new drug, there are two other issues of concern. First, how does it work? And, second, is it safe? To answer the first question regarding ivermectin, Merck began immediately. It was in their interest to do so because the answer might lead them into new directions and, possibly, even other new drugs, maybe even better ones. They learned that the mode on action for ivermectin was new. Based on research done in collaboration with a number of groups, it was determined that ivermectin opens 'chloride channels' in the nervous system, causing the parasite's paralysis. Moreover, the dose levels required to cause the effects are extraordinarily low, at least for most of the nematode species for which the drug is effective. The question of why adults of *D. immitis* are not affected while preadults and microfilariae are destroyed remains a mystery.

Safety is also always an issue, for any drug. The U.S. Food and Drug Administration must be absolutely certain the drug is safe for the host of the targeted animal. Without this information calamitous results could occur. Is it safe for the caregiver? Someone has to administer the drug. Is it dangerous for this person? If it passes from the host in excrement, will it have any sort of environmental impact? How long does it remain active? Is it biodegradable? Into which organ(s) does the drug go when the host is exposed? Does it concentrate in certain organs and, if so, does the drug hold any adverse potential for the animal, including humans, which might consume the flesh? Is the drug mutagenic, in a manner similar to thalidomide? Fortunately, in the case of ivermectin, no problems of any sort were detected and the drug came on the market in 1977. It has been tremendously successful ever since for all those who have used it in treating enteric helminths, scabies, bot flies, etc.

Onchocerca volvulus is a particularly nasty filarial worm in humans. It is, for example, the one of the leading causes of blindness in the world. Microfilariae of the parasite can also produce terrible skin itching and manifestations of elephantiasis in severe cases. Current estimates indicate the parasite infects approximately 18 million of the poorest people in the world. Various species of the black fly, *Simulium*, vector the parasite. Interestingly, as for the dog heartworm, the drug is effective against *O. volvulus* microfilariae and will not kill adult parasites, although it will cause females to cease production of microfilariae for periods lasting several months. Reduction in circulating microfilariae means vectors are no longer infected when they take a blood meal from an infected person, and community-wide spread of the parasite may, therefore, be diminished. Amazingly, a single pill given annually will do the job. In a 2004 article written for *Perspectives in Biology and Medicine*, Kimberly Collins nicely tells the story of ivermectin's transfer by Merck to the poorest of the poor, for free!

Early in the drug's development, Bill Campbell recognized the potential for ivermectin in the treatment of human onchocerciasis and sent a memo to his immediate supervisor indicating the possibility. Eventually, Roy Vagelos, Head of Merck Research Laboratories, was notified; he contacted Bill directly and instructed him to proceed with the research, which he did. The results suggested ivermectin would work as Bill thought. At this point, Mohammed Aziz took over the project and set up a test trial for human onchocerciasis at the University of Dakar in Senegal in 1981. The trials were successful. Merck was then ready to market the drug, but herein came a huge problem. According to Collin's interview with Charles Fettig, who was involved in the Merck marketing process, Fettig said, "Honestly, we couldn't find a way to price it." Fettig continued, "There's no way they [river blindness patients] can afford it." Merck estimated cost of the treatment at $3 per dose and the most a potential patient could afford was $1.

Merck found itself on those proverbial 'horns of a dilemma'. On the one hand, they knew the drug would work against *O. volvulus* and that the side effects for those receiving it were nonexistent. They also knew the drug could significantly, and in a positive way, impact the quality of life for, literally, millions of people, very poor people, in Central and South America and in sub-Saharan Africa, where the disease is endemic. On the other hand, Merck has a bottom line, and that, plain and simple, is making a profit. Research and marketing are very expensive propositions, running into the billions of dollars annually. Could they afford to begin giving drugs away for free? Would they cause other pharmaceutical companies to back away from pursuing active research programs involving tropical disease drugs, i.e., would pharmaceutical companies be dissuaded from following promising leads in fear of receiving pressure from others to 'give away' an important/effective new drug because of what Merck might have done with ivermectin? According to Collins (2004), there was even a suggestion that diseased patients might shy away from taking a free drug! After carefully considering all the options, Merck made the morally correct decision to provide the drug for free.

Initially, distribution of the drug was a real problem, but this difficulty has since been resolved. The drug is now provided at the central government level in some cases, and through nongovernmental agencies in others. It is then distributed into communities where the disease is endemic. The program has been overwhelmingly successful. Some 25 million people are treated with the drug on an annual basis and, to date, more than 525 million doses of ivermectin have been given away for free. The program has established a clear reputation for Merck as a socially aware corporation and has won Merck much well deserved recognition for its effort.

Obviously, my reference in this essay's title to the search for ivermectin in terms of a crapshoot was a feeble attempt to 'play' on words. As I mentioned earlier, Bill and I had a long and interesting discussion while we were in Philadelphia about the role of serendipity in the discovery of ivermectin. I had already told him that I had at least tentatively decided to use the word, 'crapshoot', in the title, as a way of suggesting that the search was a gamble. After I returned home, he sent me a very nice letter, in which he wrote, "I especially hope that when you come to write about ivermectin, you will reach the conclusion that, while chance was critical, it was not everything (it was, as the philosophers like to say, a 'necessary, but not sufficient cause'). You emphasized that the working title is very tentative; and, since you and your readers will be looking at the discovery of ivermectin as a whole, perhaps you would agree that the use of the word 'crapshoot' might be a little harsh!" I thought a lot about whether to include the word, or not. I knew on the one hand, that Bill and the rest of the team involved in the work were incredibly careful about the new approach they took. It was brilliant. They knew what they were after, and they knew how to find it. This part of it was not chance, or serendipity. It was nothing less than science of the very highest quality. On the other hand, Merck processed 40 000 isolates over a period of several years and found ivermectin properties just once. This means, they did not find these properties 39 999 times. Moreover, they found a microorganism that possesses the wonderful anthelmintic, insecticidal, and acaricidal characteristics in soil taken from a Japanese golf course. As I said above, if the Katasato microbiologists had taken soil from the ninth tee rather than the thirteenth fairway, they could easily have missed it.

Sorry Bill, for me the choice becomes clear again. I still think about some of the ivermectin discovery in terms of a 'crapshoot'!

REFERENCES

Campbell, W. C. 1989. *Ivermectin and Abamectin*. New York: Springer-Verlag.
Campbell, W. C. 1992. The genesis of the antiparasitic drug ivermection. In *Inventive Minds*, ed. R. J. Weber and D. J. Perkins. Oxford: Oxford University Press.
Collins, K. 2004. Profitable gifts. A history of the Merck mectizan [Ivermectin] donation program and its implications for international health. *Perspectives in Biology and Medicine* **47**: 100–109.

6

"You came a long way to see a tree"

O true apothecary!
Thy drugs are quick.

Romeo and Juliet, William Shakespeare (1564–1616)

In 1898, Sir Patrick Manson announced to the world (actually to the British Medical Association at their annual meeting in Edinburgh) that Ronald Ross had successfully proven that *Plasmodium* sp. was transmitted by mosquitoes. In the same speech, Manson said, "in virtue of the new knowledge thus acquired, we shall be able to indicate a prophylaxis of a practical character, and one which may enable the European to live in climates now rendered deadly by this pest." Of course, he was aware that there was already something available for the treatment of malaria, namely quinine, and that it had been available for a long, long time.

In the previous essay, I wrote about the discovery of ivermectin by Bill Campbell and his colleagues at Merck and the Kitasato Institute in Tokyo. In the case of ivermectin, these folks were looking for it, or at least were looking for something like it. In the process, they developed a plan and research protocol, conducted a vigilant and extensive search, made the discovery, defined the drug's structure, improved it structurally and functionally, tested it under highly controlled conditions, and then marketed it world wide. In other words, it was a carefully configured approach to discovery.

Quinine is a completely different story. In fact, I do not think it was really 'discovered', certainly not in the sense of ivermectin. Quinine just sort of appeared. One of the problems was that those in the Old World knew about the rigor and fever of a disease they referred to variously as intermittent fever, ague, or malaria. Prior to the Spanish intervention in the Americas, the greater probability is that malaria was not in the western hemisphere (except see Essay 2), hence cinchona bark as a febrifuge for malaria, at least, was not necessary. Depending on whom you believe, however, the original use in South America was for rigor. But, then, with the invasion of the proselytizing Jesuits, the value of the drug in treating the disease was discovered. It was since employed successfully in this regard, and it still is.

The best historical account that I have seen for the 'discovery' of quinine is *The Fever Trail*, written by Mark Honigsbaum (2001), a London (*The Guardian*) newspaper reporter with a penchant for malaria and malaria cures, among other things. I had a chance to sit with Mark in London in May 2005 and listen to him talk about his interests and how he got into the 'drug business'. It was fascinating.

Based on Mark's book and other historical accounts that I have read, the discovery of quinine and its value as an antimalarial drug was probably accidental, very different from ivermectin. In fact, quinine and its early use for treating malaria are shrouded in the proverbial 'fog' of history. No one knows for sure how the drug was discovered, or who made the discovery. The early writings are quite conflicting. Accordingly, there are a number of stories with regard to how the connection between malarial fevers and quinine was first made. Perhaps the most romantic account relates to the fourth Countess of Chinchon, Francisca Henriquez de Ribera, who was supposedly cured of malaria in either 1623 or 1633 by the bark from a tree, soon to be known as 'Jesuit's bark'. As the story goes, the Countess, also wife of the Viceroy of Peru in Lima, had become ill with an intermittent tertian fever. As a curative for such fevers, the Viceroy, acting on the advice of the Corregidor (Governor) of Loja, a province in Peru, prescribed the bark from a tree that grew in the Peruvian region of Loja, not far from Quito. After it was administered to the Countess, she recovered completely. The story continues that she was so thankful she obtained a large quantity of the bark and had it distributed to the poor. The bark then became known as *pulvis comitisae*, 'the powder of the countess'. Even though Linnaeus favored the Countess by naming the tree from which the bark came in her honor, i.e., *Cinchona*, he deprived her of complete glory by misspelling her name when he established the generic title for the tree – he left out the *h* after the first *c*.

Another tale regarding quinine's early use relates to the native Indians. In a remote area of Peru, not far from Quito, a group of Indians who worked in a nearby mine were apparently required to cross a deep and very cold river every day. The result was severe chills, or rigor, which 'miraculously' disappeared when they consumed powder made from the bark of cinchona trees. Jesuit priests observed the practice and began treating anyone who developed chills, ideal for the rigor that precedes the high fever associated with what we now know to be the classic paroxysm of malaria. Because of the close association between the Jesuit priests and the cinchona trees, the source of the quinine also became known as Jesuit's bark.

Many early authors indicated that cinchona bark was a long-time herbal remedy of the Indians. However, Alexander von Humboldt, a great explorer of the seventeenth century, claimed the febrifugal character of the bark was not discovered by the native Indians, but by the Jesuits. As noted by Honigsbaum (2004a), "Unfortunately, Humboldt does not say how he came by his information, and his conclusion that 'this tradition is *less probable* [his italics] than the assertion by the European authors . . . who ascribe the discovery to the Indians', suggests that he was unsure of his source."

The tree grows in a wide area of the northern region of South America, including parts of Venezuela, Bolivia, Peru, and Ecuador, at altitudes of approximately 4200–9000 feet. *Cinchona* belongs to the Rubiacea (the madder family) and has at least 23 species, although high quinine production is limited to but a handful of species.

When I heard about Mark Honigsbaum's book, I contacted him initially through his publisher. I finally obtained his home email address and arranged an interview with him in May of 2005. One of the things in which I was interested was seeing the tree. However, since I did not want to go all the way to Peru to find a tree, I thought I would ask Mark if Kew Gardens in London had one. He replied, "Yes, but they are sort of scraggly. The best one over here is in the Chelsea Physic Garden." Well, even though I thought I knew London fairly well, I had never heard of this place, so I looked it up on the web. Sure enough, there is such a garden and it is not very far from our hotel in Kensington where we normally stay while we are in London.

The first Saturday we were there in 2005, we took a taxi, with a £10 fare, to the Chelsea Physic Garden. As it turned out, it is open only on Sunday and Wednesday afternoons, and all we could see were the red brick walls sheltering the interior. We returned, by bus, the next day (£10 does not seem like much, but with the £ running at $1.91, that translates into $19.10, excluding the tip).

The trip was well worth the cost of entry, about £4. The Garden is stunningly beautiful. When we got inside, I asked a docent, "Where is the quinine tree?" At first, she looked puzzled. Then, she knew what

Figure 18. A cinchona tree, Chelsea Physic Garden, London, England

I was talking about. She responded, "It's over in the glasshouse." I said, "Great, that's what I want to see." Again, she looked puzzled, and said, "You came a long way to see a tree." I laughed and then explained what I was doing and why I wanted to see the tree. She was satisfied with my explanation. Ann and I had to look around in the greenhouses for a while before we found it (Figure 18). It really is not very unusual, as trees go. It most closely resembles coffee trees, even to the point of needing the same sort of soil conditions, temperature, and rainfall. When in flower, it produces a wonderful, lilac-like smell and is attractive for hummingbirds of all sorts. It easily cross-pollinates with other species of plants, producing hybrids, and reducing the amount of quinine in the bark as a result. More about this in just a bit.

The Chelsea Physic Garden was founded in 1673 by the Society of Apothecaries of London and, as such, is the second oldest botanic garden in the United Kingdom, with the oldest at the University of Edinburgh in Scotland. The Chelsea Garden is unusual in that, early on, it was one of just a few that was not associated with a university. Instead, London apothecaries established it so they would have plants for teaching their apprentices and local medical folks. They needed to be able to not only identify useful medicinal plants, but also to know their uses and how to distinguish them from possible poisonous ones, which they might closely resemble.

The site of the Garden is close by the River Thames in Chelsea, one of the most beautiful suburban areas of London. In fact, if you stand outside the Garden and look down the street about a block away, you can see the Albert Bridge that crosses the Thames. Another advantage for siting the Garden in that location was that it was close to the river, making irrigation water easy to acquire. Moreover, the river was preferred as a means of transport in those early days, not only because it was quicker, it was safer too. Henry VIII had a country estate close by and so did Sir Thomas Moore, later to be Henry's counselor, then enemy for refusing to give up his allegiance to the Pope, and eventually losing his head.

Hans Sloane (1660–1753) bought the Manor of Chelsea in 1712 from Charles Cheyne. In doing so, he simultaneously acquired the freehold for the Garden. Sloane was quite sympathetic to the apothecaries and had even studied at the Garden in his youth. Sloane was knighted in 1716, and later became President of the Royal Society and the Royal College of Physicians. By the way, Sloane was wealthy not only because he was a good physician, but also it was he who introduced chocolate to the U.K. and this did not hurt his wallet either! In 1722, he leased the Garden property in perpetuity to the apothecaries for the sum of £5 per year, on the promise that "it be for ever kept up and maintained a physick garden." He also required that 50 pressed plant specimens be delivered to the Royal Society each year until the total reached 2000. By 1795, the total had climbed 3700. The Garden still has medicinal plants, although they are not kept for the original purpose (Figures 19, 20). For example, both *Podophylum hexandrum*, used in the treatment of intestinal worms, and *Artemisia annua*, another plant used in treatment of malaria, are grown there.

Contrary to some conjecture, it is most likely that malaria was introduced into the western hemisphere in post-Columbian days by the early European conquerors and explorers. Honigsbaum guesses that it would have taken at least into the 1550s for the disease to spread widely among the natives. He points out it was about this time that the use of quinine in the treatment of malaria was first noted by the Jesuit priests who accompanied the European interlopers. It is a good bet the

Figure 19. Chelsea Physic Garden, London, England

Figure 20. Chelsea Physic Garden, London, England

natives prior to Columbus had knowledge of quinine and that it could be useful in treating rigor. According to Mark, it is a principle of homeopathic medicine that bitter-tasting herbs and other plants, such as the cinchona tree, often have therapeutic effects. While religious proselytizing was the primary goal of the Jesuits, many were also pretty fair scientists and physicians for their day and time, and it is a good bet they became aware early on of the power of this bark from the special tree. Mark offered that they probably obtained some of the bark, or its powder, and performed some experiments to confirm its ability to reduce fever. He went on to say that for people to be persuaded regarding the efficacy of the magic bark, stories like that told of the Countess of Chinchona's miraculous cure of malaria, or ague as it was also known then, did not hurt the assertions made by the Jesuits regarding the bark's effect on malaria. Mark then said, "They [the Jesuits] would have then been in position to oversee the very lucrative trade in the bark from the cinchona tree from Peru to, first of all, Madrid and Rome. Then, by repeating this story, they probably propagated the bark's popularity throughout all the royalty of Europe who we know by that time all had had a bout of malaria. So, what better way of popularizing the treatment in Europe than by passing the story that it worked on this glamorous member of Peru's royalty." Whether this happened in exactly the way Mark speculated is irrelevant, because in due time the Jesuits were successful in bringing the bark back to Europe. Its popularity spread throughout the continent and it became known as Jesuit's bark or Jesuit's powder.

One of the initial problems with the bark being imported into Europe in the early stages was that there appeared to be some variation in its effectiveness, which, in turn, caused some to be doubtful regarding its efficacy. Later, when chemists were finally able to isolate and quantify quinine from the cinchona bark, this variation was verified. There were two reasons for this phenomenon. First, there was a large number of species of the tree and the amount of quinine varied, one to another. Second, as mentioned above, cinchona species tended to hybridize with each other and even with species in other genera. Not only did this cause some problems for bark imported by the Jesuits initially, it was to cause some exasperating situations for later generations of cinchona exporters from the western hemisphere and for those attempting to develop cinchona plantations in other parts of the world.

Another feature that was unknown to the early users of cinchona bark was that there was more than one alkaloid present in the bark, and, in fact, there were alkaloids that were more potent for malaria than quinine. In effect, some folks were being treated with more than one antimalarial drug, which we know today is an effective way of preventing resistance from developing. One could speculate that this is a good reason quinine is still effective in the treatment of malaria 350 years after its first use, while resistance has developed against so many of the synthetic drugs. Meschnick (1997) indicated that some moderate resistance to quinine, especially by *P. vivax* and *P. malariae*, has emerged over the past several years. He also argued that quinine resistance to *P. falciparum* is not as great as for the other species because it "had relatively little exposure to quinine, and thus little evolutionary selective pressure from it."

Of course, at first the Jesuits relied on the native Indians to harvest the bark and they knew what they were doing. But early on, there is also evidence that there was a certain tree that was easily mistaken for the cinchona tree and this created even more confusion among some Europeans who were depending on the importation of the correct Jesuit's bark. There were also some unscrupulous harvesters who would export the bark of trees that looked every bit like that of the cinchona, but were not, further confusing the matter.

Perhaps one of the most confused persons in Europe regarding quinine was Oliver Cromwell, the leader of the Roundheads during the British Civil War and the executioner of Charles I. I had read somewhere that Cromwell had died of malaria after refusing to take the Jesuit's powder. I asked Mark about this and he said that Leonard Bruce-Chwatt's research into this question was probably correct in that Cromwell died of kidney disease, not malaria. He qualified Bruce-Chwatt's conclusion, however, by saying that Cromwell had probably suffered from malaria all his life and that the anemia he developed over a lifetime battle with malaria had contributed to Cromwell's ultimate demise. His hate of the Catholic Church and everything connected with it, including Jesuit's powder, thus probably contributed to his death in 1658 at the age of 59.

The 'War of Jenkins' Ear' was an event that contributed greatly to the European countries in developing an interest in the cinchona tree as a commercially important commodity. By the time they were through their era of colonial expansion, the British definitely recognized the need for a prophylactic to deal with malaria. It would be another hundred years, however, before they began formal exploration regarding the cinchona tree, etc., but 'The War of Jenkins' Ear' that occurred in 1739 definitely emphasized the need for such a process.

It is a story well told in Mark's book, so I will simply provide some highlights here and make my point regarding the impact of malaria and probably yellow fever on a major military confrontation of the day. By 1739, there were three European colonial powers in the western hemisphere, i.e., Spain, France, and Great Britain. The Portuguese were in Brazil, but this was their only sphere of influence in the New World. The Spanish were by then ensconced in Mexico, Central America, and most of South America, but the British had their eye on the major gold-producing countries of South America, one of which was Columbia. A significant port for this country was Cartagena. During those days, it was not unusual for British privateers to raid on the Spanish Main, with the Spaniards retaliating when and where possible. Such was the case when a British ship under the command of Captain Robert Jenkins was boarded by Spaniards and found to be carrying what they considered as contraband, which was, of course, duly confiscated. In what Mark describes as a "fit of pique", Jenkins' ear was cut off and the Spanish captain then spat "into what was left of it." The British captain was told to return home and tell the King that he would get the same treatment if the opportunity ever came. Jenkins sailed back to England, along with his disengaged ear in a bottle of alcohol (rum?), and described the incident to his superiors and to Parliament, which provoked a huge outrage. In actuality, it seems this is exactly the event that would give the British a reason to retaliate, which they did. Prepared to attack the fortress, sack the city, and occupy Columbia, a fleet with 18 760 fighting men sailed into Cartagena Bay. Unfortunately, the force was poorly led. They stalled long enough for the mosquito hordes from the city of Categena's underground cisterns to invade the attacking force and wreak the havoc of both malaria and yellow fever. As Mark described it in *The Fever Trail*, within a matter of a few weeks, the attacking army was decimated by disease. After withdrawing in humiliation, they attempted the same ploy in Cuba and Jamaica on the way home, but mosquitoes vectoring the same diseases doomed both efforts. By the time they returned to England, they had lost 6500 men to fever and disease and had an effective fighting force of just 3000 marines remaining.

The British were to suffer another devastating defeat by malaria, this one during the Napoleonic Wars, when some 11 000 men were killed by the disease or made ill by it at Walcheren, Belgium. This time they thought they were prepared for the problem, but the bark they had was of poor quality and they exhausted their supply in very short order.

By the mid 1800s, many tons of Jesuits' bark were being shipped out of the northern tier of South American countries. However, the supplies of cinchona bark were beginning to run low in South America. The latter countries knew it and began to protect what they had left, but not very effectively. Some Europeans, on the other hand, began efforts to find ways of getting the cinchona out of these countries to other places where the trees could be grown under controlled conditions; the supply of quinine increased by substantial amounts, and the cost of the quinine simultaneously reduced.

There were three Englishmen who were seriously involved in removing cinchona seeds and/or cuttings from these countries. Their efforts were significant in saving a lot of lives because the cinchona trade from South America was definitely drawing to a close: the reason, no conservation. The three were Charles Ledger, Sir Clements Markham, and Richard Spruce.

The first of the three was not a botanist, or a naturalist, or even a scientist. He was a merchant, the son of a merchant, who at a very early age had made his way to Peru to seek his fortune in the export of cinchona bark and alpaca fur. On arrival, according to Mark, he was immediately employed by "Naylor's, a respected British merchant house," where he spent the next seventeen years learning Spanish, about the local culture, and the difference between good and bad cinchona bark. At the end of this period, he set out on his own. One of the first things he did was to hire a Bolivian Indian, Manuel Incra Mamani, probably one of the most informed men, ever, regarding the cinchona tree. He was to be a life-long friend of Ledger and, in fact, he died from brutal treatment received in prison because he was caught carrying cinchona seeds collected for Ledger himself. Ledger had been successful earlier in collecting something like forty pounds of the seeds because Mamani knew when to collect and where the best trees were located. This quantity of seeds should have garnered Ledger a handsome profit. He had sent them to his brother, George, in London, who was to handle the sale. Through a series of disastrous mistakes, errors in judgment, and just plain ignorance, George could find no British buyers. The only buyers were the Dutch, who had an extensive cinchona plantation operation in Java, and an Anglo-Indian, who had a large plantation of cinchona trees in India. Ledger had spent £800 in securing the seeds and recovered only £170 for his efforts, plus he lost Mamani. His alpaca efforts were also a failure. He retired to Australia where he died at the age of 87, penniless and disappointed that his contributions were never appreciated as he felt they should have been. Based on my reading of Mark's book and our discussion, it is my impression that despite Ledger's obvious interest in earning money for his effort, he probably was not given the credit, either in terms of compensation or recognition that he deserved. It is interesting to note that the trees grown from Ledger's seeds produced trees that were consistently the most productive of any taken out of the cinchona region of South America. Mamani did know his trees!

The second of the three Englishmen who played a significant role in the great cinchona adventure of the nineteenth century was Richard Spruce. I asked Mark, "of the three, which is your favorite?" His instant answer was, "Richard Spruce, because he was a true naturalist. Ledger's inclinations were commercial. Spruce also writes beautifully. More importantly, he suffered from these diseases." In *The Fever Trail*, Mark describes Spruce as, "A hypochondriac Yorkshireman and moss collector, who despite his fear of disease spent fifteen years wandering the Amazon and Andes on behalf of the Royal Botanic Gardens at Kew." He was born in Yorkshire in 1817, one year before Ledger. He was a botanist throughout his life even though he had no advanced education. He came to the attention of William Hooker and George Bentham, two of the outstanding botanists of his era, when he described a new species of sedge from Yorkshire. In 1849, Spruce boarded the HMS *Britannia* and headed for Brazil. He was basically on his own, selling his collections as they were made, to various museums, institutions, and collectors. His travels took him through every conceivable sort of habitat from the edge of the Atlantic in Brazil, to high into the Andes. Finally, in 1859, the British government recognized the economic advantage the Dutch were developing in Java with their vast cinchona plantations. At the behest of William Hooker and George Bentham, the India Office hired Richard Spruce for the niggardly sum of £30 per month, about £10 more a month than he had been making working on his own.

I cannot begin to describe the travails experienced by Spruce after he entered into his new contract with the India Office. There is simply not enough space here. Spruce's charge was simple. Collect seeds and create rooted cuttings for shipment to Kew Gardens in London. One of the most serious problems he had was in locating the right kind of cinchona tree since so many had been either cut for firewood or had their bark stripped and thus died. In addition to climbing mountains to altitudes of roughly 15 000 feet, being exposed to malarious mosquitoes and other noxious insects, walking through thick jungles in the worst heat, rain, and humidity, riding mules and horses on trails that were so narrow and steep that they should not even be considered as trails, dodging government soldiers and rebels who were engaged in a bloody civil war, contending with a mysterious rheumatoid illness that had kept him bedridden for nearly two months, Spruce and his companion, James Taylor, made it to a place in Andean Ecuador that became known as Spruce's Ridge, so that they could do their work. John Cross, who had been dispatched by William Hooker of Kew Gardens to assist, joined them there. Spruce had contracted a really severe case of ague while on the Orinoco River in 1854, probably falciparum malaria, so he understood the imperative of his task. The quinine he had been given then had undoubtedly spared his life and he had carried the Jesuit's bark with him constantly ever more. Despite the many hardships and difficulties, they were successful in their effort. An important aspect to this contribution is that Spruce probably recognized the significance of their work and, while not contrite about it, was quite self-satisfied nonetheless. After all, hundreds of thousands of lives were saved in India alone because of the quinine produced by seeds and cuttings sent out of Ecuador by Spruce and his partners.

According to *The Fever Trail*, only two tons of the red bark were shipped from Limon, Ecuador, in 1859 and none in 1860. The practice of cutting so much cinchona had brought a close to the trade of Jesuit's bark from Ecuador, a practice that, in part, kept it from becoming a prosperous country. The cost for 100 pounds of the bark in 1859 was $43. Within 30 years, the price had dropped to $10, mostly because of the seeds and cuttings sent by Spruce to Kew Gardens in 1859 and the trees grown on Indian plantations as a result.

Spruce stayed on in South America for several years until, while traveling in the Andes, he was "struck by a paralysis that kept him bedridden for nearly three months. He could hardly sit up straight and he suffered terrible pains in his abdomen." He returned to England via Southampton, and immediately went to work at Kew Gardens where he prepared several of his monumental monographs on botany of the Andean South America, including the descriptions of 300 new species of plants! His life savings of £700 was lost when his bank went under, but thankfully he was saved by the action of several friends. While at Kew, he suffered another attack of malaria, this one probably due to *P. vivax*, but copious quantities of quinine reduced the impact and he recovered quickly. He died of influenza at the age of 76 in 1893 at his home in Yorkshire.

The third player in the British trio involved in the search for the cinchona 'fountainhead' during the middle nineteenth century was Sir Clements Markham, who was also to become the longest serving president of the Royal Geographical Society. His greatest fame is associated with the two Scott expeditions to the Antarctic, but Mark says in *The Fever Trail* that his greatest personal success was his involvement in getting the cinchona tree out of South America. At the age of fourteen, as a naval cadet, he was privileged to visit Peru. This resulted in a life-long attachment to the people of Peru, past and present. He vowed at the time, in fact, to write a history of the Incas. He left the Navy early and secured enough money from his father to return to Peru. It took him ten months to get there because of various stops made along the way. On arrival, he discovered his father had died in the meantime and he returned home in haste. He required immediate employment and was lucky to quickly find a position with the East India Company, which gave him the time to write his history.

It was about then that the British finally awakened to the importance of the cinchona tree and the need for quinine in the treatment of malaria. William Hooker, Director of Kew Gardens, also recognized the need to get cinchona out of South America and into India where plantations could be established for growing the tree. After much negotiating and cajoling, the East India Company was finally convinced that an expedition was required to meet this goal; after all, the Dutch were already way ahead of them in this regard. Cinchona bark from South America was becoming limited for export and the price of quinine was rising. The Court of Directors of the East India Company finally relented and agreed to send an expedition to resolve the problem, but who was to lead it? Markham convinced the Board that he was the right person, after all, had not he been there already, and had not he written a history of the Peruvian Incas, and was he not familiar with the geography and cinchona forests (he really wasn't, according to Honigsbaum)? Although he was relatively unfamiliar with cinchona, he got the job, at the age of twenty-one! He asked that four expeditions be formed and that he would lead one of them. He succeeded in getting three. The luckiest coincidence of his life was in securing the help of a little known botanist to lead one of the expeditions as well. This person was none other than Richard Spruce. Spruce's success meant automatic success for Clements Markham, and, "Now we know the rest of the story," to paraphrase one of the most respected radio commentators in the U.S.A. Markham's reputation was made.

After reading Mark's account of Markham's actions on learning of his appointment, I will say that Markham did everything in his power to insure success. He spoke with every expert he could find, he read Ledger's accounts of his exploration in the area, and he arranged with Hooker at Kew for the transport of cinchona cuttings. He also secured the services of John Cross and John Weir, both excellent botanists, to tend the plants after they were collected and to take them safely back to Kew. As Mark noted, however, Markham was politically incorrect – he talked too much. Everyone in the key countries knew he was coming, and why. Despite this propensity for exuberant talk, Markham managed to secure several hundred of the Cinchona plants and then get most of them back to Kew, under the watchful eye of John Weir. Markham went on to a wonderful and productive career, and was even knighted for his many contributions. At the age of 85, while reading by candlelight in his Eccleston Square home in London, the candle fell on the bed lighting up his bedclothes. He cried out for help, and even though the fire was quickly extinguished, he died without regaining consciousness.

At the end of our discussion in London, I asked Mark, "What do you think it takes for a man to travel six thousand miles via a sailing vessel in the nineteenth century, then wander for many months/years in the jungles of Columbia, Argentina, Peru, Bolivia, Ecuador, and/or Brazil, knowing all the while he could be killed and maybe even eaten by local natives, robbed and/or beaten by thieves, savagely attacked by any of several dozen different blood-sucking insects, drown in a flooded river, or freeze to death in a sudden blizzard up at 10000 feet in the Andes, just in search of a tree, or its seedlings, and/or its seeds?" I repeated, "What does it take for a man to do something like this? You've got Markham, Spruce, and Ledger, what was it about these men?" Mark responded, "I think they all had different motivations. In Spruce's case, he was passionate about botany. There was nothing more wonderful for him than to be in the Amazon or Andes. For him, it was his life. Every day was an adventure for him. Whatever the hardship, there was always reward waiting for him. In Markham's case, he had already become passionate about the history of South America. He was also deeply motivated by Christian humanitarianism, classic Victorian 'do-good' principles. He saw it as his duty as a Christian, a Victorian philanthropist, to spread a cure that saved lives. When he was given this opportunity, it was a double, because on the one hand he could do something he was passionate about in a moral sense, but it was also an opportunity for him to make his name, to become part of the Establishment. For Ledger, it was commercial, because he saw that there were things there that were valuable. He was constantly scheming for ways to make a profit. The tragedy of his life was that he never really profited from anything he tried. He was not successful because he could never get to the right botanist, chemist, or horticulturist in, say, Java, where the Dutch were setting up these huge plantations necessary to raise the right species of the cinchona trees and cultivate them so there would be sufficient quantities to be harvested commercially. They were able to turn quinine into a lucrative pharmaceutical product. The thing all three had in common was the desire for adventure. Some men have always been driven by the need for adventure and excitement, and the thrill of discovery, whether it's an archeological remains, or a pot of gold, or a new medicine."

Ann spoke again, "You are a composite of these three by doing what you did to write *The Fever Trail* and *Valverde's Gold* (2004b)." Even though Mark demurred, she was right. I seriously doubt that Mark would hesitate to do anything that any one of these three explorers had done in their relentless search for the cinchona tree. I too went a long way to find the tree, but I chose London, not Peru, to see it. It's just too hot and humid down there for me and, besides, I love riding on the big double-decker buses as opposed to the hard backside of a Peruvian mule.

REFERENCES

Honigsbaum, M. S. 2001. *The Fever Trail*. London: Macmillan.
Honigsbaum, M. S. 2004a. Cinchona. In *Traditional and Medicinal Plants and Malaria*, ed. M. L. Willcox, G. Bodeker, and P. Rasanova, pp. 21–41. Boca Raton: CRC Press.
Honigsbaum, M. S. 2004b. *Valverde's Gold*. London: Macmillan.
Meschnick, S. R. 1997. Why does quinine still work after 350 years of use? *Parasitology Today* 13: 89–90.

7

Infectious disease and modern epidemiology

Life is as tedious as a twice-told tale,
Vexing the dull ear of a drowsy man;

King John, William Shakespeare (1564–1616)

Ross, Bailey, Hairston, Bradley, Michel, McDonald, are all names that will be recognized immediately by any epidemiologist/modeler working on eukaryotic parasites. Almost all of these people studied and published prior to 1971, beginning with Sir Ronald Ross who, I am sorry to say, rather feebly attempted to predict the prevalence of malaria in mathematical terms. Most investigators in this area would agree that the seminal publications for the modern epidemiology of helminth parasites were those of Crofton (1971a, b). His efforts were truly the 'seeds' for what followed. Harry Crofton, most unfortunately, died very young, not long in fact after publishing his two papers in 1971. I have often wondered what would have followed had he lived a longer life.

Early in my career, I had the pleasure of spending almost an entire year (1971–72) at Imperial College in London with Desmond Smyth, who was trying to teach me the intricacies of in vitro culture. He succeeded, but in doing so, I also learned that this sort of research is tedious, very expensive, and, frankly (for me, at least), terribly boring. It was about that time that I turned my complete attention to ecological pursuits. While I was in London during that year, I had heard about a young parasite ecologist who had just finished his Ph.D. with June Mahon at Imperial and then gone off to Oxford to do a postdoc. His name was Roy Anderson. In the spring of 1984, I was invited to give a seminar at the Institute of Parasitology at McGill University in Montreal. I chose to talk about the really excellent doctoral research just completed by one of my Ph.D. students, Dennis Lemly, on black spot disease that occurs in centrarchid sunfishes. My wife and I were invited by our hosts to a very nice dinner at a local French restaurant in Montreal the night before the seminar. As it turns out, Roy was in town to consult with Marilyn Scott, another excellent parasite ecologist. This was my first meeting with Roy and I must admit to being somewhat intimidated; after all, by that time, he was one of the true 'stars' in the world of mathematical parasitology, modeling, and epidemiology, and I was not sure what to expect. The dinner that night couldn't have gone better, and the seminar the next day was, I thought, very well received (mainly because Dennis really did a good piece of work). Subsequently, I of course kept track of Roy's travels, and travails, after our meeting in Montreal. When I knew I was going to write this book, I also knew I wanted to do an essay on modeling and epidemiology. As I wrote in his biographic sketch earlier, this is when I contacted Roy to see if he would be willing to help out.

I wanted to write about modeling and epidemiology not so much for the mathematics involved, about which I know very little in practical terms, but to see if I couldn't learn something about Roy's thinking in developing these models and in his general attitude and approach to the epidemiology of infectious disease. Roy has authored or coauthored several hundred papers since receiving his Ph.D. in 1971, and has directed the research of countless graduate students and postdocs. His influence in this field has been enormous and his successes have been rewarding to him professionally and personally; not always, however, without cost, to wit his experience at Oxford.

Roy's most significant contributions began, at least in my opinion, in 1978, when he and Bob May published two papers in *Journal of Animal Ecology*, followed by two more in *Nature* in 1979. The first 1978 manuscript (Anderson and May, 1978) was directed at developing mathematical models to help explain the regulation and stability of host–parasite interactions at the population level. A central conclusion of the first paper is that three factors tend to help stabilize the dynamic quality of host–parasite interactions. These include: overdispersion, or aggregation, of parasites within host populations; nonlinear functional relationships between parasite numbers per host and death rate of the host; and "density dependent constraints on parasite population growth within individual hosts." In the second paper (May and Anderson, 1978), they examined the nature of factors that tend to destabilize host–parasite interactions. Again, they reached three conclusions with respect to destabilizing influences. These include: "parasite induced reduction in host reproductive potential; parasite reproduction within a host which directly increases parasite population size; and time delays in parasite reproduction and transmission."

I asked Roy to describe how their approaches differed from mathematical modeling schemes developed by earlier parasitologists, i.e., McDonald's (1965) work on the schistosomes. He responded, "First of all, I came to the problem from my perspective as firmly interested in population ecology, primarily, and not parasite biology. With an ecological perspective, you should always think in terms of dynamic species interactions. You think of host populations, not just humans, which have their own dynamic . . . I had been part of a very vibrant ecological community [in London], which was my main source of intellectual stimulus at that time, but I was also a parasitologist in the sense that I had training, done third year projects, and had received a Ph.D. in this area. It was a melding of those two fields that turned out to be so productive. So, I had a conceptual framework of how I wanted to go about approaching the problem. Then, from Bob May's end, he added a considerable degree of mathematical rigor that enabled us to explore a very broad range of biological assumptions in a quite generic framework. The most important of those was, in fact, that most parasite distributions, i.e., numbers of parasites per host, are highly heterogeneous, with most parasites in a few hosts. We were able to embed into the model these distributional properties in what are called hybrid structures that have certain probability elements embedded in a deterministic framework. This allowed us to explore a whole range of problems and hypotheses, beginning with the ecological notions of density dependence, stability, resilience, perturbations, cyclic fluctuation in abundance, and so on, for these populations."

He continued, "We could explore a hundred and one different life cycles of all sorts of complexity, ranging from very simple, like direct transmission, all the way up to the most complex of the digenean cycles. The result was that we had a framework that allowed us to explore all these bits of biology and that really was quite fascinating. Then, very shortly after the 1978 papers were published, we broadened our scope to look beyond the traditional parasites, like helminths and protozoans. In some sort of haphazard way, we invented this dichotomy of micro- and macroparasites. This was in the 1979 *Nature* papers [Anderson and May, 1979; May and Anderson, 1979]. We used microparasites to mean viruses and bacteria that multiplied directly within the host. Using Bob's rigorous approach, we could explore the properties of these models analytically with mathematical tools. We could also explore more complex structures, with many different populations representing complex parasite life cycles and many forms of heterogeneity. In those days, Bob did not like computers, but I was heavily into this sort of technology, so we could explore very complex problems with stratifications of age, sex, and space. It was a melding of biology and the mathematics, coming at it from an ecological and evolutionary perspective."

I then asked, "Did you have to make large adjustments in your thinking when you made the jump out of the traditional parasitology to include viruses and bacteria?" He answered immediately, "No, not at all. You should remember that studies on infectious diseases at that time were quite Balkanized. You had departments of parasitology, of bacteriology/microbiology, of virology, etc. Despite this, however, the ecological concepts and the evolutionary issues faced [in any of these areas] were identical. They are all parasites. That's why we were seeking a minimal number of frameworks to explore the problems . . . In those days, quantitative PCR [polymerase chain reaction] was not available, and you couldn't easily quantify the amount of virus in a person. You could just know that they were infected or not, because you had immunological or serologolical tests, which meant plus or minus. In today's world, with quantitative PCR, you can determine how much virus or bacteria there are in a person. One can now construct a single generic framework for all of the infectious agents."

The development of new molecular technologies has opened all sorts of new doors for those investigating epidemiology and infectious disease, just as new avenues were opened thirty to forty years ago with the development of computers. I asked if the creation of modern modeling was driven by technology, i.e., the computer, or was it the other way around. In other words, "Did the computer open the door for you?" Roy responded, "In those early days, the computer really was not the key thing. When I went to Oxford to do postdoctoral studies in the biomathematics department, I learned quite a few mathematical techniques. It was, therefore, possible to explore the properties of these models largely by analytical methods alone. You didn't need a computer. If you wanted to add in all the 'bells and whistles', you know, the fine details of complex life cycles and all the pervasive heterogeneities of real environments, you could resort to the big computers. I was also computer literate and could write programs in the languages of the day, such as Fortran and Pascal. I was also able to use large mainframes to deal with more complex problems. Today, we can do all of that with the smallest laptop, but, in those days, one had to use the mainframe. My postdoc supervisor, Maurice Bartlett, was using an analog computer in the basement of the department. That machine took weeks to compute the passage of an epidemic through the U.K. population, while today it can be done quickly on a laptop. I think on the conceptual side, the idea was to capture the biology as rigorously as possible in small sets of nonlinear equations. The main difference in people like McDonald and Crofton, and us, was that our approach was to try and make the mathematics as rigorous and well defined as possible. The assumptions made were absolutely transparent and explicit, and there was a hard mathematical framework on which you could use analytical methods to explore the properties. Very quickly, by using these tools, with which neither McDonald nor Crofton were familiar, for example, we entered this very subtle world of complex nonlinear dynamics. There, we began to make connections with a much broader applied mathematics community, not just in biology and medicine, . . . but, more broadly, those dealing with thermodynamics, aeronautics, and all that as well."

Since Roy had mentioned Harry Crofton, I decided to ask a question that had really been on my mind for several years. So, I inquired, "Harry Crofton introduced the notion of frequency distributions for parasites within host populations, correct?" and Roy agreed immediately. I then continued, "And he did it within the context of overdispersion." I then queried again, "This is a minor question, but I want to ask it anyway. The first time I talked about overdispersion with a very well known and highly respected plant ecologist, he about 'crapped out' because he told me that they used the terminology in just exactly the opposite way. He immediately chastised me and, through me, Harry Crofton, for changing things around the way Crofton had done." The plant ecologist explained further, "When we [plant ecologists] talk about overdispersion, we are actually talking about a regular distribution." I asked Roy, "Do you know why Harry did it that way?" Roy was quite animated in his response. He said, "Well, the plant ecologist is wrong! If he goes to the formal statistical literature, he would find that these probability distributions are discrete ones and categorized by the relationship of the variance to the mean. If the variance is equal to the mean, then the distribution is Poisson, or random. If the variance is less than the mean, it's a positive binomial, which means there is less heterogeneity then you would expect by chance and this is called underdispersed." I responded by saying, "Then these guys are wrong and Crofton was correct." Roy replied, "Yes, definitely. This terminology has been around a long time if one examines the appropriate statistical literature. In fact, one of the first mathematical people to look at what biological circumstances could generate a negative binomial was my supervisor at Oxford, Maurice Bartlett. His papers were published in the 1960s, and I became aware of them then. Also, although this is no discredit to Harry Crofton, because he was the first to publish it, in my Ph.D. thesis and my undergraduate project done in 1968, these distributions were discussed there. They were already in wide use by ecologists."

All mathematical models are based on certain assumptions. After all, these models are, in reality, nothing more than sophisticated hypotheses that have been framed within a mathematical context. In order to develop the models for their 1978 publications, there were several things Anderson and May had to do. One of them was to define parasitism. I should note that defining this concept is not an easy proposition. Anyone who has ever done any parasitology would immediately understand what I am saying. If one examines any of a number of current parasitology textbooks, or even the primary literature, the range of definitions is quite striking. In most cases, these definitions are biased by the background, training, and interest of the person doing the writing. In fact, in a number of books, authors have justifiably claimed that a precise characterization of parasitism, as for life itself, is impossible to fashion to the satisfaction of all. Some folks, therefore, have been content with a carefully crafted perception of it instead. This is, for example, exactly what Crofton did in his 1971 publications that appeared in *Parasitology*.

In the first paper of Anderson and May (1978), they took a similar approach to that of Crofton (1971a, b). They really did not define parasitism. Instead, they too described it, not unlike Crofton. For example, early in the paper, they argue that, "to classify an animal species as parasitic we . . . require that three conditions be satisfied: utilization of the host as a habitat; nutritional dependence; and causing 'harm' to its host." In their conclusion, they stated, "We regard the inducement of host mortalities and/or reduction in host reproductive potential as a necessary condition for the classification of an organism as parasitic." This is how they define harm. If one compares their description with that of Crofton, they are essentially the same.

I have been personally intrigued by the various definitions and/or descriptions of parasites/parasitism ever since I began teaching parasitology to undergraduate students more than forty years ago. When I came across the characterization of harm by Anderson and May as requiring "inducement of host mortalities and/or reduction in host reproductive potential *as a necessary condition* [my emphasis] for the classification of an organism as a parasite," I felt compelled to ask Roy the following question. "How do you account for all of the parasites that don't do this, the ones we normally consider as parasites? I mean, if you look at the geohelminths, for example, most of them are not 'killers'." As I look back on his response, I guess I should have seen it coming. Roy replied emphatically, "If they are not 'killers', they are not parasitic. If there is no evolutionary or reproductive cost in having the organism within the host, then they are not parasites. They are commensals." I then asked, "What about morbidity? How do you factor this into your scheme of things?" He fired back at me, in a friendly way, "Well, we defined that very carefully. It has to have an effect on the net growth rate of the population, which was measured in terms of either mortality, behavioral relation to mortality or vulnerability, or reproductive efficiency." Roy continued, "To clarify this issue in my own mind, I first turned to the literature, ranging from viruses to helminths. I began to quantify from the laboratory experiments whether one could get good ideas in carefully designed experimental studies, where you had uninfected and infected controls with varying burdens, and how the burden of infection related to the likelihood of death or reduction in fertility, weighted against the controlled experiments. Every time somebody had done a well-designed and controlled experiment, then the organism caused a detriment to the population growth rate, defined as birth rate minus death rate. But also, both of us [both Roy and Bob May] believed it was fundamental irrespective of models. We believed this from a philosophical point having come to the conclusion from an ecological perspective. When two-species interactions are divided into symbiont/commensal, predator/prey, insect/parasitoid, and host/parasite, that was the only definition that could apply. You couldn't call it a parasite if it had no effect on the host. It was then a commensal." I pressed him again by saying, "If an organism makes an animal sick, but doesn't kill the host, or reduce the host's reproductive capacity, then it isn't a parasite?" Roy's answer was both well taken and reasonable. He said, "If it makes the host sick, its probability of being eaten by a predator, for example, in a Serengeti game park, is much greater than the nonsick animal and, therefore, its intrinsic mortality rate is invariably higher during that period. If you take the old wildebeest with the high blood parasitemia or intestinal nematode burdens, then the hyenas or lions are going to take them out first because they don't move so fast any longer. In the real ecological world, their per capita natural intrinsic growth rate is going to be lower. Their fitness is decreased. Another example comes with birds, and a lot of this work has been done lately. If you look at birds ringed [tagged or marked] at birth and then follow them through life by mark–recapture, those which have high parasite burdens turn up less often in the recaptures, and their mortality is higher."

I then said, "You are looking at parasitism in a less typical way." He responded, "Yes, it's an ecological perspective. If I was even to take something as trivial as *Hymenolepis diminuta*, the rat tapeworm, and place the rats on a poor diet throughout their lives, then the infected individuals will, on average, die earlier than uninfected ones. There are small, almost immeasurable, fitness costs, and to me, and to Bob, if the organism does not produce a cost, then it is a commensal."

I then turned to another topic of interest regarding the epidemiology of parasitic organisms. I approached it as follows. "Throughout your publications in dealing with transmission biology of the geohelminths in humans, you emphasize the role of genetic predisposition. But there are many examples showing that genetic predisposition is not a factor. For example, how do you account for the results of a number of investigators, i.e., Croll and Ghadarian in Iran, and Kightlinger and his group in Madagascar, in dealing with this issue?" Roy interrupted by saying, "If I could interject here. The experimental work done with Gerry Schad, for example, raised this question. Is there predisposition to infection? In experiments, or in the field, with humans, if you cleared the worm burden, then during the reinfection process, those individuals who originally had heavy worm burdens would reacquire higher than average worm loads. We were the first to do those field studies, for hookworm, *Ascaris*, *Trichuris*, and schistosomes, and show that, in each and every case, when you did the work properly and had large enough sample sizes, there was strong predisposition to either heavy or light infection. So, that's the epidemiological observation. Causation is a totally different question. For example, it could be that the child lives in a hut in which there is poor hygiene or sanitary conditions, or the hut is near a tree in the shade, which gives hookworm larvae better life expectancy in the soil. So, the predisposition can come from many different sources, of which genetics is only one. However, I would argue that in today's world of inbred mice, and transgenic mice, evidence shows very, very clearly that, even for the most mundane of infections, genetics is the dominant feature. We didn't have transgenic mice back in the old days." Then, rhetorically, he asked, "Why do experimenters in the laboratory working with pathogens, ranging from viruses to helminths, use inbred mice?" He answered his own question by saying, "The reason they do that is because if they used outbred strains, you get greater heterogeneity in the results. That's true for all pathogens that I know about. A second observation relates to the human genome. The largest chunk of the human genome is devoted to two things, the recognition of nonself and the immune system. So this says that in the evolution of primates, and ourselves as one example, these two factors have been the major driving forces in our evolution. And it tells an important story about the early days of parasitology, i.e., why did we use inbred strains of mice?"

I replied by saying, "Okay, I agree, but let's reverse this and ask, doesn't genetic variability in the parasite also play a role in this business of transmission?" He replied, "Oh yes it does and increasingly today, with modern technology, we know a lot more. In fact, today I was on the telephone talking with someone in Vietnam about this new avian influenza. We know that certain strains are more transmissible than others. We know that for HIV, in quite a bit of detail, there are quite a large number of genetic quasi-species in the infected person, and that these are typically only some subset of those genetic forms that are transmissible. If you think of transmissible pathogens, you have two dominant selective forces. One is the ability to survive in the host in the face of immunological attack and the environment chosen, so that the pathogen can get to reproductive maturity. The second dominant selective pressure is getting from one host to the next. And these two selective pressures can be very different. Sometimes, with HIV, for example, evolution is dominantly concerned with surviving in the infected environment and, therefore, those that can do that inside the human host and can transmit tend to be a small fraction of the total gene pool. Most of the viruses have been selected heavily over many years for invading three or four cell types, thereby circumventing the immunological attack of the human host. Only a small subset of the viruses is fit for transmission. And I don't think any of the helminth or protozoan parasites are any different. So, we have this complex interplay of pathogen or parasite evolution typically on this very fast timescale, and HIV is a very good biological example. Every mutation across the entire genome occurs multiple times every day because the rate of replication by the virus is so great compared with the human generation time of about thirty years. So, the virus is operating on an evolutionary framework of minutes to hours and the human is operating in an evolutionary framework of many decades. These two different timescales themselves present problems. The evolutionary consequence of these differences is in long-lived animals and trees, and the evolution of sophisticated immune systems with memory."

One of the practical outcomes of their modeling efforts had to do with the treatment of infectious diseases, primarily the geohelminths, but schistosomes and several others as well. Their predictions indicated that if the parasite population densities could be driven below some threshold, the population biology of the parasite would begin to be subject to chance effects at very low densities and that the parasite would become locally extinct. To reach achieve this goal, they suggested three strategies. The first was mass treatment, repeated at intervals determined by the life expectancy of the adult worm in the human host, not unlike the approach of that used by the Rockefeller Foundation in the southeastern U.S.A. during the first part of the twentieth century to knock down the hookworm 'epidemic' that was occurring there. This simply involved the treatment of everyone with an appropriate drug. As I have suggested elsewhere, and as Peter Hotez has pointed out, this is okay if the drug is effective *and* long lasting in its effect. However, it does not work for hookworm, a helminth for which repeated drug therapy is necessary. Roy told me that, "repeated drug therapy is necessary, perhaps at intervals of one year for *Ascaris*, a little longer for hookworms, and perhaps every three years for schistosomes." The second strategy is selective treatment. In this instance, one must isolate stool examples from the population under study, or treatment, and determine who is the most heavily infected, then treat just that portion of the population. Ideally, this should work, but the cost can be prohibitive because of the labor required to do the stool exams on a repeated basis. The third method is targeted treatment. In this case, knowledge of age/sex/prevalence curves must be obtained first. If we know what they are, then without stool exams, those individuals who should have the parasite receive treatment. Others in the population can be disregarded. This strategy is fine if we have this sort of information. In the case of the geohelminths and schistosomes, this segment of the population should include the school-age children, and, in fact, is being used in many parts of the world at the present time, apparently, according to Don Bundy, with some success. A note of caution is required here. In some places in the world, hookworm prevalence can be seen in convex curves, i.e., school-age children are most likely to be infected and carry the heaviest infections, but not always. Peter Hotez and John Hawdon, and their colleagues in China, have provided strong evidence that prevalence and intensity of infection with hookworm continues to rise with age, especially in the interior of the mainland. It is imperative, therefore, to know which group to target before designing experiments and determining delivery strategies.

I then remarked, "There is another aspect to this host/parasite genetic problem. If you are using targeted treatment, rather than selective treatment, then it really doesn't make any difference regarding genetic predisposition, does it? I mean, who cares who it is among the children's populations who has the largest worm burdens because you are treating all the children?" Roy responded in the affirmative, "Yes, absolutely. But there are two caveats in this regard. Usually, the reason for adopting a particular control strategy is largely determined by economics. The easiest thing to do in many poor resource settings is not to diagnose, but to select a group of kids, or use some other attribute, such as age, or just being in school. In this way, you don't have to do the lengthy diagnostics such as looking at a fecal sample, or, with viral or bacteriological problems, of doing an immune test. The cost of diagnostics is too high. A serological test might run as high as $50 in the latter case, or somewhat less for a fecal examination, but it is still very considerable in resource-poor settings. So, the aim of targeted therapy is to deal with the pragmatics of cost issues. If diagnosis was simple and very cheap, and these technologies are emerging, then we could be more selective in who we treat. The cost of diagnosis must fall below the cost of treatment before it can be successfully used. The second caveat has to do with side effects. If the therapy has a side effect, then I would use the drug in rare cases. We are very fortunate with the helminth parasites that this is not the case, in general. With cancer treatment, almost all therapies have quite considerable side effects. Some antibacteria and antiviral drugs have significant side effects as well. HIV is an example. Therefore, there is a tendency only to treat when the viral load is very high." Instead of targeting a group of children, or a group of people suspected of having a given parasite, there are situations in which selective treatment is the best course of action. He continued, "So, both diagnostics and the side effects of treatment help dictate our treatment strategy."

I decided to pursue the idea of predisposition with a quotation from Don Bundy, who said, "The resistance mechanisms involved in predisposition cannot primarily depend on acquired immunity because the evidence for predisposition suggests that it is the individuals with the greatest prior experience of infection who subsequently reacquire intense infection." I asked Roy if he agreed with Don. He responded by saying, "I would put a lot of caveats around that. As I said earlier, predisposition can be caused by many factors. We know that immunity to geohelminths is not an easy task for the human immune system via its gut interface. In many cases, immunity develops over a long period of time. So, predisposition may just indicate that the kid is living in a nonhygienic household, or, as I said earlier, next to a tree in which moisture is retained in the soil so survival of the hookworm larvae is increased. There can be many factors. Or the kid may have behavioral patterns which predispose anyway."

I then asked, "Do you believe that immunosuppression by geohelminths is having an effect of the geodistributional patterns of TB and AIDS?" His answer was interesting. He replied, "I think it's a much broader problem than that. I don't know the answer to that question. This is an opinion, not based on an analysis. In a funny sort of way, President Mbeke in Africa made the point that many of these diseases are diseases of poverty and I think the factor that's most important for most tropical parasitic infections is poverty, that is to say, poor education, poor nutrition, etc., etc. The latter tends to predispose because the immune system requires a lot of nutrients to be sustained and these people are getting just enough to do the job." His response was similar to that of both Bundy and Hotez. Poverty is not the root cause of tropical parasitic disease, but it enhances its dimensions in so many different ways and, clearly, lousy nutrition is one of them!

I then asked if he felt that resistance to anthelmintic drugs was on the increase? He said, "The quantitative evidence in terms of detailed molecular epidemiological studies is not available at the moment, except for one area, and that's probably the antifilarial agents. There is reasonable evidence that the volume of treatment is linked to the likelihood of detecting resistant strains. More broadly, I think the question is still open. Selection among human helminths works quite slowly, in a sigmoidal fashion over time, from initiation of treatment in a defined community. It increases slowly at first, before entering an exponential phase of rapid growth, then leveling out. In fact, all genetic selection operates sigmoidally. I just don't know of any consistent international surveys that say there is a relationship for the intestinal nematodes or schistosomes, as yet, between treatment levels and resistance. However, I *have no doubt that it will be there, absolutely none whatsoever* [his emphasis], because the golden rule in population genetics is, the stronger the selective pressure you apply, the more likely there will be resistance." I said, "So we are going to see it?" He replied without hesitation, "Yes, we are."

What about vaccines as a way of controlling some of the geohelminths in humans? I asked, "Do you think an effective vaccine is possible for any of the geohelminths?" I asked if he was aware of the field trials about to start in Brazil by Peter Hotez and his group under the auspices of the Gates Foundation. He said that he was, and his response was quick. He said, "Absolutely, and I'll tell you why I am enthusiastic. In a summer job, just before I left school, I worked for a company that was taken over by Glaxo and there was a chap named David Pointer, who developed the first helminth vaccine for a lung nematode, *Dictycaulus*, in cattle. It was an excellent veterinary helminth vaccine, nearly 100% efficacious. If you can do it for cows, you can do it for humans. What is hindering us from doing it in humans are, of course, all the subtleties and complexities of modern legislation about safety. You know, helminths are big beasts. They reproduce sexually. There is a lot of genetic heterogeneity, so you are not dealing with a fixed set of antigens. You are dealing with a lot of different, and evolving, antigens. If today's rules and regulations applied when Salk was doing his polio vaccine, and the smallpox vaccine was developed ten years later, we would have neither of those vaccines." My response was simple, "It sounds kind of scary!" He replied, "Absolutely. So it's our obsession with safety that is in the way. Now, as regards Peter, I wish him every luck, and I have no doubt that scientifically it is possible, but his real problem is going to come with licensure and regulation."

My next to last question, "Where do you think we are going to be ten years down the road in dealing with these problems, the geo-helminths in particular?" Roy responded, "I think the Gates Foundation has made a huge difference. They've got vast resources in the charity. In the modern world, parasitic organisms have decreased in their sexiness, as it were, to WHO and other bodies, largely because of the emergence of HIV and tuberculosis. These diseases have in some ways drawn attention away because of the sheer magnitude of the HIV problem. There are also a lot of the emerging viruses, such as SARS, due to our increasingly globally mixed society where you can go from Bangkok to London in nine hours. So, Bill and Melinda Gates have made a difference because they have the money, the largest charity in the world at present, and they are interested in diseases of poverty such as those caused by the helminthes. They believe in making a difference on the ground in developing countries with the disease problems. The World Bank has done a bit, but to be honest, despite Don Bundy's very best of intentions, it's the money that Gates brings to the table that helps. So, via the Gates funding profile with regards to hookworm and schistosomiasis control in five African countries, and extensive funding for filariasis control programs, progress should be made. The free ivermectin program by Merck for onchocerciasis in certain African countries should also be effective. But it's not just the free drug program, it's infrastructure support to deliver it." I then chimed in, "You know, I was so darned impressed with Don Bundy when I talked with him. He's got so much passion for controlling these geohelminths." Roy came back, "Oh yes, I agree and it is a great challenge, but one where success is possible. You've got to work to the Ministries of Health and Education and so forth, and get top-level government approval. It helps so much if you've got a huge chunk of money. In the five-country program for the schistosomiasis control program now supported by a significant grant from the Gates Foundation, Alan Fenwick, of my Department at Imperial, keeps a tight control of the money so it doesn't get into the wrong hands and is used to treat the children. He provides the infrastructure for the delivery to schools of the free praziquantel supplied by the manufacturers, and you've also got to track what you are doing because one of the quite dangerous things that could happen is that some of the drugs could possibly be stolen in a poor country without being delivered to the schools and then sold by the thieves. So, it's repetitive education and delivery that's the real problem after you secure the money."

I had one more question to ask, a kind of philosophical query of Roy. I was interested in his approach to science. I said, "It seems to me that it included first the identification of a question that you wanted to answer and then, second, a search for a system to test your hypothesis. Is this basically correct?" Roy's answer was in the affirmative, "at least with some questions. You touched on one earlier, i.e., what's the demographic impact of pathogens with different biological properties in natural communities? You know, I became very irritated with a young ecologist who claimed that it was insects, parasitoids, plant herbivores, and predators that were doing all the damage and were the really important controlling factors in community structure. I feel quite strongly that infectious agents are very, very important, so it was finding a system in which to analyze that and then a mathematical framework that permitted even more rigorous analysis than had been the case before. So, I'm very much a theoretical person. I am a concepts/ideas person, and I always need to work with the pragmatic experimentalist or field worker. And it's this combination that I enjoy most." I commented, "Well, it sounds like you feel good being back home at Imperial and that this probably makes a big difference." He replied, "Keeping your feet on the ground here with graduate students and friends means that it is a nice haven. Most of the academic staff here are long-standing friends, so it's a very stimulating and friendly place for academic pursuits, research on parasite ecology and epidemiology!"

He closed by saying, "You know, I am in part a helminthologist at the core, but more broadly I find biology absolutely fascinating in all its nooks and crannies. So good luck with the project, and it's a great task." I was pleased with the outcome of our interview, and I promised him a pint or two in my local pub, 'The Goat', the next time I was in London. He accepted enthusiastically.

Between the time of our interview and now, as I sit writing this essay, I reflected on Roy's comments about the young ecologist and his naïve perspective on the nature of things as they regard organisms that have an impact on our communities. I recall one of my own earlier experiences teaching parasitology at the W. K. Kellogg Biological Station. The Director at the time and my good friend, George Lauff, and I were talking one day about what should be taught there. He related that one of the Station's faculty members had questioned the place of parasitology in the curriculum of a field station, saying it should be taught instead on the vet school campus in East Lansing. I was outraged at such a suggestion and told him so. Fortunately, he agreed with me. However, thirty years later, George is retired, I am gone from there, and so is parasitology. It is not being taught. This is such a tragedy, because it has happened at so many field stations and in universities and colleges across North America! Some biologists claim that at least 50% of all species now extant have a parasitic phase at some point in their life cycles. I agree. If we accept the description of parasitism in the sense of Anderson and May, i.e., that an organism must cost its host either energetically or reproductively to be a parasite, then I would still place the percentage at around 50%. Fortunately, not all parasites are destructive in the sense of causing host mortality. However, morbidity is costly and it can lead to host death. In some cases, the parasite actually depends on morbidity that will lead to host death to complete a life cycle, or lead to dissemination of a parasite's reproductive propagules.

I really enjoyed my three hours with Roy and the several hours I spent in reading his papers and preparing for the interview. He is a very intense man, with a strong perspective on his broad approach to ecology, epidemiology, and mathematics. Although his career has stretched well beyond what I suspect he thought it might have become in his early days as a graduate student at Imperial College, I could see, and hear, a strong sense of commitment to his roots as a parasite ecologist. Most importantly, I could feel the intensity of his commitment to this important niche in the field of parasitology. He may now hold a very high post in Her Majesty's Ministry of Defence, but my bottom dollar would be to bet that, deep down, he still thinks of himself as a 'gut-scraping' field parasitologist!

REFERENCES

Anderson, R. M. and R. M. May. 1978. Regulation and stability of host-parasite population interactions: I. Regulatory processes. *Journal of Animal Ecology* **47**: 219–247.
Anderson, R. M. and R. M. May. 1979. Population biology of infectious diseases. Part I. *Nature* **280**: 361–367.
Crofton, H. D. 1971a. A quantitative approach to parasitism. *Parasitology* **62**: 646–661.
Crofton, H. D. 1971b. A model for host-parasite relationships. *Parasitology* **63**: 343–364.
McDonald, G. 1965. The dynamics of helminth infections with special reference to schistosomes. *Transactions of the Royal Society of Tropical Medicine and Hygiene* **59**: 489–506.
May, R. M. and R. M. Anderson. 1978. Regulation and stability of host-parasite interactions: II. Destabilizing processes. *Journal of Animal Ecology* **47**: 249–267.
May, R. M. and R. M. Anderson. 1979. Population biology of infectious diseases. Part II. *Nature* **280**: 455–454.

8

The 'unholy trinity' and the geohelminths: an intractable problem?

We have unmistakable proof that throughout all past time, there has been a ceaseless devouring of the weak by the strong.

First Principles, Herbert Spencer (1820–1903)

Obviously, the 'unholy trinity' mentioned in the title of this essay is a play on words. However, when I make the connection between the so-called 'unholy trinity' and the geohelminths, any parasitologist would know the three parasites to which I refer. These would include *Ascaris lumbricoides*, *Trichuris trichiura*, and the hookworms. Generally, the two species of human hookworms, *Ancylostoma duodenale* and *Necator americanus*, are lumped together and considered as one, mainly because their biology is so similar, and because the disease they cause is so nearly the same. While their geographic distributions are essentially sympatric in today's world, *A. duodenale* probably had an Asian origin. Charles Wardell Stiles of the U.S. Bureau of Animal Industry in Beltsville, Maryland, first described *Necator americanus* in 1906, but its origins are not of the New World. It most likely evolved in Africa and was imported into the western hemisphere, along with malaria, yellow fever, schistosomiasis, and several other diseases, during the slave trade.

Estimates regarding the numbers of people infected with the 'unholy trinity' vary, but all would agree that these parasites, collectively, have perhaps the greatest impact on DALYs (Disability-Adjusted Life Years) on a worldwide basis when it comes to the helminth parasites. According to David Crompton's (1999) paper in the *Journal of Parasitology*, the number infected with *A. lumbricoides* is >1.4 billion, with hookworms greater than a billion, and *T. trichiura* about 750 million. That's a lot of people! Most would agree that all three are on the rise. To confirm this assertion, all one has to do is check Norman Stoll's (1947)

Presidential Address to the American Society of Parasitologists, "This wormy world". He placed the number for *Ascaris lumbricoides* at 644 million, *T. trichiura* at 355 million, and hookworm at 457 million. In other words, each of the 'unholy trinity' has doubled, or more, since 1947. To me, this seems astonishing! Why the increase and are they still on the rise? One of the best people to answer these questions and others regarding control of the 'unholy trinity' and the geohelminths, is Don Bundy. So, I went to see him.

More than likely, all of these worms have been with humans for a long time, at least 5000 years, and probably a lot longer than that. Most agree that both *Ascaris* and *Trichuris* were acquired by humans from pigs. As Don pointed out, "This is largely based on the very close genetic relationships between *A. lumbricoides* and *A. suum*, and between *T. trichiura* and *T. suis*, as well as the very close personal living relationships between pigs and humans historically. I find this idea persuasive, but then we humans have 98% of the DNA of chimps and 88% of the DNA of rats, so perhaps genetics doesn't say that much after all."

Elsewhere (Essay 9), I have discussed the history and life cycle discovery of the human hookworms, so I need not repeat it. Here, we'll briefly review *A. lumbricoides* and *T. trichiura*. Many people include, correctly, *Strongyloides stercoralis* as a geohelminth, but this species is not always included as part of what I am calling the 'unholy trinity'. However, its life cycle can be virtually identical to that of the hookworms, so I will refer it too, but only in passing.

The life cycles of the geohelminths took some real time to resolve. There are several reasons. First, the idea of alternating generations in the nineteenth century was new and I think a lot of those folks in that era had trouble in really comprehending how a tiny egg could develop into a twenty-centimeter worm, for example. Second, I also think the idea of a life cycle was a difficult concept to grasp. Under similar circumstances, I am confident it would be for me as well. Third, even though most of these parasites had been assigned scientific names, too often, new ones would be superimposed on the old ones. This made the literature very confusing, to the point that parasitologists of that time frequently were uncertain about the identity of the parasite on which they might be working. Fourth, the life cycle of any of the human helminths posed a special problem in that the host could not be killed and necropsied to check for larvae or developmental stages. Fifth, I honestly believe that most of the nineteenth century parasitologists did not have a reasonable idea about how many different helminths could actually occur in the gut of a human. Finally, microscopy in the nineteenth century was still very primitive and this too posed a serious handicap.

It is most likely that *A. lumbricoides* was the first helminth parasite known to humans. Thus, a twenty-centimeter worm is hardly something that could be ignored when passed in a stool, or out through one's nose. While several names were attached to this parasite in the eighteenth and nineteenth centuries, the designation given by Linnaeus in 1758 is the one that eventually stuck. The dioecious state was described by Redi and eggs were described by Tyson. The life cycle was ultimately resolved by Salvatore Collandruccio in 1886, although Giovanni Batista Grassi claimed the experiments were his and published them without giving any credit to Collandruccio. Both had tried to infect themselves using eggs, but apparently the infection did not take in either. However, Collandruccio gave 150 ascarid eggs to a young child who subsequently expelled 143 *A. lumbricoides*. He did the critical experiment, but Grassi took (stole) the credit.

Development of ascarid larvae inside the eggs was described several years before the life cycle was actually worked out by Collandruccio. The pattern of internal larval migration and its significance was initially misunderstood. Francis Stewart, an Englishman who worked in Hong Kong, was the first to attempt to deal with the issue. His experimental model first included *A. suum* and pigs. Having no success, he switched to using both *A. suum* and *A. lumbricoides* in rats. This time, he was successful, to some extent at least. After feeding eggs of both ascarids to rats, he was able to detect larvae in the lungs. Based on these results, he concluded that the rats were intermediate hosts for the ascarid nematodes and published these conclusions in the *British Medical Journal* in 1916. It is of interest to note that an editorial regarding Stewart's findings followed immediately in the same journal, but it made no mention of rats as potential intermediate hosts. Instead, it alluded to the possibility that the larvae were transient visitors in the lungs and were on their way to the gut, not unlike the larvae of *Ancylostoma duodenale*, as described by Looss several years earlier. Four years after the editorial in the *British Medical Journal*, Stewart changed his mind and concluded *A. lumbricoides* was monoxenous.

Shemesu Koino provided the final evidence for a single host, or monoxenous, life cycle in 1922. He swallowed 2000 *A. lumbricoides* eggs and began seeing larvae in his sputum after just three days (Grove, 1990). Fifty days later, he ingested an anthelminthic and quickly passed 667 immature worms in his stool. By the way, he also became very ill several days after ingesting the eggs, a clear case of overexuberance I would say.

Trichuris trichiura, while not a minute worm, was not recognized and described until the fourteenth century. For many years, the worm was frequently confused with *Trichinella spiralis*, *Enterobius vermicularis*, or *A. lumbricoides*, even by some of the virtuoso parasitologists of the nineteenth century, including Kuchenmeister, Virchow, and Leuckart. It too went by several different names until 1941, when the American Society of Parasitologists definitively determined that the Rules of Zoological Nomenclature required the worm to be designated as *Trichuris trichiura*.

The resolution of the life cycle for *T. trichiura*, however, was creatively accomplished using species of helminths that were similar to the one that occurred in humans, but in other hosts. Casimir Davaine, for example, used eggs of *Trichocephalus* (= *Trichuris*) *affinis* to show that this parasite would develop in sheep. Leuckart did the same thing with *T. crenatus* in swine. Calandruccio examined his own feces for something like six months, to make sure he was not infected, before ingesting eggs of *T. trichiura*. About four weeks later, eggs of the parasite appeared in his stool. Oddly, history repeated itself when his colleague Grassi published these results as his own, "with minimal acknowledgment to Calandruccio, by Grassi" (Grove, 1990). Grassi was a very good parasitologist, but his 'hijacking' and publication of results generated by other investigators was certainly unappreciated. Calandruccio even publicly castigated Grassi for this behavior, but no response and no defense whatsoever ever came from the latter parasitologist.

Perhaps the most complicated life cycle of any helminth parasite is that of *Strongyloides stercoralis*. As might be expected, its resolution was just as complicated; perhaps confusing is a better way of expressing it. We now know, for example, that parasitic males are nowhere to be found, and that females are parthenogenetic. We also know that the life cycle can be direct (homogonic), with filariform larvae penetrating the skin and then migrating through the lungs, where they use the bronchial escalator to be coughed up and then swallowed. There can also be internal and external autoinfection. Then, if external environmental conditions are suitable, both free-living males and females can develop. This pattern (heterogonic) can be sustained through several generations. When environmental conditions are no longer conducive to the free-living cycle, infective filariform larvae will be produced and percutaneous infection will occur. Can you imagine the confusion in working out such varied patterns in this parasite's life cycle? I can.

As for *Trichuris trichiura* and a number of the other parasitic helminths, *S. stercoralis* was known by several names before the International Commission on Nomenclature settled the issue in 1915. A French physician, Louis Normand, first described larvae of the parasite in 1876. He observed these worms in French soldiers who had developed severe diarrhea while serving in Vietnam. Because it has such a weird life cycle, it took quite a while to figure out just exactly what was going on. However, Grassi, in 1882, this time without Calandruccio's help, correctly suggested its parthenogenetic character. In the same year, he also indicated the possibility of internal autoinfection. Percutaneous infection by filariform larvae was demonstrated by Looss in 1904 when he experimentally infected himself. After surgically performing a tracheotomy and an esophagostomy in dogs in 1911, Fulleborn was able to demonstrate that filariform larvae of *S. stercoralis* followed the same internal migratory route of hookworms. One of the really interesting characteristics of the parasite is the longevity of infection in humans, presumably because of internal or external autoinfection. There are, for example, several reports of former American prisoners from WWII retaining infections for up to fifty years after returning home.

The pathologies produced by the geohelminths are similar in some ways, but different in others. For example, both species of hookworms and *S. stercoralis* all cause 'ground itch' when filariform larve enter the skin. *Ascaris lumbricoides*, the two hookworms, and *S. stercoralis* all pass through the lungs where the antigenic qualities of the cuticle and molting hormones can induce localized hypersensitivity, which may lead to localized edema and the development of pneumonia. In the gut, however, we see characteristic qualities of pathology that are unique to each species. In the case of *A. lumbricoides*, adults are in contact with the gut wall, but cause no tissue damage. Because of their size and occasional propensity to roam, they are known to appear sometimes in unusual places, i.e., the bile duct or pancreatic duct, which may cause blockage and even become lethal. There are even occasions when an adult may move into the stomach and up the esophagus, making an uncomfortable appearance in the mouth or even the nostril. They are also known to 'ballup' in the intestine and cause intestinal blockage as well.

Of course, we all know about skin pallor and iron-deficiency anemia caused by hookworms. Since this parasite is always associated with poverty, it is particularly dangerous in developing countries, especially in reproductive-age females and children. The anemia is usually exacerbated by low protein diets. *Trichuris trichiura* is also dangerous, especially in high numbers. In those with heavy infections, there is the potential for rectal prolapse and severe diarrhea or dysentery. Adult *S. stercoralis* is not a tissue-dwelling parasite, but because of internal and external autoinfection, its numbers can become exceedingly high and may even lead to severe diarrhea and death.

In the 1970s, there was radical change in approach for the epidemiological study of parasitic helminths, including the 'unholy trinity'. Almost all of this change was related to Crofton's two (1971) papers, plus those of Roy Anderson and Bob May that were published in the late 1970s and early 1980s. I will not repeat these points here since they are covered thoroughly in Essay 7 and in my interview with Roy Anderson. As he described the transition, a number of conceptual aspects of what could be considered as then extant ecological theory were simply shifted to host–parasite systems and, in a short period of time, became fixed in modern epidemiological doctrine as applied to infectious disease.

By the time Don Bundy finished his Ph.D., he had developed a strong interest in the modern epidemiology of human intestinal worms, including the 'unholy trinity'. He said that, "I was also going through a mental process of thinking that I really needed to care more about people, and more about the disparities between 'north and south.'" I asked if he thought that his growing up in Singapore and learning about 'auntie Par' might have colored his thinking in this regard, and he responded emphatically, "Oh, I'm sure that it did! One of the things I did during my undergraduate years was that I led an expedition to the northern Sahara, specifically Morocco. This was in the 1970s and, in those days, Morocco wasn't the great country it is today. It was a poor country and there was a great deal of inequity. I was there ostensibly to collect parasites from dead animals. It really did strike me at the time that, here I was, living a good life and employing intellectual energy in London, when people were combating disease and poverty as part of their normal lives. I felt then that I should be doing more about that. It became an issue in my way of thinking. I wanted to work on parasite control.

"I had just completed my Ph.D. with Phil Whitfield. Since now I was going to be a parasitologist, I wanted to work on human parasites as an issue. And, so, I looked around for opportunities. About that time, there was an advertisement for a position at the University of the West Indies in Jamaica, which was originally established as part of the colonial university system. It was British based, and twinned with London University." Since it was still operating within the context of the old colonial system, he was interviewed for the position in London. He indicated that some colleagues were encouraging him to take a postdoctoral fellowship, but he said, "No, this is what I want to do, this lectureship is in a place where I knew they had some problems." He continued, "I was pretty naïve. I had never been to the West Indies. I had no idea what was going on there." I asked, "So, you really made a conscious decision to follow this course, leading to where you are right now," and he responded in the affirmative. I then realized the full extent of his purely humanitarian commitment. My admiration and respect for Don suddenly increased by at least an order of magnitude, and I already had quite a bit of both.

He continued, "My goal of working in this area of epidemiology and parasitology was not an obsession. It was simply a very strong feeling that there was something to be done and that I had been given the skills to do it and that I should go and get on with it." So, he got the job and set out for Jamaica on a banana boat. It took two weeks to get there, a long ride.

According to Don, the University of the West Indies was really a good place to be, with high standards. The main campus is in Jamaica, with another in Barbados, and a third in Trinidad. It services fourteen countries in the West Indies, essentially the English speaking territories of the Caribbean. It was really quite an exciting experience for him because he was called upon to lecture in everything from beginning zoology to protozology and, of course, parasitology. In other words, it was heavy on the teaching side. He enthusiastically remarked, however, "It was all new to me and I was eager and excited to do it! And the students were an absolute delight. It was really fun to teach them. They were dedicated to learning."

He began working with geohelminths while he was in the West Indies and, in particular, *Trichuris trichiura* in children. The experience he obtained during those early years was important in his career, and largely gained by his own effort. There were no other parasitologists in the University and he was forced to make his own way. "I had to learn all the techniques on my own and, in some cases, invent the techniques. The whole expulsion of *Trichuris* was a technique that we put together. The *Ascaris* expulsions were pretty easy, nice big worms, but the *Trichuris*, that was a different kettle of fish altogether. And you can get hundreds of worms with *Ascaris* expulsion. We developed a whole battery of techniques for quantifying infection. For measuring eggs in soil, there was no standard procedure. We developed techniques for expelling worms and, with whipworm, since they are in tissues in the colon rather than the small intestine, expulsion occurs over a matter of days. There was a fantastic young pediatric surgeon there. Dr. Venugopal was his name. One of the things that he became engaged in was colonoscopy of children, because most of the children we worked with had bloody stools. Their colons were hugely damaged by infection with these worms. If we could treat them, the problem would go away, just like that," he said, with the snap of the fingers. "Within a matter of a few weeks, these kids would be thriving. We were trying to understand what was causing the colitis and bloody stools. It took 10–15 years to figure it out. As it turns out, the colitis caused by *Trichuris* was very much like that caused by Krohn's disease. There is a whole range of inflammatory bowel diseases. They are extremely debilitating. In this case, you've got the front end [of the worm] laced into the gut wall and it's pumping out all kinds of antigens in the gut wall. So they are causing a hypersensitivity reaction that pervades the whole gut. Why wasn't more damage being done with so much exposure to antigen? It wasn't until we could repeat the work of Leslie Drake that we could answer this question. We showed the physiology of how the worms' front end secreted a protein, a pore-forming protein, that essentially allowed the membrane punctured by the worm to reform around the worm." I asked at this point why the host was not responding to this protein in some way so that its action could be blocked. Don responded by saying that "the immune response seemed to have no consequence for the parasite, and the host was responding with hypersensitivity, with mast cell secretions causing the colitis." Before we left the topic, Don felt strongly that he wanted to mention "Ed Cooper, a young pediatrician who became interested in this issue and really contributed a lot to our program and success while I was down there. He always helped me so much with the clinical dimension of the infection, and we are still working together to this day."

The time in the West Indies was exceedingly important to Don's work 'down stream'. Because they were able to refine techniques and collect data, real data on worm burdens and distributions, he felt they were then in position to begin looking at the theoretical suppositions that Roy Anderson and Bob May were generating at the time regarding the epidemiology and control of the geohelminths. He concluded this part of the interview by saying, "Jamaica and the West Indies were about getting the evidence phase together and looking at what the theory had predicted. Then we examined the empirical evidence to see if it supported the theory. By and large, it did, particularly that the distribution of worms in the population was overdispersed and that all of the data fit to our expectation."

Don left the West Indies after seven years. I asked why he left and he replied, "Roy offered me a job at Imperial College in London. I had by then put in place a number of the things I wanted to do. I took it because I felt isolated in the West Indies and I felt I could make much more of a contribution by going back." He was recruited by Roy Anderson who was by then the Head of the Department of Biology, having left King's College and moved very rapidly through the academic ranks at Imperial. "I [Don] took the Imperial College job on the understanding and under the expectations that I would continue to work abroad, that I could maintain research programs abroad. That is exactly how my life as an academic worked from then on. I was very lucky in getting grants from the Wellcome Trust to work overseas. So, I have spent substantial parts of my life on the road."

In the latter part of the 1980s and into the early 1990s, Don focused his geohelminth research in Montserrat, a small island in the eastern Caribbean. He was attempting during this period to discover the practical applications of his earlier findings. The island was small enough that they could treat all of the school children. They were testing the hypothesis that "if you treated the school kids, there would be an impact on the overall level of infection, in other words, targeted chemotherapy [of Anderson and May]. On Montserrat, the Minister of Education and the Minister of Health was the same person, which helped a great deal. We did the study with John Horton over a period of several years. The hypothesis was, if we affected the school children, we would impact preschoolers and adults as well. It was targeted chemotherapy. The results were very successful because we saw a decrease in infection among the [untreated] preschoolers and the adult population, simply because the worms die in the most heavily infected group. In other words, they weren't getting reinfected. And, of course, we have now shown the same thing in many other places. The paradox of our research on Montserrat was that we got the level of infection down and then the volcano erupted and virtually the entire population left."

In my opinion, these results were most revealing because the approach taken was a test of the 'new' epidemiology. In dealing with the control of any of these geohelminths, the 'unholy trinity', or otherwise, there were several problems that required resolution. There was diagnosis. Was it necessary to do stool exams on everyone and first determine who was heavily infected? Then, there was the treatment strategy: i.e., to reduce prevalence and intensity of infection, was it necessary to treat everyone? The Rockefeller Foundation in the southeastern U.S.A. during the 1920s did not use diagnosis for their hookworm problem, but they did employ mass treatment. Something worked because prevalence of the hookworm plummeted. At the time, and even today, many attributed success to mass treatment and a huge education program.

However, there are some differences of opinion that also make a great deal of sense. For example, Peter Hotez says it was because the economy in the south improved significantly during those years and that the treatment programs were more or less superfluous. In other words, although poverty did not completely disappear, it declined enough to significantly impact the geohelminths.

At this point, I should insert here a commentary from Don about Peter Hotez's assertion regarding the impact of the Rockefeller program for hookworm disease in the southeastern U.S.A. during the early twentieth century. Don is quite emphatic that the Rockefeller group played a very significant role in controlling the hookworm problem. In a letter written to me after both interviews, Don said, "First, while it is undoubtedly true that the long-term reduction in worms in the southern U.S. is associated with improved economic conditions, there is clear evidence that those areas where Rockefeller implemented programs showed faster reduction than elsewhere, and indeed showed statistically significant improvement in such factors as school attendance, growth, and, ultimately, labor growth." He then cited a recently published paper (Bleakley, 2003), "that goes back to the original data and uses a smart technique to progressively compare where intervention was initiated at different times. What particularly impresses me is that the scale of the impact is (remarkably) the same as that shown by Miguel and Kremer in their randomized trial in Kenya, where there is no doubt at all that the only intervention was deworming and the trial design was as good as it could be. Second, I would also draw attention to the experiences in Japan, South Korea, and urban Indonesia, where worm prevalence fell precipitously, while at the same time there was both extensive active control and economic development. In these places too, there is good evidence of the direct effect of the program as well as the long-term benefits from economic development, which, through improved sanitation and health care, have kept the prevalence down. Third, rather than say that economic growth led to parasite control, many would make a causal connection in the other direction – that parasite control boosted school participation and learning, and that a more educated workforce laid the foundation for economic growth. This is not merely my view. There are several classic analyses of the economic miracle of southeastern Asia that show – and this is not controversial – that the single most important factor in the spectacular growth of the economies of Japan, South Korea, Thailand, and Indonesia since WWII was universal basic education. Indeed, this is why education for all, including female education, is so strongly targeted now by the development agencies, including the [World] Bank. The evidence from the Rockefeller program from 80 years ago, now fully supported by modern pysychometric work, has shown that deworming is a key and cost effective factor in educational participation and achievement – this too is uncontroversial. Thus, it is now argued that school health, including deworming, as part of an education for all program, leads to a better educated population and that, in turn, leads to economic growth (which itself leads to lower worm burdens). The Japanese are so convinced of this link that they established the Hashimoto Initiative to promote school-based parasite control – a program worth $5.3 billion and very active in Asia and Africa. No one would argue that the Japanese would be putting this kind of money into a program unless there was a strong evidence base!"

With the concepts and mathematical models generated by Crofton, Anderson, and May, i.e., overdispersion and targeted treatment being the 'centerfolds' of modern epidemiological theory, the notions of both diagnosis and mass treatment became moot. Moreover, new drugs by the 1970s had become available and they were not only cheap, they were effective. Although they were effective, however, there was still a serious caveat, at least in the case of hookworms. With these parasites, the drugs had to be administered three or four times a year. Why? Because reinfection occurred among children. As Peter Hotez said, "Yes, the drugs are effective for hookworms, but the kids do not build any immunity. If they remain in an endemic area, they will become reinfected and they must be treated over and over." This is one reason Hotez directed his attention at a vaccine for these tiny bloodsuckers. However, with the other geohelminths, if we have a good idea of who requires treatment (without the cost of diagnosis), and if we have a drug that can be targeted at a certain group, and if the drug is inexpensive, then there remains but one problem, and that problem is delivery. Health care delivery can be expensive, but is it always necessary to use professionals? Can the cost of delivery be cut if another system can be developed? This is where Don Bundy, and others like him, came into the picture.

I cannot help but note here that there are nonetheless conflicting opinions regarding the frequency of treatment of hookworms versus that for the other two members of the 'unholy trinity'. Accordingly, there is also a difference of opinion regarding the usefulness of drugs in the treatment of hookworm disease. Peter Hotez, for example, is adamant that drug therapy is virtually worthless for hookworms because it is not effective over the long term and must be repeated every few months. In contrast, Don Bundy says that hookworm, among the 'unholy trinity', "requires the least frequent treatment." He bases this assertion on the work of Anderson and May. According to Bundy, "The 'reproductive rate' of hookworm is lower than that of either *Ascaris* or *Trichuris*, so that lower control 'effort' is required to push down the prevalence and intensity. Studies reliably show that suppressing hookworm prevalence can be achieved with more infrequent treatments than for the other two genera." In other words, we have an honest difference of opinion between Bundy and Hotez regarding the usefulness of drug therapy for the control of hookworm disease in the world.

Don continued, "Around 1990, a lot of 'stuff' came out. Interestingly though, just last year, economists, some from the Word Bank and some from MIT and Harvard, produced a study showing what they called 'externality'. This term refers to the consequences of actions that are external to the group you are trying to improve. In Kenya, they looked at a population of kids in school, and showed that hookworm treatment of kids in school not only improved their school attendance, but it also improved the attendance of kids that were not treated. It had a significant impact on the treated schools and on those schools that were nearest to the treated schools, but not the ones that were further away. In other words, the predictions of the parasitologists regarding this phenomenon were being borne out. This piece of work, the idea of externality, was of profound importance in getting other things from the development and economic communities. Even the present White House administration in the U.S.A. has incorporated the idea of externality into their thinking in regards to development planning in the Third World. Fifteen years ago, we raised the idea that you could use schools in the delivery of treatment programs for the geohelminths. The idea of using school public health programs as a way of dealing with worms brought a lot of interest. I should emphasize here that there were several foundations that got behind this idea too. One was the Edna McConnell Clark Foundation under the leadership of Joe Cook, a really fantastic guy. He's retired now, but we owe him a huge debt. The Wellcome Trust was involved. Another was the Rockefeller Foundation and then a group called the James S. McDonald Foundation. The latter bunch was concerned with cognition, and mental development. They became interested because we tried to make the case that if we were going to talk about school health, using schools as an existing infrastructure for the delivery of treatment, then what were the benefits for those kids? We had already shown that there were benefits in clinical trials, for example the work with Ed Cooper in the West Indies on trichuriasis. But, beyond these clinical consequences for treatment, were there consequences for physical and mental development? If you could say yes, then there would be justification for recruiting school personnel to work in these school health programs. Giving a kid a 5–10 cent pill has almost no delivery cost, because the infrastructure is already in place. We started looking at the intellectual benefits in the early 1990s, and produced our first paper on that issue. That was a very important turning point.

"These foundations and the UNDP [United Nations Development Program] got together and supported the Partnership for Child Development, which was an organization I started while at Imperial. The idea was to look at the reality of the experience of treating school children and they supported the programs. The two principal ones were in Ghana and another in Tanzania, where we would deliver treatment to school kids, and school kids only, and then look at the broader consequences."

I asked Don, "One of the problems I think you would have concerns about is the political instability like in many countries of sub-Saharan Africa. How do you deal with these situations?" He responded, "They may be unstable at the top, but what we are doing is working with the schools. I mean the schools are almost always there. What I had to do first though was deal directly with the political directorate. There is a Minister for Health, one for Education, and another for Finance. They must be fully on board. They must be convinced that this is something worth doing, because what you are asking is for the teachers to deliver the treatment." This was the key to delivery, using the school systems and the public health systems within the schools.

When Don went to Oxford with Roy, he took the new program with him. After the program was under way, they confirmed that it was practical, the costs were trivial, and the payoffs were very substantial. The demonstration projects were very important in bringing new people "on side", as he expressed it. "One of the things that became very clear to us during these demonstration projects was that you could address schistosomiasis in the same way you were working on the geohelminths. However, this ran into big political problems in the parasitology community." I was immediately puzzled and asked him why? He responded, "Because the schisto community was a separate and rather 'aristocratic' group. They worked on schistosomes, which were not the same at all as the rather 'trivial' intestinal worms. It was a most extraordinary thing. You would have thought they would embrace the new approach, but they saw this as a challenge to their centrality. I had to spend endless time in addressing this issue. The perception was that the geohelminths were *just* [his emphasis] worms and that schisto was a special case. They also believed that one had to do diagnosis before treating with drugs. If you look at the literature prior to 1990, that distinction was clearly made." He then noted that this was not the Rockefeller approach for the elimination of hookworm disease in the southeastern United States in the early part of the twentieth century. "They treated everyone. This, plus a huge education program, did the trick. You know there is hookworm around, so treat everyone. You don't need to do a diagnosis first. In the case of the schistosomes, it was mainly the physicians who were taking the lead on diagnosis first. They believed it would be unethical to treat without doing a diagnosis first. So, we did some analysis and what we found was that you miss lots of the infections with diagnosis." I asked if this attitude and resistance have changed. He responded in the affirmative. "Yes. We won that battle, but it was a tough fight that went on for quite a long time. In part, we won because we showed that it could be done." Don added, "The cost of diagnosis is on the order of five to ten times the cost of treatment, so the cost of inclusion of a screening step typically made programs too expensive to do at all, so no one got treated. The bottom line is that screening approaches are simply not cost-effective.

"I wanted to mention my colleague, Andrew Hall, who worked with me throughout this period of time on actually running these programs in Ghana and Tanzania. It was Andrew who pushed forward with the technical program on the ground while we were having these interesting tussles regarding the approach that should be taken." He added, "One of the problems for the delivery of drugs via school teachers is that the dose of drugs is dependent on the child's weight. If you want to provide a weight scale for every school in Africa, then you are talking about a very big bill indeed! And, how long will they keep working properly? This was an obstacle. I know it sounds trivial, but in a poor country, it's very real. So, Andrew Hall came up with the very simple idea of using a 'weight' pole. You measure the height of a child and, using a height/weight curve, you can calculate the correct dose to be given. Now, WHO distributes a tape that you can just put on the wall. Trivial things like that were major problems, but they were solved. Another thing that came up was how do you determine what parts of a given country actually have worms. You can't afford to send out teams to do an analysis all over the place. One of my former students, Simon Booker, became very interested in using GIS. Going on to a decade now, he has refined the system so that we can take data from a country using the remote sensing satellite to look at patterns and then use the information to predict parasite distributions. And, it turns out, this is a remarkably efficient way of doing it." I reminded Don that this was basically the same way that Sir Patrick Manson had determined that mosquitoes were the vectors for *Wuchereria bancrofti* in 1878, simply by plotting the distribution of elephantiasis on a map and then trying to figure out which blood-sucking insect matched the distribution pattern for the disease.

With the establishment of credentials and his success with the Partnership for Child Development, Don was invited to join the World Bank in Washington, D.C. The World Bank is the world's largest funding agency for development programs. They recognized the importance of educational and health programs in development in developing countries. It is one of the few agencies to work in both of these areas. So, the World Bank wanted Don because of his expertise in health and education. He's been with them since the late 1990s. I asked how he liked it. He responded, "I like it a lot, and I'll tell you why. It's because one of the key things that happened when I first came to Washington was that we started going around and talking to the other agencies about their public health programs. Included were WHO, UNICEF, and UNESCO, with WHO a leader on public health, and UNESCO and UNICEF leaders on education. The four of us agreed on a new program of cooperation, i.e., Focusing Resources on Effective School Houses, or FRESH. It's been absolutely terrific. We agreed that there are four things that we all did that were the same thing. First, school health meant that we had to have policies in government that allowed it to happen. Then, there had to be improved sanitation. Third, there had to be health education to re-enforce sanitation. Finally, there had to be health services, which, in this case, would mean delivery of medicine in schools to treat the diseases. We took our FRESH concept to the World Education Forum [WEF]. In 1990, the world had declared through the WEF that all children should have the opportunity of going to school. However, ten years later, there were actually more children out of school." He explained, "This happened, in part, because the population had grown, but also because there had been no concerted effort to get those kids into school. So, in the year 2000, there was a major meeting of all the agencies to address what had gone wrong and what are we going to do now? This meeting, The World Forum for the Education of All, was held in Dakar, Senegal. The Education for All declaration in 2000 (by 181 countries and essentially all the main development partner agencies) recognized that children would only go to school and learn while they were there provided they were healthy and well nourished. The declaration specifically endorsed the need for school health programs using the FRESH framework, and thus opened the way for deworming to become an established and normal part of any education program in a low-income country where worms are common. I hope you see that this is not like WHO on its own declaring some disease or other is important – it is a specific requirement that a program must include this component if it is to become a success, and a requirement that has global and interagency endorsement. It is a real big deal! The truth, somewhat unpalatable to the majority of public health parasitologists who come at the issue from a health perspective, is that countries and most development donors did not see worms as a priority from a health perspective (which is why there were so few deworming programs previously), but now see deworming as a developmental priority largely because of the impact on education. My guess is that it will take our community at least another ten years to recognize this paradigm shift."

Don emphasized, "Andrew Hall helped me to manage the Program for Child Development until I left for the World Bank at the end of the 90s. Since then, Lesley Drake has been in sole charge and has done a fantastic job of shifting the focus of the Program towards operations while at the same time maintaining the strong scientific core."

Don focuses on work in low-income countries – those with less than $965 Gross National Income (GNI) per capita – which are eligible for support from the World Bank's International Development Association. This IDA support is currently running at some $11 billion per year, of which some 20% is in the form of grants and the rest is essentially zero interest loans with a long payback time. About a third of this money is for health and education programs. The World Bank has a technical team assigned to each of the eligible countries and Don works with this team to assist the government in deciding what are the best programmatic investments for the country. When a country decides to include a school health component, then he works with the country's technical experts and other development partners to help design the plan for the school children, which nowadays is likely to address malaria, HIV/AIDS prevention, and worms. An important part of this work is in ensuring that the development partners work together in supporting the country's vision – which is why the FRESH consensus framework is so important. For school health programs, you might see coordination among the UN agencies, especially UNESCO, UNICEF, UNAIDS, WHO and the World Food Program, bilateral aid agencies representing countries such as the U.S.A., U.K., Norway, and Ireland that have shown a special interest in the health of school children, and among organizations that deliberately have no direct links with governments, but instead represent civil society, such as Save the Children, the Partnership for Child Development, and Education International. In designing worm control programs, the countries have received particularly important assistance from Lorenzo Savioli's team at WHO Geneva, as well as WHO's regional teams. Don's job is on the one hand to help ensure that programs are based on sound, evidence-based policy decisions and strong technical designs, and on the other hand to ensure that the money is moved in a responsible fashion. As Don said, "We have a responsibility to the institutions in the countries that they implement programs that are cost-effective and serve the primary aim of reducing poverty, and at the same time a responsibility to the people of the recipient countries, as well as 184 countries that are shareholders of the World Bank, that the funds are used responsibly and for the intended purposes."

This is a huge job that he has. I was tremendously impressed with his sense of duty and his regard for the children in these developing countries. I really had very little knowledge of the World Bank and what it does around the globe. But I do now. Moreover, I have a great deal of confidence in their objectives. I asked Don, "Will you solve the worm problem?" He responded, "Not this year, but maybe in twenty years they will be able to put an appreciable dent in the problem." I thought to myself, is this a realistic prediction? Don is clearly an optimist, but he is also in a position to know. He said that Sri Lanka had instigated the school delivery system for treatment about 15 years ago and that there had been a really significant decline in the prevalence of geohelminths since then. Maybe, just maybe, he is correct. I truly hope so!

REFERENCES

Bleakley, H. 2003. Disease and development: evidence from the American south. *Journal of European Economic Association* **1**: 376–386.
Crompton, D. W. T. 1999. How much human helminthiasis is there in the world? *Journal of Parasitology* **85**: 397–403.
Grove, D. I. 1990. *A History of Human Helminthology*. Wallingford: CAB International.
Stoll, N. R. 1947. This wormy world. *Journal of Parasitology* **33**: 1–18.

9

Hookworm disease: insidious, stealthily treacherous

Let's talk of graves, of worms, and epitaphs.

Richard II, William Shakespeare (1564–1616)

According to the most recent estimates, there are almost a billion men, women, and children infected with hookworm. Among the most affected peoples in the world are the poorest populations in sub-Saharan Africa, the Americas, and Asia. For centuries, hookworm has been a major problem among the Chinese. According to Peter Hotez, between 1990 and 1992, parasitologists in various parts of China undertook what I consider to be one of the most intense surveys of the problem ever conducted. They collected more that 1.4 million stool samples from 2848 study sites in 726 counties in every province in the country. Of these, 17% passed eggs of either *Necator americanus* or *Ancylostoma duodenale*. This extrapolates to approximately 194 million cases of hookworm disease in just that country, and it does not include those who have had the disease and lost it for one reason or another. The enormity of the problem is exacerbated by Norman Stoll's estimate in 1962 that each day hookworms suck enough blood to cause the total exsanguination of 1.5 million people. The number of infections by hookworm has nearly doubled since the publication of Stoll's paper, so we can safely assume the amount of blood lost on a daily basis has, likewise, doubled!

Hookworm disease has been associated with humans since we changed from being hunter-gatherers to farmers. Ancient writings were undoubtedly describing the signs and symptoms of this disease, among other ones, when they referred to problems with ground itch, pneumonia, anemic pallor, and diarrhea. The ancient Chinese referred to the infection as, "the able to eat but lazy to work disease." Today, in the south and southwest of China, the locals still talk about "*lan huang bin*, the lazy yellow disease, and *huang zhung bing*, the yellow puffy disease", both of which are attributable to hookworm.

Hookworm was discovered 1838 when Angelo Dubini saw it in a dying peasant in a hospital in Milan, Italy. While performing autopsies over the next few years, he encountered hookworms again several times, but always just females and never males. Then, in 1843, he found a large number of worms in a man dying of pneumonia, but this time there were males and females present. He was then able to complete a description, and he named the parasite *Agchylostoma duodenale*. By 1853, while working in Egypt, W. Greisinger and T. Bilharz asserted that there was a definite connection between hookworm and severe anemia, also known as 'Egyptian chlorosis'. In 1860, a Negro slave in a Benedictine monastery in Bahia, Brazil, was seen by O. Wucherer. The slave was suffering from what he termed "hypoaemia." At the slave's death, Wucherer was able to perform a partial autopsy and discovered hookworms in the man's duodenum. He made an immediate connection with Greisinger's 'Egyptian chlorosis' and confirmed it was due to the hookworm infection.

Not long after, an interesting epidemiological situation regarding the hookworm problem was about to be manifested in Europe. Hookworm disease was really not a serious public health concern in Europe in the nineteenth century, but that was about to change. Construction of the St. Gotthard Tunnel connecting Italy and Switzerland was begun in 1880. Soon, there was a severe outbreak of hookworm anemia among the miners. When they finished work on the tunnel, the miners dispersed throughout Europe and wherever they went, the anemia went with them. I recall that W. W. Cort, one of the world's early experts and researchers on hookworm disease, began working on this nearly ubiquitous nematode in California gold mines in the early twentieth century. This surprised me initially, but after thinking about it for a while, it really made good sense. As in the European tunnel, many of these mines were warm, almost hot, depending on your personal thermostat. What better place than a warm, 'shady', confining habitat than a mine for the transmission of hookworm! It is no wonder that this parasite could become established and then reach near epidemic proportions in European mines.

Initially, transmission of the hookworm was believed to be the result of hand-to-mouth passage of eggs, based on experiments conducted by Leuckart and Leichtenstern. However, Arthur Looss, who had studied under Leuckart in Leipzig, was to change this conclusion. In

1898, while feeding cultured hookworm larvae orally to guinea pigs in Egypt, he accidentally (serendipity!) allowed some of the medium to drop on his hand. It quickly produced an itching sensation and then erythema. Looss immediately dropped more of the culture on the back of his hand, and then waited. He periodically checked his own feces; after a few weeks, he observed hookworm eggs. He then learned of an Egyptian boy who was about to have a leg amputated. About an hour before the surgery, he placed drops of culture medium containing third-stage hookworm larvae on the lad's bad leg. After the surgery, Looss took pieces of skin from the leg and found larvae penetrating the skin. He repeated the work with *A. caninum* and discovered that hookworm larvae entered the skin of dogs in the same way.

It is of interest to note that the symptoms experienced by Looss were not unlike those of William Walter Cort nearly thirty years later, when he was isolating snails collected in Douglas Lake, Michigan. Cort's accidental confrontation with schistosome cercariae led him to discover the cause of swimmer's itch. Clearly, both Cort and Looss were the benefactors of serendipity.

Initially, however, efforts to confirm Looss' work failed. Then, C. A. Bentley, working under very primitive conditions on a tea plantation in Assam, India, produced data that sealed the skin transmission hypothesis of Looss. Bentley first showed that rich soil from plantation contained '*Ankylostoma duodenale*' larvae. Next, he demonstrated that application of the soil itself was sufficient to produce ground itch. Then, he took skin in which ground itch lesions had just developed and scraped it; he discovered infective larvae and empty larval sheaths. With no microtome, he could not do any tissue sectioning. So, he devised an experimental plan to resolve the issue. He obtained soil and into one part he placed feces containing eggs of the parasite. The other received feces without eggs. He dried both for eight hours. Then, he took a portion of each sample, applied it to his wrists, and covered both with a bandage. Ground itch developed on the wrist exposed to the contaminated sample, but not the other. Except for the internal migration of the L_3 stage, this was the *coup d'état* with respect to the discovery of hookworm life cycle.

Subsequently, Looss, in 1905, demonstrated that larvae of *A. caninum* were picked up in both the lymphatic system and the venous side of the circulatory system following percutaneous penetration. He was able to show that the larvae passed through the lungs and the trachea on their way to being coughed up and swallowed into the gastrointestinal tract. He also noted distinctive morphological changes to the larvae during their passage through the lungs. Schaudinn, Fulleborne, Miyagawa, and others shortly thereafter confirmed these findings.

So, by early in the twentieth century, the entire life cycle of hookworm was known, as was the etiology of the disease caused by these parasites. As more and more was learned, the seriousness of the problem became apparent worldwide. Just as important though, was who was prepared to do anything about it. The first concerted effort to deal with a bad situation was made by the Sanitary Commission of the Rockefeller Foundation. This group stepped forward and went to work, first in Puerto Rico, where it was estimated that one third of the annual deaths on this small island were being caused by hookworm disease. Then, they switched into the southeastern part of North America where prevalence of hookworm disease was very high, ranging from roughly 25% in Tennessee to 53% in Mississippi between 1910 and 1914. Rockefeller had become interested in the hookworm problem indirectly. His Baptist minister, a man named Gates, was to interact with Charles Stiles, a very reputable parasitologist with the Bureau of Animal Industry. Stiles, an expert on hookworms, explained the problem to Gates, who then took it to Rockefeller's foundation. He had convinced the Reverend Mr. Gates that this was a major cause of anemia in Puerto Rico and in the southeastern U.S.A., and a program of eradication was initiated.

Epidemiologically, there are four important features of hookworm disease. First, the extent of endemic disease among the locals is related in part to the quality of the soil. If the soil is warm, moist, and shaded, there will be a good chance for hookworm. If not, then hookworm will not be present. This clearly places hookworm disease into the tropics and subtropics. Second, the prevalence curves are mostly convex when compared with age class, irrespective of the sex. Thus, it is low among younger age classes, followed by a rise and peak during the mid teens, followed by a decline. For example, in the southeastern U.S.A. during the early twentieth century, in the 0- to 4-year-old group, prevalence was approximately 2.5%. The highest prevalence was 17% in the 15–19 year age class, followed by a decline with age (Keller and Leathers, 1940). I should emphasize here that the age-related convex prevalence curve for hookworm does not hold for all localities. In some areas, prevalences rise steadily, and level off when the twenties are reached. The proportion of those infected, therefore, does not decline. In contrast, in China, the tendency is for prevalence to continue rising, throughout life. As noted by Brooker *et al.* (2004), these variations could have far-reaching implications for the treatment of hookworm disease and for strategies of reducing infection in various communities by targeting treatment at just school-age children. Third, the prevalence of hookworm infection in the stools of nearly 17 000 Caucasians living in the coastal plain of North Carolina in the early twentieth century was 20%; in a sample of 6300 African-Americans from the same area, it was 4%. Based on these data, one would assume the racial difference must be genetic. I can think of no other explanation. Finally, whether in the southeastern U.S.A. early in the twentieth century, or China at the present time, a consistent companion of hookworm disease is poverty. Inevitably, accompanying destitution is inadequate sanitation and poor, if any, health education. Without debate or question, hookworm is the benefactor of all these conditions.

Whether genetic or not, no matter the age, and despite the depth of poverty, the order of the day for the Rockefeller Foundation was to reduce the prevalence of hookworm. Their efforts were three-pronged. First, do a diagnosis; if infection was found, use a drug to treat the infected people. Initially, they employed something called oil of chenopodium but, subsequently, they switched to thymol. Second, and simultaneously, they made a really serious effort to improve sanitation. This was accomplished in two ways. First, build privies and, second, educate the locals with respect of the need to use them. I recall my first general parasitology course in college, taught by Robert M. (Doc) Stabler. He told us that in the 1920s and 1930s, billboards were constructed all over the southeastern U.S.A., urging the people to use privies and among other things, to "Cut One for the Youngsters". On the same billboard was the photograph of a pile of feces seen through a disgustingly large hole. The idea was clear. Cut a small hole for younger children and encourage them to use it. It seems that a great many of the early privies were 'one-holer's' and youngsters were afraid to sit on them for fear of falling in and suffocating in the mess at the bottom. More than 100 000 privies were constructed throughout the southeastern U.S.A. over the course of just a few years. Finally, according to Peter Hotez, the Rockefeller group also gave serious attention to helping local municipalities by building a strong public health infrastructure via county health departments.

By the late 1930s, prevalence in all of the southeastern states had fallen radically. In Mississippi, for example, it had declined from 53% to about 20% and in Tennessee from 25% to less than 7%. Today, infection with hookworm has virtually disappeared from the continental U.S.A. I might add though, so has the abject poverty of the early twentieth century. In other words, there is no reason to have hookworm here anymore.

At this point though, I had an interesting exchange with Peter regarding the success of the hookworm control program in the southeastern U.S.A. He reminded me, at the turn of the twentieth century, the nonurban area of the southeastern region of the U.S.A. was just like a developing country of today. "You had endemic typhoid fever. You had endemic malaria, with huge outbreaks in the summer and fall. You had yellow fever in the ports. Yet, we were able to get rid of all those problems without a hookworm vaccine, without a typhoid vaccine, and without a malaria vaccine, in fact, without even much of a malaria control program. We had all these serious problems, but we managed to eliminate all of them. What really did it was economic development. The local residents, for example, improved the quality of their dwellings by using hardware cloth to cover the windows. They developed air conditioning. They paved highways and roads, and covered parking with lots of concrete or asphalt. They improved sanitation. It was not a single intervention that did it. It was all of these things coming together at the same time under the umbrella of economic development."

He continued, "That's what makes parasites so successful. Parasitic diseases are, for the most part, diseases of poverty. There are a number of exceptions, like some of the zoonoses. Let me give you an example of what I mean. We were working fifteen years ago in an area of China about an hour or two from Shanghai. At the time, there were reports of hookworm at 70–80% in the local population. At present, because of aggressive economic reform in the area, including the building of factories, invasion by Kentucky Fried Chicken, McDonald's, and other sorts of fast-food outlets, and shops, the prevalence of hookworm has slipped to 15%. The solution to the hookworm problem was not chemo- or immunotherapy, but economic development. This is the ultimate answer, i.e., economic development and poverty reduction! In most places where hookworm disease is endemic, however, this is not going to happen in our lifetime. For now, it will only happen near the major cities in Asia and perhaps South America, but it isn't going to happen worldwide. The interesting thing is that you cannot put your finger on any single component of economic development as the major contributor for hookworm reduction. Even with improved sanitation, in the absence of economic development, you are looking at only a 4% reduction! So, it's not just sanitation. It's the other things that go along with it. You can't look at isolated variables. They all have to occur together. So, what happened in the southeastern U.S. in the early part of the twentieth century is that everything occurred at the same time." Accordingly, Peter would contend that the efforts of the Rockefeller Foundation were mostly coincidental to the localized eradication of hookworm disease. If one looks at the profound changes in hookworm, typhoid fever, and malaria over the first forty years of the twentieth century in the U.S.A., combined with all of the socioeconomic changes that were simultaneously occurring, then it is easy to see his point.

If we accept the notion that economic development is not going to be a 'cure all' for endemic hookworm disease throughout the world, however, then there are two other modes of attack that can be employed. The first of these is via chemotherapeutic means. At the present time, there are four effective drugs (for hookworm) on the market, although each has its own limitations for one reason or another. These include pyrantel pamoate, levamisole, albendazole, and mebendazole. The latter two are most frequently the drugs of choice because they can be administered in a one-dose tablet and do not require the patients to be weighed first. Despite their proven efficacy, however, there is nonetheless variability in effectiveness. Another ugly situation that could well be upon us soon is the development of resistance to these drugs. The story of resistance developing among helminths of veterinary importance is well known and, according to Roy Anderson (see Essay 7), there is every reason to believe something similar will happen for the chemotherapeutic remedies used by humans in the treatment of hookworms. Peter also kept emphasizing that a problem with these drugs is that they require delivery every 4–6 months. This means that to keep prevalence of hookworm low within a certain age group of our population, drug treatment is necessary at least twice a year. Moreover, chemotherapy must be continued for several years running.

Traditionally, the delivery of drugs in developing countries has also been a major problem. Based on the modeling efforts of Anderson and May (1979) and May and Anderson (1979), however, a number of epidemiologists, including Don Bundy at the World Bank (see Essay 8), have developed what appears to be an excellent strategy to resolve the problem of delivery. Based on these models, there are three approaches for treating helminth parasites, including hookworm and the other geohelminths. The first is referred to as mass treatment, and this occurs when everyone in a given village is treated, usually without diagnosis. The second is referred to as selective treatment. In this case, residents in a village would all be diagnosed and only the relative few with the heavy infections would be treated. The third approach is called targeted treatment. In general, enough is now known regarding

the transmission dynamics of most parasitic worms that one can predict who, or what group, will be infected. Also of critical importance is
knowing who would be of greatest risk for developing iron deficiency
anemia from hookworm: in this case, school-aged children and women
of reproductive age. Then, the idea is to treat just these groups. Forget
about the others. Moreover, a diagnosis to confirm the epidemiologic
'speculation' is not required. The models further predict that if infections can be brought below some threshold in a given community then
the population dynamics of the parasite will be disrupted, even to the
point of eliminating the parasite locally. With any of these strategies,
however, there is the problem of delivery. Rather than training and
hiring expensive health care personnel, Don Bundy and other epidemiologists believe the best way to deliver the drugs is use teachers in the
schools. Since the pills cost just a few cents, and administration of a
delivery program by teachers costs nothing, the plan seems feasible.
The only other potential problem might be the necessity to cajole a
Minister of Education, or a Minister of Health. Several such programs
are currently under way in a number of developing countries, and they
appear to be working (Essay 8).

   The other mode of attack for hookworms involves the use of vaccines. Peter Hotez has entered into what he hopes will be the culmina-
tion of a lifetime's work in this area. Even as am I writing this essay,
field trials for his new hookworm vaccine are under way in Brazil. The
first question I asked Peter when we began his interview was, "Why
did you choose a hookworm vaccine rather than chemotherapy as a
way of dealing with this parasite?" He responded, "There were several reasons. First, one of the problems with hookworm, unlike many
infectious pathogens, is that they do not create a naturally induced
immune response. Actually, it turns out that they do in some cases,
but in most they don't. So, what happens is, you get infected and you
are administered a benzimidazole anthelmintic drug such as mebendazole. Here, you encounter the first problem with chemotherapy. Mebendazole doesn't work very well, as sometimes only half the patients will
respond and lose their infections. Let's assume though that most of
the recipients do respond. The worms will always come back in 4–6
months. With a lot of infectious organisms, like measles or polio, if
you are exposed to the wild type or an attenuated strain of the virus
and you survive, sometimes you are immune for years, even a lifetime
in some cases. Hookworm is just the opposite. No matter how many
times you are exposed and successfully treated, you are vulnerable to
reinfection. So, no matter that there are good drugs around, and that

they are cheap, they are ineffective for public health control because you have to be able to go into these endemic areas, which are sometimes present in rural and remote areas of Third World countries, and administer these drugs up to three times a year." I asked, "Why is it you can't find an effective drug?" He replied, "But these drugs are effective, they just don't prevent reinfection. If someone were to come here from Zanzibar and be treated with one of these drugs, they would be just fine because there is no transmission here. But, if they returned to Zanzibar, they would be reinfected in 4–6 months. So, we are looking for a prophylactic approach, something that will prevent infection, or reinfection, after treatment."

One of the things I was curious about after reading some of Peter's papers, and with up-front knowledge of the parasite's transmission ecology, it occurred to me that his strategy for development of a vaccine could come from any of several directions. So, I asked him, "Why did you choose to go after the enteric form of the parasite rather than, say, the L_3 stage?" His response was interesting. "We actually chose parallel approaches. We have approached the L_3 and the adult worm, both of them. In fact, we think you need both to make an effective vaccine. The vaccine will actually be comprised of two antigens, one that is produced by the L_3 just after it penetrates the skin and before it enters the circulatory system. The other targets the adult. The vaccine that is in clinical trials now is the one for the larval stage. The reason we chose the larval stage first was that Cort's team, including Drs. Foster, Otto, McCoy, and others, found, back in the 1930s, if you take a laboratory dog and give it small innocula, either orally or subcutaneously, of live L_3 *A. caninum* over a several month period, you could render the animal virtually resistant to challenge infection. In other words, 70 to 80% of the L_3 would not become established in a challenge infection, whereas the naïve dog would perish from exsanguination." I asked if the resistance lasted for a long period and he answered in the affirmative. He continued, "Industry actually took this finding a step further. They actually used that principle that Cort developed to make a commercial dog hookworm vaccine. This was done in the 1960s by a Scottish parasitologist named Tom Miller, who ultimately moved to the U.S. to become the principal scientist for a company that made a dog hookworm vaccine. The larval source had been attenuated by ionizing radiation. It didn't matter if it was UV irradiation, gamma irradiation, or X-irradiation, but he ended up using the latter. This vaccine, using attenuated larvae, was ultimately marketed commercially back in 1972, first in Florida and the next year along the entire east coast of the U.S. You could actually buy a bottle of irradiated hookworm larvae from your veterinarian. There were two problems though. One was that production costs were high because you needed someone to harvest kilogram quantities of dog feces to isolate the eggs and rear L_3 stages of the parasite. The greater problem was that the L_3 stages had to be alive. Once the larvae died on the shelf, they wouldn't work any longer. They would give two infections of larvae and that would produce about an 80% protection in terms of worm burden reduction. There was also a marketing problem between the veterinarian and the pet owner because it wasn't a sterilizing immunity, although it reduced the worm burden by 80%. The pet owner would bring their dog back to the vet after treatment and have a stool exam done on the pet. When hookworm eggs were seen in the stool, the pet owner would erroneously conclude the vaccine had failed. All of these problems worked against the success of the vaccine. By 1974, it was off the commercial market.

"We looked at the system, knowing that you couldn't make a vaccine with irradiated larvae, but if you could understand the antigens that were associated with those irradiated larvae, then you could develop a good hookworm vaccine. So, when I restarted my lab after my residency was completed, we began to look at the larvae. I had the idea that I was going to see what the larvae secrete, isolate those proteins, clone them, and make hookworm vaccine out of the cloned products. Gerry Schad started me up in the life cycle again. But the L_3s didn't secrete anything. There wasn't anything there." He wondered, "What in the hell is going on here?"

At this point you could feel the frustration in his voice. Peter continued, "I was really stumped. I had obtained a grant for this and everything looked so promising when I started. And then there was nothing there! John Hawdon, however, was working in Gerry's lab on his doctoral thesis. He had found that if you take an ultrafiltrate of serum from dogs and put the larvae in, that they resume development, because they are in an arrested state as third stage larvae. If you add some of this serum ultrafiltrate with some glutathione, they actually start esophageal pumping. They begin partial development. So, I said to John, why don't you come to my lab as a postdoc and let's do this thing together?"

I asked John Hawdon about this during our interview in Mobile. He said that Peter had been, "taking a brute force approach to the problem. He was grinding up worms and running them over red agarose columns and trying to isolate proteins for his vaccine. I had this [hookworm] system where I knew I could get the L_3 larvae to feed. I had narrowed down the stimulus as far as I could. I found that I could use a low molecular weight fraction of serum, coupled with glutathione. And we could stimulate development beyond the nonfeeding filariform stage. Eventually, I found that glutathione with a methyl group added to cysteine worked just as well. So, it's some kind of single receptor thing and that will work just fine. There is a synergism between the glutathione and something in the serum. We have been trying to isolate the serum factor since then. We've got it down to a single small fraction, but even with mass spectrometry, we haven't been able to figure out what it is. We have some ideas, but aren't there yet. We are at the point of getting 90–100% of the population to start feeding. The next step is to find out how they do it. What are the molecular steps involved in the transduction? What are they releasing? That was what was interesting to Peter. He thought that whatever they were releasing might be a good vaccine candidate." And, in fact, it was.

When John arrived at Peter's lab, they isolated L_3s, exposed them to serum and glutathione, and incubated them for 24 hours. "We ran the medium out on an SDS-polyacrylamide, and when we did this we found the protein we knew should have been there." According to both Peter and John, this was the breakthrough for the vaccine side of things. As Peter said, "We had to trick the larvae into 'thinking' that they were in the host before they would make the protein. The first protein they identified was a protease. Then, we cut bands from the gel, the ones in highest concentration, and sent them to the protein people at Yale."

While he was working with Peter at Yale, John was learning his molecular biology from a woman named Elizabeth Ullu, a highly respected *T. b. brucei* biologist. She and her husband, Chris Tschudi, helped John produce a DNA library for hookworm, the first of its kind. About this time John managed to obtain an NRSA postdoctoral fellowship from NIH to work on kinase in hookworm infection. John continued, "We looked at PKA [protein kinase A], which was the first thing we pulled out of our library. We found a couple of other kinases, but nothing with which to work because you are sort of stuck because you can't do genetics, you can't do knockouts, to characterize these molecules. But we made the primers from the protein sequence of the cut bands and used them to clone this molecule that was secreted and I named it Ancylostoma-Secreted Protein. I presented the results here [at an ASP meeting]; so, it's ASP-1. We found this molecule and it was interesting because it had two domains that were highly related, almost a duplication kind of thing. Then we cut another band out of that gel and, as it turned out, it was closely related to the first one, but there was only one domain [ASP-2]. It had a high level of similarity with the carboxy-containing terminal domain of the first molecule. So the two carboxy terminal domains lined up great. We published the first description in 1996 and the second in 1999. About the same time, this gene started showing up everywhere. It shows homology with a whole bunch of things in different taxa, ranging from a plant pathogenesis factor to the venom of hymenopterans.

I asked Peter about the source of the protein. He responded by saying, "It took us years to figure that one out." He continued, "One of the problems with larvae is that they have such a thick cuticle and you can't fix them properly. You have to do immuno-electron microscopy and it is only in the last year that we got the immuno-EM pictures in collaboration with Sarah Lustigman at the New York Blood Center. It turns out that a lot of these proteins are made in the glandular esophagus and then they are secreted via the lumen of the esophagus and out the mouth. Then, there are also 'channels' that go from the esophagus directly out to the cuticle." I asked him about the function of the proteins, but he said, "We don't know. We have the X-ray crystal structure of these proteins now. The two proteins, ASP-1 and ASP-2, are in a unique family of proteins that are found in plants, insect venom, and in mammalian testes. We don't know the function of those proteins either. There is some similarity to a protease, which would make sense. There is also some similarity to certain types of chemokines."

After months of bench work, he and John Hawdon finally obtained enough protein that they were able to obtain the amino acid sequence. They had been able to ultimately clone the gene and publish their results in the *Journal of Biological Chemistry* (Hawdon *et al.*, 1996). When they attempted to immunize mice with the cloned proteins, there was a hint of some success. "So, we went back into the literature and found a paper published by a fellow named Kerr at Johns Hopkins in 1936 in the *American Journal of Hygiene*. He had worked out a model in which he immunized mice with larvae and then followed with a challenge. They found that the number of larvae reaching the lung would be reduced. Hookworm larvae do not complete their life cycle in mice, but we could still use Kerr's system to assay the efficacy of the vaccine. I had begun talking with vaccine people to get a better 'handle' on the approach we should be taking." Peter even scheduled a meeting with the Nobel Laureate, Jonas Salk, out in La Jolla, California. He said that he had been invited to a meeting in San Diego and that Salk "had always been one of my heroes anyway, so I wrote to him and asked if he would meet with me," which he did. He said that Salk sat with him for nearly an hour and a half, going over his data and asking questions about his research. It was during this conversation that Salk told him he needed a mouse model. That is when they discovered Kerr's work from 1936 and that is why they adopted it and began using it in their search for a vaccine.

"By this time, I was a young Assistant Professor at Yale. One of the problems with hookworm is that it is not a high priority for NIH. I had my NIH grant, but I never had two of them. To be a big time player, you need two, or three, or four of them. So, Yale isolated me off in a tiny lab in the Department of Epidemiology and Hygiene. I had a technician and John as a postdoc, and that was how I got started. The other thing that I thought was very important at the time was that I didn't have a field connection. I wasn't working with anybody in a hookworm endemic area. That really bugged me. So, I made some false starts in India and even made a trip to Guatemala without much success in terms of building a long-term collaboration. But, about that time, the Institute of Parasitic Diseases in China was opening up to the western world. I managed to finance a trip to China to see the Director of the Institute who was actually working on hookworm in Shanghai. This was really an eye-opening experience. Back then, in 1994, this was like the old China where I had known that a competition for NIH grants to work with overseas tropical institutes was coming up and I needed to find some collaborators. One day when I was with the Director of the Institute, I asked him if there were any Americans working there that we could write another grant with? He responded, no, we really don't have any other Americans around. But on his desk, there was a picture of George Davis from the Academy of Natural Sciences in Philadelphia. I asked him, what about that guy, the one in the picture? He said, 'oh, that's George Davis. He's in Philadelphia.' So, I said okay and, when I returned home, I called Gerry Schad at the University of Pennsylvania. I had actually heard a lecture Davis had given at an ASP meeting, so I at least knew of him. I called George and set up a meeting in Philadelphia for lunch where I approached him about writing a grant together. He immediately was positive and agreed to do it."

Peter explained, "The Tropical Medicine Research Center Program is unique at NIH. The Principal Investigator had to be overseas. But, obviously, they can't write the grant without help from western scientists. So, George and I went back to Shanghai with our laptops and for three weeks we sat at the Institute with their staff. When we received word that the grant was to be funded, we knew this would give us a field opportunity in China and also access to all their clinical material. We were now hooked up with the central Institute of Parasitic Diseases, and China is very centralized even though they have all of the provincial institutes as well. It's one of the nice things about a communist system. This created an opportunity for close collaboration with Shu-hua Xiao, and we started doing experiments together on hookworm. At the same time, I met another guy named Zhan Bin who ultimately became the head of our molecular biology discovery group for the Human Hookworm Vaccine Initiative.

Peter continued, "This was now getting up to 1999. One of the problems we had was that NIH would not fund our particular line of research." I asked, "Why?" Peter said, "Because they felt it was something that industry should do. It was really not hypothesis testing. It was industrial research and that's when you should be collaborating with a company. But try and talk with a company about a hookworm vaccine. There is a market with 740 million people infected with hookworm [some say the number is greater than 1.3 billion], but they are all in developing countries. The people who get hookworm there are not even in the middle class. It's not like there are just a few people infected, there are millions and millions. The problem is they don't have any money to pay for a vaccine. It was about that time I heard rumors that Bill Gates was going to start funding programs for disease control in developing countries. I gained access to the Gates Foundation through Jeffrey Sachs, an international development economist at Harvard. He actually provided me with an introduction to the Foundation. They said they would be potentially interested, but there were a number of problems related to policies on intellectual property and other factors that made it difficult to go through a university such as Yale. I started thinking about a small nongovernmental organization that I could recommend. At the time, I knew the head of the Sabin Vaccine Institute, H. R. ('Shep') Shepherd. I had met him by trying to bring him to Yale. We became friends even though I was not successful in getting him to Yale. Shep was not a professional scientist, but a very successful businessman with a science background and a respected philanthropist. He also had a real appreciation for fostering young scientists, a kind of modern-day Basil O'Connor [referring to the founding Director of the National Infantile Paralysis Foundation, now known as the March of Dimes]. Knowing the kind of person I needed, Shep introduced me to Phillip K. Russell, who was a Major General in the Army and the highest ranking physician in the military at the time. He had just retired from the military after service as the Commandant of Walter Reed Army Institute for Research and the U.S. Army Institute of

250   *Parasites and Infectious Disease*

Research on Infectious Diseases at Fort Dietrich. And he was interested in doing something with the Gates Foundation. I approached Phil with the idea of going to them about my hookworm vaccine. At first, I think he thought I was coming from 'outer space', but in the end I won him over. This was a big step because Russell knows more about the vaccine development process than just about anyone. In due time, we submitted a grant and was funded. So I went from being one of the poorest investigators in terms of grant funding at Yale to being the Principal Investigator on an $18 million grant!"

After isolating the antigens ASP-1 and ASP-2, spending years building a team of protein expression and fermentation experts (headed by Gadam Goud), re-engineering the proteins in yeast in order to produce soluble and properly folded molecules, and finally preliminary trials in the laboratory, they were ready to proceed with human testing. This required establishing the equivalent of a biotechnology company with an academic setting in order to develop and support a biologic product. Project management is critical for such an enterprise and he was fortunate to recruit Maria Elena Bottazzi to take leadership of this phase. Known as the Human Hookworm Vaccine Initiative, Peter sometimes refers to this as his "guaranteed money-losing company," producing hookworm vaccine for the world's poorest people. The project was established jointly in the labs of a newly created Department of Microbiology and Tropical Medicine at George Washington University, together with program management at the Sabin Vaccine Institute, which was just a few blocks away in the Foggy Bottom section of Washington, D.C. (Hotez *et al.*, 2005).

Ultimately, human testing in the field requires a hookworm-endemic region of the developing world. According to Peter, several criteria were involved in selecting a country in which to proceed. First, the country had to have endemic hookworm disease, with a high rate of transmission. Then, the country had to have the wherewithal to develop, produce, and distribute the vaccine. The latter criterion automatically excluded a large number of developing countries. It meant that only 'middle-income' countries would work, e.g., China, India, South Africa, Mexico, and Brazil, for example. For a variety of reasons, the latter was selected. According to Peter, "Brazil is a good 'poster child' for this effort. They have a serious hookworm problem, plus the ability to make their own vaccine. So, what we've done is sign a memorandum of understanding between the Sabin Vaccine Institute and two Brazilian institutions, Instituto Butantan, a Brazilian manufacturer sponsored by the State of Sao Paolo, and the Oswaldo Cruz Foundation (FIOCRUZ) for clinical trials. The agreements state that if we can show that the vaccines are efficacious, we will transfer the technology to them provided they agree to produce and distribute it in their country. In turn, Brazil has now started partnerships with Portuguese-speaking countries in sub-Saharan Africa, e.g., Angola and Mozambique. Brazil will also export this vaccine to these countries. We think we will be able to do the same thing with China and India, perhaps Indonesia." A team from the Sabin Vaccine Institute, headed by David Diemert and Kari Sto-ever, is now putting together essential components needed to conduct safe and ethical clinical trials of the hookworm vaccine in Brazil. For this, they are working closely with [Jeff] Bethony and a FIOCRUX team headed by Rodrigo Correa-Oliveira. It is a complex series of trials for which they have strong clinical trial and statistical support by a team from the London School of Hygiene and Tropical Medicine (headed by Peter Smith, Laura Rodroguez, and Simon Brooker). Downstream, the vaccine may be augmented with a second antigen from adult stages of the parasite, an antigen discovered originally by Alex Loukas and his colleagues, now at the Queensland Institute of Medical Research in Australia.

I then asked Peter how the vaccine was going to be administered, i.e., oral or injected? He responded, "You deworm the individual, then follow with the vaccine via injection." It is to be a targeted group, i.e., school-aged children. "It's a practical matter. Now that Don [Bundy] and others have set up this school infrastructure for deworming, we have a perfect segue for us to go in and deliver the vaccine to a captive pop-ulation, because we can follow on the heels of the deworming efforts. If you were to give it as an infantile vaccine, we think that it would be too much. These infants are already 'pin cushions', because they get so many vaccines. Adding another vaccine on top of those already given could be damaging." Peter is now working with the World Health Organization, the United Nations Millennium Project, and other inter-national agencies to identify mechanisms for incorporating the hook-worm vaccine into existing and control programs for tropical diseases (Hotez, 2006; Sachs and Hotez, 2006).

I was curious about the Rockefeller and Gates Foundations. I asked Peter to compare them for me. He said that the Rockefeller Foundation's approach was oriented toward "treating and controlling the disease. It was about using existing control measures, i.e., chemotherapy. We have gone about as far as we can with that sort of control. We need a new generation of controls, this time prophylactic measures, i.e., an AIDS vaccine, a malaria vaccine, a hookworm vaccine." I changed the subject somewhat at that point by asking Peter if he thought there would ever be a malaria vaccine that really worked? He responded, "Well, you know they have this new malaria vaccine, called RTSS, that's been developed by Glaxo-Smith-Kline, the military, and Walter Reed. The paper just came out in *Lancet* a couple of months ago. It shows you can get a 58% reduction in the number of cases of severe malaria, and a 30% reduction in overall infection. I think this is the first breakthrough. It's the first proof of principle." As he was citing the results, I kept thinking to myself, "I wonder if this is true? Is this really the breakthrough? Why am I in denial of the idea that it is a breakthrough? I guess my pessimism stems from nearly thirty years of promise, of hearing it's just around the corner." Peter must have been reading my mind, "I know, we've been waiting around a long time, but it was just a lot more complicated than we thought." He reminded me of the time in India "when we went from 75 million cases down to 50 000 when we were spraying with DDT, and then we saw resistance and malaria prevalence went back up. This is the same thing that worries me about existing control methods for hookworm, about benzimdazole anthelmintic drugs. They are going to 'douse' the world in benzimdazole, and we know that cattle and sheep both can become benzimdazole resistant." Based on what Peter said, and on what Roy Anderson related in my interview with him, I can easily see a resistance problem coming down the road, and sooner, not later.

I ended our talk by asking him about genetic variability and the role it might play in the epidemiology of hookworm disease. He responded by telling me about an, as yet (as of then), unpublished tidbit of fascinating information. "In both Brazil and China (Hainan), Jeff Bethony in Brazil and Shu-hua Xiao in China have discovered an interesting subpopulation in endemic areas of hookworm disease. About 15% of the population naturally makes IgE antibodies against ASP-2 antigen. In these individuals, there is a 62% reduction in risk of acquiring heavy hookworm infection. This means that there is a small segment of the human population that has natural protection against hookworm infection." I mumbled, "Genetics?" Peter spoke up immediately, "Could be. You would think so. Now the question is, how do you make the rest of the population respond in the same way by giving them the vaccine?"

While I flew back home from Washington, D.C., I could not help but think about how Peter's life had been so carefully structured from reading de Kruif's *The Microbe Hunters*, Chandler's *Introduction to Parasitology*, and, finally, Craig and Faust's *Clinical Parasitology*, all before leaving high school. I also was impressed that he picked Yale because there was a parasitologist there (Curtis Patton) and that he went to Rockefeller to do his Ph.D./M.D. because of another parasitologist (Bill Trager).

Virtually all of Peter's professional focus has been on parasitology. He chose to work on hookworm in part because of Stoll's remarkable 1962 paper in *Experimental Parasitology*, and partly because no one else was working on it. He decided to develop a vaccine because he could clearly see the need for one rather than working in chemotherapy. The thing that really impressed me though was his 'bulldoggedness' in pursuing each of these goals. It seems as though almost nothing could deter him. When he saw something worth doing, it's almost like he determined he would do it no matter what. He would find a way to success and he did! This pugnacious attitude has served him well. If he is successful with the vaccine, it will have served a lot of people very well.

Gerry Schad and John Hawdon were also a great team. Based on what John told me, Gerry had a great impact on his career. Peter presented a second huge opportunity for John, not only in the area of molecular biology and genetics, but in epidemiology as well. John and Peter still work on projects together from time to time, but they have mostly gone their separate ways in recent years. That's okay. They achieved their own goals and, at the same time, served the needs of the other, a great way to do successful research.

REFERENCES

Anderson, R. M. and R. M. May. 1979. Population biology of infectious diseases: Part I. *Nature* **280**: 361–367.
Brooker, S., J. Bethony, and P. J. Hotez. 2004. Hookworm infection in the 21st century. *Advances in Parasitology* **58**: 197–288.
Hawdon, J. M., B. F. Jones, D. R. Hoffman, and P. J. Hotez. 1996. Cloning and characterization of *Ancylostoma*-secreted protein. A novel protein associated with transition to parasitism by infective hookworm larvae. *Journal of Biological Chemistry* **22**: 6672–6678.
Hotez, P. J. 2006. The biblical diseases and U.S. vaccine diplomacy. *Brown World Affairs Journal* **12**: 247–258.
Hotez, P. J., J. Bethony, M. E. Bottazzi, S. Brooker, and P. Buss. 2005. Hookworm: the great infection of mankind. *PLOS Medicine* **2**: e67.
Keller, A. E. and W. S. Leathers. 1940. The results of recent studies of hookworm in eight southern states. *American Journal of Tropical Medicine* **20**: 493–509.
Kerr, K. B. 1936. Studies on acquired immunity to the dog hookworm *Ancylostoma caninum*. *American Journal of Hygiene* **23**, 381–406.
May, R. M. and R. M. Anderson. 1979. Population biology of infectious diseases: Part II. *Nature* **280**: 455–461.
Sachs, J. D. and P. J. Hotez. 2006. Fighting tropical diseases. *Science* **311**: 1521.
Stoll, N. R. 1962. On endemic hookworm, where do we stand today? *Experimental Parasitology* **12**: 241–252.

10

The spadefoot toad and *Pseudodiplorchis americanus*: an amazing success story of two very aquatic species in a very dry land

I shall not see the shadows,
I shall not feel the rain;

Song, Christina Georgina Rossetti (1830–1894)

Even though they are buried several feet below the desert's scorched sand for most of the year, spadefoot toads, *Scaphiopus couchii*, definitely feel the rain when it comes. Remarkably, they feel it even though they cannot see it and they are not touched by it.

The beginning of Richard Tinsley's interest in the spadefoot toad, and *Pseudodiplorchis americanus*, a monogenetic trematode infecting the toad's urinary bladder, began in the late 1960s while he working on his Ph.D. at the University of Leeds in England. It was then, very early in his career, that he came across a publication written in 1940 for an obscure journal (*The Wassman Collector*) by L. O. Rodgers and Bob Kuntz. In this one and a half paged paper, Rodgers and Kuntz had described *P. americanus*, but nothing about the biology of the extraordinary interaction between this parasite and its host. Richard said, however, holding a now yellowed Xerox copy of the paper in his Bristol office, "There is enough in here [in the Rodgers/Kuntz paper] to show that there is an exciting story to be told. First of all it's a monogenean. Then, with my interests in this group of vertebrates, I knew that its host was completely terrestrial. But, I knew that the toad returned to water for a very short time each year. It is absolutely obvious that can be the only time a monogenean parasite can be transmitted to new hosts. At least part of the solution of the transmission problem by the parasite can be seen in this drawing [in the Rodgers/Kuntz paper]. It shows all the offspring as fully developed in the uterus of the parasite. So, it was all set up for a study!" In other words, the larval stage inside the eggshell in the uterus was ready to emerge as soon as the egg was released from the monogene. There would be no delay required for embryonation outside. This was an exceptionally astute observation, a noteworthy one considering he knew nothing more about this southwestern U.S. monogene than what appeared in the tiny paper, and that certainly was not very much.

During the first ten years after receiving his Ph.D., Richard was totally occupied with *Xenopus* spp. and its monogenean parasites, not with the spadefoot toad. Much of his early research effort required frequent trips into remote areas of Africa, including Ethiopia, Kenya, Uganda, Rwanda, Congo, Ghana, Sierra Leone, and South Africa. The outcome, alongside parasitology, included descriptions of two new species of *Xenopus*. However, the idea of working on *Scaphiopus* and *Pseudodiplorchis* had stayed in his mind since seeing the Rodgers/Kuntz paper. Then, in 1979, the Nuffield Foundation announced the establishment of a new fellowship program, one that would make funds available to support college professors with heavy teaching loads to take sabbaticals. The Foundation was trying to encourage opportunities to engage in unbroken research for up to a year. He remarked as we sat in his University of Bristol office, "That was just what I needed, relief from the heavy teaching load that I had at Westfield College."

A truly serendipitous event was to occur about the same time. He told me that he was on his way to an executive committee meeting of the British Society for Parasitology one spring day in 1979, but that he arrived too early. The meeting was up in Regent's Park in London and he had about 45 minutes before it started. The Society of Zoology offices were in the vicinity, so he thought he would stop by and browse their library as a way of killing some time. While there, he picked up the recent copy of *Copeia*, a journal that publishes both ichthyological and herpetological papers. In that particular issue, as pure luck would have it, there was an extensive article dealing with "the terrestrial ecology of spadefoot toads" in Arizona. His fire for the toad and the urinary bladder monogenean was immediately rekindled. The next morning, he wrote the senior author of the *Copeia* paper, Rodolfo Ruibal, at the University of California-Riverside, inquiring if he had seen any parasitic worms in the toads' urinary bladders during the study. A short time later, Ruibal replied in the affirmative, but that he was not a parasitologist and could not confirm if they were *P. americanus* or not. Richard believed (hoped) they were and proceeded immediately to make application to the Nuffield Foundation to support a year's leave from teaching.

His proposal to the Foundation was written as though the urinary bladder monogene was present, a good supposition, but without any empirical evidence to support it. Not surprisingly, he was successful with the grant. However, he hoped his time in Arizona was not going to be a 'wild goose chase' because he still was not sure the parasite was present. As it turned out, luck was on his side. In our discussion, he remarked, "I had been incubating this idea [work on the spadefoot toad and its polystomatid monogene] for about ten years, but it was the serendipity of seeing this wonderful paper on spadefoot toads in the library of the Society of Zoology that afternoon in 1979 that soon took me to the Southwest Field Station in Arizona," and launched him into nearly twenty years of research roughly 6000 miles away from London and Bristol. He continued, "I suppose that I eventually would have seen that particular article at some point later on, but having the funding available simultaneously was extremely important."

Richard continued his story, "Everything worked exactly to order. A lot of it was luck. Sometimes, for example, the rains in the desert southwest of the U.S. don't come on schedule. They are not entirely predictable. My technician, Celia Earle, went with me on the first trip. When we arrived that first night from London, we rented a car and drove from Phoenix down to the Station, close to the Arizona–New Mexico border. By chance, it rained. The roads became covered with spadefoot toads, which I had never seen before. We stopped the car and caught several. They all had spades on their feet, so I knew we had the animal I was looking for. The next morning, I dissected some of them and found parasites. They were *Pseudodiplorchis*," and he was in business.

The only serious problem Richard had with work in Arizona over time was funding. It was erratic. It would have been so much better for ecological research if funds for successive years' fieldwork had been known in advance to allow for the design of long-term studies. Instead, in many years, Richard was forced to do research not knowing if he would return the following year. Ultimately, however, these isolated studies developed into an established record of population ecology extending from the early 1980s to the late 1990s. Overall, he said, "The work was done on a shoestring basis." In my humble opinion, it was a very good run. So, what did he learn about this amazing host/parasite relationship?

It is widely known that polystomatid flukes are distributed among a range of frogs and toads. Those monogenes found in completely aquatic hosts reproduce continuously. In this way, he said, "these monogenes resemble fish parasites." However, inevitably, those found in mostly terrestrial animals exhibit a reproductive behavior matching that of their vertebrate host. In the case of *P. americanus* in the spadefoot toad, reproduction is directly tied to the annual, 'monsoon' period of about two months in the southwestern U.S.A. He said, "However, the opportunity for parasite transmission is actually limited to less than 24 hours in a whole year, a restriction greater than for any other helminth of which I am aware."

The life cycle of *P. americanus* is direct, like that of other monogenes. However, it is somewhat unique in that not only is oncomiracidium development completed while the egg is still *in utero*, the egg hatches in the uterus as well. There is then live birth, making the parasite ovoviviparous. This also means that development of parasite eggs in the uterus has proceeded for several months while the toad lies dormant, buried deep in the desert earth. The presence of these fully developed oncomiracidia in the figure drawn by Rodgers and Kuntz (1940) was the clue recognized by Richard ten years previously. For Richard, it immediately suggested that a potentially exciting story would be told by this parasite and its host.

Because of the huge restraints placed on both the host and the parasite in the extreme desert habitat of the southwestern U.S.A., there are a variety of special adaptations exhibited by both the parasite and the host. The first years of research focused on unraveling these adaptations. However, when one notes that prevalence of the parasite is approximately 30% and its overall mean intensity is in excess of six worms per host, *P. americanus* does quite well in the harsh desert. So does the spadefoot toad. Both host and parasite are, for example, beautifully geared toward rapid reproduction and growth. They both must be ready for the onset of the monsoon rainfall (if it comes) each summer.

How does the toad know it is raining when it is buried down to a meter below the desert's surface? Then, how does the parasite know it is time to release the contents of its uterus to the outside? In the case of the toad, Richard says the stimulus for leaving hibernation (= aestivation) is the "low frequency vibration" created by raindrops striking the desert floor. Then, when the toads enter newly formed ponds to mate, the parasite reacts to intense "sexual excitement", clearly a response to a hormonal cue (Tinsley, 1999), and discharges its parasite larvae *en masse*. As Richard put it during our conversation, "host sexual activity is infallible as a cue for stimulating release of the oncomiracidia." He continued, "The key is absolute precision. An hour too soon while the toad is still traveling to the pond, or an hour too late in the morning after the toad has departed the pond, and a whole year's production of larvae would be wasted." Oncomiracidium release from the toad must perfectly coincide with the toad's appearance at the ephermeral pond. This opportunity is more fleeting in females than in males. In both sexes, pond visitation is nocturnal. After release of her eggs, however, the female departs the pond and will not return. Thus, a female may stay only for up to four hours. Males 'hang around' longer, a maximum of seven hours. This also means that males end up with heavier infections than females.

Once oncomiracidia emerge from the toad into the pond, they swim continuously. Since they do not feed, they must rely entirely on stored energy reserves and the presence of cilia on their surface to move. Survivorship of the free-swimming larvae is surprisingly good, with up to 50% of the parasites remaining motile for at least 15 hours (Tinsley, 1999).

Entry by oncomiracidia into the toad is via the nares. Since the nares of a given toad are generally kept above the water line in the pond, this means that the oncomiracidia are briefly exposed to the air, but they still manage to locate the opening and enter the toad. Now comes the very peculiar, and even dangerous, migration route inside the toad. After 24 hours in the nostrils/sinuses, the larval parasites migrate into the oral cavity where they reside for up to a week before heading for the lungs. While in the oral cavity, the parasites begin to consume blood from host tissues. The lung excursion is apparently unique among monogeneans, since only *P. americanus* and *Neodiplorchis scaphiopodus* are known to take it. Their stay in the lung is relatively brief, lasting only 7 to 14 days. After about three weeks inside the toad, the parasites move to the buccal cavity in preparation for another arduous trek, this time to travel down to the cloaca through the stomach and intestine, and up into the urinary bladder.

If dormancy of the toad occurs before the monogenes enter the genitourinary tract, they will stop any further development and enter dormancy as well. They are able to remain in this arrested state for up to a year, according to Richard. When the parasites enter the digestive system and encounter the extreme environment there, they are prepared. In the first place, they know their way through the gut. It is quick, being completed in a matter of just a few minutes. While in the respiratory system, the juvenile worms also begin to accumulate two types of membrane-bound vesicles in their syncitial outer tegument. During gut migration, these vesicular contents are released to the outside, creating a covering that shields the surface tegument from the otherwise certain digestion. I use the word, otherwise, because Richard and his students have conducted a number of different sorts of experiments to show that death and digestion are certain outcomes unless the parasite is completely 'equipped' for the gut journey. All of this preparation for migration through the gut suggests a cue must stimulate movement, but the nature of the stimulus is, as yet, unknown.

Once through the gut and into the cloaca, the parasites move quickly into the urinary bladder where they take up residence that will last until the monsoon rains come the following year. On arrival, they immediately begin to consume blood and continue to grow until hibernation starts. Interestingly, monogene reproduction begins at about four weeks after infection of the toad. This means that the parasites are reproductively active before reaching their final destination in the toad.

The fact that they consume blood suggested to me the potential for pathology. Naturally, Richard and his students had tested this idea too. I asked him about this possibility during our conversation in Bristol. He responded, "They have the potential to induce it because they feed on blood. Remember, for the toad parasite, the host doesn't feed for eleven months out of the year, so it's a sealed container with its nutrients in it. When the parasite takes blood from the host during dormancy, the host must make up the difference from somewhere. The loss is made good from the fat reserves in the toad laid down for its own survival. So, during periods of activity, these hosts must accumulate enough lipid to see them through nearly a year of starvation. Part of the energy reserve is being taken away as the parasites take blood for their own use and the toads must compensate for this loss. We've been able to measure the drain on host blood, which has the potential for causing anemia. We've been able to see how this translates into depletion in fat reserves. Of course, this is a density-dependent phenomenon. So, the heaviest infections are potentially life threatening. For an average burden [about six worms] during the course of the eleven month hibernation period, it works out that the parasites take about 7% of the reserves above what the toad needs for itself. It's a straightforward equation. If the host goes into hibernation with enough reserves for itself, along with 7% extra, there is enough to sustain the parasites' basal metabolism and provide for their reproduction. If the host does not have the surplus, it may die." I asked him if this was the same as a threshold? He replied in the affirmative, "But it's a difficult threshold. The length of hibernation may vary each year because it is dependent on when the monsoon rains come during the summer. It might have to wait an extra month, which requires a correspondingly greater reserve of lipid. All of this work has been verified in the laboratory by maintaining animals under different conditions and measuring hematocrit and fat reserves to see what would happen." The bottom line is that life in the desert for the toad and its parasites is very precarious!

I then asked him about any sort of immune response on the part of the toad? Again, he replied in the affirmative, "and it is temperature dependent. If we maintain toads in the lab at 15 °C, then the animals will not lose any parasites, but if we raise the temperature, then they begin to lose their parasites. This means that infection levels remain stable during the winter in the desert. At these low temperatures, the host immune response is more or less suppressed. The parasites take relatively little blood and their growth and reproductive preparation are minimal. There is a critical period as it is warming up when host immunity is becoming more and more effective. Parasite numbers begin to drop, but those that survive take increasing amounts of blood and produce more fully developed infective stages. So, there is a complicated balance here as well."

Seasonal changes in temperature are an important feature in regulating the population biology of *P. americanus*. Richard has estimated that about 97% of the worms that successfully invade a host population fail to survive to contribute to their first opportunity for transmission a year later. He has reviewed the nature of the factors contributing to this massive attrition (Tinsley, 2005) and concluded that, "The dominant regulation of *P. americanus* infrapopulations . . . is a highly effective immune response." He continued, "There is a very delicate balance between parasite pathology and host immunity. There is evidence, circumstantial in this monogenean system, that acquired immunity has an important role in parasite population ecology. However, during each year's breeding season, all mating toads become infected with, on average, about 100 worms, and in hosts with substantial protective immunity it takes about a month for these worms to be reduced or eliminated." Richard explained that recent studies have shown that lung epithelium accumulates scar tissue in response to the short-lived presence of larval monogenes and that this may prejudice respiratory efficiency, especially since this is the time that toads are actively pursuing prey. So, "One of the additional costs of infection is that the host's ability to survive the following year's hibernation may be compromised. Indeed, this may affect even those toads that have effective immunity." He concluded, "Despite host resistance that prevents recently invading worms from reaching maturity, the pathology incurred during initial pulmonary infection may have long-term consequences."

The complicated relationship between the host, parasite, and temperature, raises an interesting question regarding the geographic distribution of the parasite. Richard said he had not tested any ideas along this line, but speculated that, "as you move south, temperatures should be too high for parasite survival. Further to the north, the parasite should survive better, but the season is too short for the parasites to produce enough offspring for the parasites' life cycle to be sustained. Alternatively, there may be different 'strains' of the parasite adapted for local conditions of temperature and corresponding interactions with host immunity."

I asked Richard about the role of serendipity in his research. He said that the only serendipitous event in his research, at least that involving the spadefoot toad and *P. americanus*, came on an afternoon in the spring of 1979 when he showed up early for a meeting in Regent's Park, and he paid a visit to the Society of Zoology library. However, this connection was possible only because of Richard's knowledge of the literature, especially a very obscure journal, and the ideas that had formed ten years previously on the potential for research investigation. This launched him into twenty years of research roughly 6000 miles away from London and Bristol. He said, "I suppose that I eventually would have seen that special article at some point later on. But having the funding simultaneously was very important." Considering the breadth, length, and depth of his work in the desert southwest of the U.S.A., I view the absence of serendipity as a strong indicator of a well-conceived lifetime of research.

Richard emphasized that he also learned a great deal about research during the "Arizona years". His fascination with the system has included the potential for obtaining very precise empirical data for a naturally occurring host–parasite system "in the wild". He added, "There is great scope for field experiments, including the use of enclosures and plankton sampling equipment, to obtain direct measurements of transmission success." He explained that studies of this sort required comprehensive preparation beforehand, including the transport of large quanties of equipment night after night in hopes that the appropriate conditions would be found in some part of the desert that would permit him to conduct his studies. All field investigations had to be designed and completed during the hours of darkness before toads leave the water and each transmission episode comes to an end. He found that his schedule had a major influence on his training for research. Thus, he said, "There is an absolute need to have every-thing ready in advance, especially the ideas for a range of different possibilities for field studies. The decision on which line of investigation to pursue depends on the conditions occurring on a given night."

He expressed regret that many of the results he had generated from these very intense efforts still awaited publication. He lamented, "The fieldwork was so productive, for a few weeks each summer, that it would generate far more results than could be prepared for publication, alongside teaching and other work, before the start of the next summer's fieldwork, and this would provide even more data for publication." Despite the excitement of this program, Richard explained that the backlog convinced him several times that he should give up the annual trip to Arizona and concentrate on writing up what he already had in hand. It was not long, however, that he developed "withdrawal symptoms" and applied for additional funding to go back into the field. Finally, he did give it up in 1992 and, for the next three years, the Sonoran Desert of Arizona was subjected to its worst drought in history. In fact, many researchers at the Station had to abandon their projects because of the lack of rain. This turn of events encouraged Richard, however, to design a three-year project to document the effects of the extreme conditions on parasite and host populations. He was hopeful that the new research would provide insight into the nature of factors regulating infection levels under these new conditions. He discovered that in some places, the absence of rain caused local extinction of the parasite. In contrast, the toad populations were relatively unaffected because of their longevity, up to a maximum of seventeen years. The parasite, on the other hand, was significantly impacted since their life span rarely exceeds three years, and two or three successive recruitment failures eliminated the parasite. He noted that the failed transmission was continued for a few years after the drought ended when there were torrential rain storms and all infective stages were washed away by local flooding. It made him aware that this 'aquatic' parasite was not only vulnerable to excessive drought, but to excessive rain as well.

A major project in recent years has been concerned with determining the age and survivorship of *S. couchii*. Because the toads feed actively for only about a month each year and then 'slow to a crawl', visible 'growth rings' are produced in their bones. His age studies were directed at correlating age and infection by *P. americanus*. His analysis revealed that worm numbers in the toads are strongly age dependent. Infections in toads begin to occur at about three years when they first enter ponds at sexual maturity. They then rise in abundance before leveling off at about six to eight years of age, and then declining in the oldest hosts. Richard believes that this is circumstantial evidence to support his thinking about the effects of acquired immunity of the population biology of the parasite. He said the the last few years of data provided convincing evidence that, "even in an extreme environment, like the desert, and for a helminth parasite transmitted exclusively in water, the major controlling factor for infection was not the external environment. Instead, it was clear that the major controlling factor for regulating parasite densities was acquired immunity."

Richard's enthusiasm for monogeneans and ectothermic vertebrates is clear. He spent a great deal of time toward the end of our conversation in extolling their research advantages as opposed to many of the host/parasite models used in investigating systems of medical/veterinary importance. His rationale is convincing for any number of reasons. Importantly, though, he notes that, "In many cases, these host species would not normally be infected in nature, and the characteristics of the parasites have become modified by repeated passage to the point where they may no longer be infective to the natural host."

At the very end of our productive discussion, I asked Richard about his current research. He responded by describing a newly initiated study on *Xenopus laevis* and *Protopolystoma xenopodis*, both of which he worked while a graduate student at Leeds. He explained that *X. laevis* has been introduced into a range of new locations on several continents. There is even a feral population in Wales and, moreover, *P. xenopodis* infects it. Laboratory results indicate that *X. laevis* develops a very powerful acquired immunity to *P. xenopodis*. This fact, plus the very low ambient water temperatures in Wales, would suggest that the parasite's presence is highly unusual. In fact, Richard indicated the parasite should have become locally extinct soon after introduction. Field data that he has generated to date indicates, however, that, "The parasite survives at very low infection levels and principally in juvenile *X. laevis* experiencing their first infection. This indicates the parasite may persist by exploiting naïve offspring produced each time the toads reproduce." Complicating the biology of the parasite is the observation that environmental conditions in Wales are such that toad reproduction has occurred just five or six times in the past 25 years. So, how does the parasite 'bridge the gap' between successive appearances of naïve recruits into the host population?

Based on recent findings in Africa (Jackson and Tinsley, 2005), natural populations of hosts and parasites were shown to have significant heterogeneity infectivity/susceptibility characteristics, which Richard believes may point to the possibility "that 'strains' of the parasite in Wales have been selected for compatability with local genotypes of the host. Additionally, low temperatures may provide the key to parasite persistence. Thus, on the one hand, parasite developmental rates are likely to be very slow, extending the intervals between generations. On the other hand, the suppression of host immunity at low temperatures may permit survival of worms for far longer than is known from laboratory experiments, at temperatures equivalent to those in Africa."

Richard was quite excited about this new project. He views it as a natural experiment, "carried out over a forty year time frame." He has been looking at this *Xenopus* population for twenty-five years and all of the toads are individually marked, allowing for a 'longitudinal' character to be employed as well. "Indeed," he remarked, "the maximum longevity of *Xenopus* in this Welsh population exceeds twenty years, so the host–parasite system provides the basis for a relatively rare, long-term interpretation of population ecology", and towards this end he also intends to look at relationships within the population using mtDNA analysis. He concluded our interview by saying that this new project represents his real love for parasitological research where there is likely to be a complex interaction between parasite, host, and environmental factors in determining the persistence of infection under conditions which should have led to extinction. He said, "The key to understanding the interaction will be the close integration of fieldwork studies with experiments carried out under controlled conditions in the lab, exactly as with my most enjoyable previous projects."

As usual, my stay in Bristol was informative, interesting, and fun. Talking with Richard is always this way.

REFERENCES

Jackson, J. A. and R. C. Tinsley. 2005. Geographic and within-population structure in variable resistance to parasite species and strains in a vertebrate host. *International Journal for Parasitology* **35**: 29–37.

Rodgers, L. O. and R. E. Kuntz. 1940. A new polystomatid monogenean fluke from a spadefoot. *Wassman Collector* **4**: 37–40.

Tinsley, R. C. 1999. Parasite adaptation to extreme conditions in a desert environment. *Parasitology* **119**: S31–S56.

Tinsley, R. C. 2005. Parasitism and hostile environments. In *Parasitism and Ecosystems*, ed. F. Thomas, F. Renaud, and J.-F. Guégan, pp. 85–112. Oxford: Oxford University Press.

11

The schistosomes: split-bodied flukes

Humanity has but three great enemies; fever, famine, and war; of these, by far the greatest, by far the most terrible, is fever.

Sir William Osler (1849–1919)

As I have written elsewhere in this tome, there were two important contributions that allowed those working in the field of parasitology to make breakthrough discoveries in the nineteenth and early twentieth centuries. The first was the microscope. It was the Dutch rug trader, Antonie van Leeuwenhoek, who began to develop and refine this technology in the seventeenth century. Robert Hooke, an early British microscopist and a contemporary of the resourceful Dutchman, created what was ultimately to evolve into one of the most powerful of all biological concepts, in fact, one that still is being cultivated today. While looking through one of his primitive scopes one day, Hooke noted that the structure of a piece of cork he had sliced was divided internally into what he called "cellulae". With this observation, the cell theory was borne.

The contribution of van Leeuwenhoek provided the way for technology to eventually take us inside cells and build on Hooke's observation. The cell theory, along with Darwin's evolutionary theory, unquestionably did more to alter the biological landscape than any other conceptualization. There was, however, a widely held idea that had to be purged before significant biological research could progress. Thus, at some point, the scientific community had to eliminate a tenacious and long-lasting roadblock, namely the notion of spontaneous generation, part of the foundation for the stubbornly held thesis of early creationism.

The belief in spontaneous generation is an old one, espoused continuously since the time of the ancient Egyptians and Chinese. It was not until the seventeenth century, however, that new views began to appear, for example, the idea of preformation, which held that inside an egg was a fully formed being. If an appropriate stimulus were provided, then development would proceed. The downfall of spontaneous generation actually began in 1668, with Francisco Redi, who did not believe that 'worms' spontaneously appeared in dead meat as a consequence of putrefaction. He thought that the worms were deposited there, as 'seeds', and developed into worms, which used the flesh as a nutrient supply. In his initial experiment, he killed three snakes and left them in the open. Very quickly, the flesh was full of maggots. Gradually, as the flesh was consumed, he saw the maggots transformed into what we would today call pupae. Soon, from pupae emerged flies. His conclusion was that flies were dropping something on to the dead flesh, something that would eventually develop to maggots, pupae, and flies, in that order. In his next experiments, he placed a dead snake, some dead fish, and the flesh of a cow into glass flasks, some of which he sealed and some of which he left open. He waited. Maggots soon appeared in the open flasks, but not the closed ones. His hypothesis was confirmed through this simple, but brilliant, experiment. Even with this hard evidence, however, Redi continued to hold the view that intestinal worms developed by spontaneous generation.

The next crucial idea came in a helminthology book published in 1700, written by Nicholas Andry, who championed the idea that enteric helminths in humans came from seeds. One of the serious problems in his thinking, however, was that each person was actually predisposed to infection with specific types of worms, depending on the nature of the 'humors' with which they were born. Those fortunate individuals born without humors were to remain without worms.

As time passed, the microscope became a favorite research tool of the biologist, and new microscopic creatures were being seen everywhere and in virtually everything. The idea of heterogeny then emerged. This notion held that tiny animals were introduced into the human body and then developed into specific kinds of worms.

While the concept of a life cycle had not yet been conceived, the first attempt to complete a worm life cycle using an experimental approach was nonetheless carried out in 1790 by a Dane, Peter Christian Albildgaard. It seems that he had observed the nonreproductive larval [plerocercoid] stages of worms in the body cavities of sticklebacks that bore a striking resemblance to reproductive worms in the gut of mergansers and other fish-eating birds. His hypothesis stipulated that the two kinds of worms were related to each other, a huge 'leap of faith' at the time. He designed a simple experiment. He obtained a pair of ducks and fed them large numbers of worms from the sticklebacks. After three days, he killed both ducks. In the intestine of one, he found 63 reproductive worms and a single reproductive worm in the other. This was the first successful experimental evidence for the idea of alternation of generation. At the same time, it was a hugely important step for the rebuttal of spontaneous generation, although it received little, if any, attention. Pasteur, in the 1880s, is generally given credit for disproving the idea of spontaneous generation although it seems to me that his experiments were little more than simplified repetitions of Redi's work from more than 200 years before.

All of us now accept a central theme in the idea regarding alternation of generation. Nowhere among the parasitic helminths is this idea better illustrated than with the life cycle of a trematode. Although cercariae had been first described by Muller and assigned a generic name, *Cercaria*, it was Nitzsch who noted the potential for transformation of these 'organisms' when he observed they could shed their tail and develop into what he referred to as "pupa" (undoubtedly metacercaria). However, he did not make any sort of connection to idea of alternating generations. Not long after, just prior to 1820, another German, Ludwig Bojanus, observed motile sacs in the viscera of newly dissected snails and identified cercariae as emerging from these sacs. According to Grove (1990), he even "wagered" that these cercariae were embryonic distomes, but he too overlooked the real significance of his observation.

The concept of alternation of generation was formalized in 1821 by von Chamisso who employed it in describing a marine tunicate in the Mediterranean Sea that was tied together in chains, with alternating generations next to one another. Twenty-one years later, another Dane, Johannes Steenstrup, finally provided a proper conceptualization regarding alternating generations as it applied to digenetic trematodes. His view, simply stated, was that some animals, including trematodes, developed through a succession of distinct morphological stages on their way to becoming sexually mature adults. Throughout the rest of the nineteenth century, the life cycles of a vast array of parasitic flatworms and nematodes were elucidated, mostly in Europe. Between 1879 and 1882, the life cycle of *Fasciola hepatica* was resolved, independently, by Rudolph Leuckart in Germany and A. P. W. Thomas in England. This was the first for a trematode.

The history of discovery for the schistosome life cycle is one of much confusion, extensive contradiction, a great deal of controversy, and retaliatory spite, especially among the Europeans who were the first to engage in an effort to resolve the problem. An interesting feature of the discovery of the schistosome life cycle is that success came first to the Japanese, not the Europeans. Moreover, the latter parasitologists had a 53-year head start, i.e., Bilharz discovered adult *S. hematobium* in 1851, while Fijiro Katusurda did not observe adult *S. japonicum* until 1904 (translated 1905). After seeing adult *S. japonicum* for the first time, it took the Japanese just nine more years to publish the full life cycle of this parasite (Miyairi and Suzuki, 1913). There are several good reasons for this quick success as opposed to that of the Europeans. These I will detail shortly. Before continuing, I should note that while there are several very good books on the history of parasitology, by far the best, *A History of Human Helminthology*, was written by D. I. Grove and published in 1990. I have made extensive use of it as a source for what follows regarding the history of discovery related to the schistosome life cycles.

Rather than examine the success story first, I would rather describe some of the accomplishments, and blunders, made by the Europeans as they stumbled along for 64 years before finally discovering how everything worked. The first thing we can say definitively about the discovery of *S. hematobium*, by Theodor Bilharz in 1851 in Egypt, was that it was purely serendipitous. He was doing autopsies in Cairo and seeing lots of intestinal helminths, some new and some already described, when he came upon a delicate, thread-like worm in a portal vein. In subsequent autopsies, he frequently encountered it. He quickly deduced that the worms were distomate, and nonhermaphroditic, a first in terms of sexual orientation for the distome worms that were beginning to turn up with greater and greater frequency in those years. He named the parasite *Distomum haematobium*. Over the next 66 years, four other generic names and a couple of species names were used to identify the parasite first seen by Bilharz. The International Commission on Zoological Nomenclature settled the issue in 1917 by assigning it to *Schistosoma*.

The life cycle of the new group of parasites was just two years less perplexing to resolve than their generic name, even though Bilharz had successfully identified both the egg and the miracidium in the same year he first saw the adults. The eggs were described as terminally spined, the miracidia had cilia, and swimming occurred with a rotating motion. In his 1864 textbook, the acerbic and aggressive Spencer Cobbold made the first suggestion that a molluscan host could be involved in the life cycle. Nonetheless, a great many respected parasitologists championed the idea of a direct life cycle, including John Harley, an assistant physician at King's College and the London Fever Hospital. Harley became involved with the debate when he was sent urine samples from a South African physician and colleague that contained eggs with terminal spines. In 1866, Harley received a fascinating piece of information from another South African physician named Rubidge. In his letter to Harley, Rubidge reported the following. First, boys bathing in nearby freshwater rivers inevitably suffered from hematuria. Second, those who bathed in the sea rarely exhibited hematuria. Third, girls seldom bathed or swam in the freshwater rivers and rarely exhibited hematuria. In fact, Rubidge wrote to Harley, "I have not myself observed a single well marked case in this sex." Any modern-day epidemiologist with an interest in schistosomiasis would recognize these observations as virtually identical to any of a number of studies done in recent years in Africa. Harley, however, 'stuck to his guns' regarding the direct transmission. He believed that eggs were somehow directly inserted into the skin of the boys who swam in the freshwater rivers of South Africa.

The argument regarding the parasite's entry into humans continued unabated until the successful resolution of the *Fasciola hepatica* life cycle by Thomas and Leuckart in 1882. This triumph focused attention on the snail as a potential intermediate host/vector for schistosomes. Several investigators directed their efforts at exposing snails to miracidia, or dissecting snails in endemic areas, in hopes of finding intramolluscan stages of the parasite, or both. Failure to resolve the schistosome life cycle was the 'name of the game', until 1915. The European who eventually succeeded was named Robert Leiper. He succeeded where others had miserably failed for two reasons. First, he had recently been in Asia and was familiar with the success of two Japanese investigators regarding the *S. japonicum* life cycle, although he did not focus much attention on their success in his subsequent writings. Second, on his return from the Far East, he was posted to Cairo by the British War Office with explicit instructions to solve the problem of *S.* (*Bilharzia* at the time) *haematobium*, at least with regards to its life cycle. Lord Kitchener, the top Brit in Egypt, was concerned that too many British soldiers were being affected by this parasitic scourge and believed that resolution of its life cycle was an absolute necessity. After arriving and setting up their laboratory, Leiper and his colleagues undertook field collections of snails near a small village about ten miles north of Cairo. They were successful in isolating three types of fork-tailed cercariae from a species of *Bulinus* (misspelled as *Bullinus* at the time). They then used these snails as a source of cercariae to expose a variety of birds and

mammals. After an appropriate wait and necropsy, the elusive adult
S. (Bilharzia) haematobium was found in one rat and one mouse. They
repeated the experiment using rats, mice, and monkeys. All became
heavily infected. *Voilà*, at last there was success for the Europeans. They
knew that snails were the required intermediate hosts. Several others
repeated their work and the results were confirmed.

One of the early problems not recognized by Bilharz, or any of
the other European parasitologists for that matter, was that a second
species of *Schistosoma* was also present in Africa and, in some places, it
was even sympatric with *S. heamatobium*. Other reports soon appeared
that the second species was also present in the New World, a puzzling
discovery at first. Bilharz saw eggs of *S. mansoni* in 1851, but did not
make a distinction from those of *S. haematobium* even though he had
made drawings of both egg types. The first to openly suggest the existence of two species was Sir Patrick Manson. A colleague of his at the
London School of Hygiene and Public Health, Lois Sambon, then proceeded to formalize the recognition of the new species by naming it
for Manson. The criteria she used to described *S. mansoni* included the
presence of a lateral spine, the very different clinical manifestations
of infection by the two species, and the geographical distribution, i.e.,
S. haematobium was strictly African, while the new species was African
and occurred in the New World as well. The proposed description of
a new species drew the immediate and intense wrath of Robert Looss,
then one of the preeminent parasitologists in the world and a 'self-
styled' expert on the schistosomes. The vitriol extended by Looss (1909)
toward Sambon in his 1907 paper in the *Annals of Tropical Medicine and
Parasitology* was just slightly greater than Sambon's 1909 reply.

An interesting exchange between Looss and Sambon was recorded
in *A History of Human Helminthology* by D. I. Grove (1990):

> Looss to Sambon: "Among scientific workers, it is a good custom that
> anyone who believes he has made a new discovery also takes the trouble
> to prove it; it is not customary among scientists to assert something
> then call for the help of others to establish it."(Looss, 1909)

> Sambon to Looss: "I never for a moment placed myself on the same level
> in the latter respects (as a helminthologist) with the celebrated professor
> [Looss] of Cairo, but at the same time I would say that I have paid some
> attention to the subject, and cannot abandon my independence of
> judgment, or my right to give expression to my views." (Sambon, 1909)

And so it went with the scientific deliberations of the time as they
pertained to the schistosome life cycle. Perhaps if they had spent less

time dealing with nonissues, they might have been more successful than the Japanese. At any rate, other investigators quickly supported Sambon's hypothesis regarding two, separate and distinct species of *Schistosoma*.

As I have already noted, *Schistosoma japonicum* was the first of the three species to have its life cycle resolved. There were many reasons for this success, but three stand out among the others. First, the Japanese were familiar with the European literature. They were thus not required to make the same mistakes. Second, and more importantly, *S. japonicum* is not host specific. It is a classic zoonotic parasite. This permitted them a wide choice of experimental hosts to use during their studies. Third, when two Japanese collaborators began their life cycle search, they almost immediately found the correct snail host.

The discovery followed a clear sequence of events. First, Naganori saw the egg of *S. japonicum* as early as 1888 during an autopsy of a person suffering from "ascites and peripheral oedema." When histological preparations were made of the granular nodules on the surface of the liver, he found round eggs, but no spines on them, at least none that he could see.

Then, in 1904, Fujiro Katsurada conducted a thorough investigation of twelve patients suffering from a series of symptoms that were strikingly similar in character to those described by Naganori (Grove, 1990). Their illness was characterized by hepatosplenomegaly, diarrhea, anemia, weight loss, ascites, and peripheral edema. In five of these patients, he discovered the round eggs described by Naganori sixteen years earlier. After examining the eggs, the disease with which they were associated, their location within the body, and hatched miracidia, he concluded the parasites were most closely related to *S. haematobium*, "but not exactly the same" (Katsurada, 1904, 1905). In the same year, Katsurada found eggs of the same type he had recovered from a human, but this time in a cat. He also found adult worms in the portal system of the same cat. Later that same year, he described the parasite as *Schistosoma japonicum*. Others soon confirmed his findings.

Grove (1990) refers to the work of A. Fujinama and H. Nakamura as fundamental to resolving the life cycle of *S. japonicum*. The argument at the time was still whether the parasites gained access to their vertebrate hosts directly or by a percutaneous route. So, they designed a very simple experiment. They obtained seventeen cows from a source where the parasite was known to be absent. An area known to be endemic for the parasite was selected as the site for study. One group of six cows was fed food and water that had been boiled; when not feeding, the cows' faces were covered so they could not feed on native vegetation. Daily, three of these cows were forced to stand knee-deep in a rice paddy. The other three were made to stand knee-deep in a stream known to receive significant runoff from near-by rice paddies. A second group of seven cows had their legs washed in soap and alcohol, and then their legs were oiled and covered in a waterproof device. Four of these cows were permitted to feed and drink from the rice paddies and three were allowed to eat along the stream fed from rice paddies' runoff. Two other cows were allowed to feed and drink like those in the first group, but one cow was kept in a barn and the second cow was placed in the runoff stream for five and a half hours on one day only. The results were exactly as one would predict if percutaneous infection was the route of entry for the schistosome. They repeated the experiments using dogs and rabbits and the results were the same. Moreover, they showed that penetration of the infective stage and 'partnering' of male and female worms was accomplished in 23 days; furthermore, they determined that egg production by females began within eight weeks of initial exposure to the parasite. Other investigators quickly confirmed their results.

There was one last question to be answered, what happens to an egg once it is shed in human feces? The answer had eluded researchers since 1851. In 1913, K. Miyairi and M. Suzuki published results of a study that revealed the final secrets held so closely by these devastating parasites. The first thing they did was to locate an ox that was shedding eggs of the parasite, thereby providing them with a constant egg source. Next, they acquired some human feces along a rural roadside in an endemic area, and then scoured the area for snails, which they found (and even though they did not know it at the time, the snails they collected were species of *Oncomelania*, the snail intermediate host for *S. japonicum*). On return to the laboratory, the snails were isolated into separate containers. Newly emerged eggs from the ox were then placed with young, isolated snails. On emergence from their shells, miracidia immediately targeted the snails, became attached, and penetrated. After entering the snails, the miracidia shed their ciliated coat and transformed quickly into sporocysts. Within 12 days, a second generation of sporocysts was present (although they mistakenly referred to the second generation as rediae). After 32 more days, split-tailed cercariae were visible inside the daughter sporocysts. They attempted to infect mice with these cercariae, but all attempts failed. They returned to the field and collected the same kind of snails that had been used in their lab experiments. Some of these snails were shedding cercariae of the exact type they had obtained from experimentally infected lab snails. These cercariae from naturally infected snails were then used for exposure to mice. Ironically, one of their experimental mice was killed by a cage-mate three weeks after infection. On dissection of the dead mouse, they found adult male and female schistosomes. They repeated the experiment over and over, and each time the results were the same. For the first time, the entire life cycle of a schistosome had been completed experimentally, from beginning to beginning!

An early research effort, even before the parasites' life cycles were resolved, dealt with the pathology of the schistosomiasis. Over the years that followed, pathophysiology of the disease became a serious focus. A natural transition from these research areas led into the immunobiology of schistosomiasis and a number of fascinating discoveries were made. Some of the early work also directed attention at treatment, with relatively recent research culminating in development of praziquantel, a broad-spectrum anthelmintic also effective for the schistosomes. Other efforts have been directed toward developing a better understanding of the epidemiology of the disease caused by each of the three primary species.

In the time since the discovery and naming of the first three schistosomes, several new species in this group have been identified from a wide assortment of vertebrate animals, ranging from crocodiles to birds and mammals. With the advent of modern molecular tools, e.g., RAPD, PCR, RFLP, sequencing, etc., present-day investigators have begun to delve into evolutionary relationships among these new species and their hosts and, in turn, how they may be related to the three primary schistosomes in humans. A great deal of fascinating information has been gleaned from these efforts and more is unquestionably on its way. A recognized leader in this area of research has been David Rollinson and our conversation naturally drifted in that direction, the evolutionary biology of schistosomes.

Based on data generated over the years, it is known that there are four species groups that comprise the genus *Schistosoma*. In Africa, the first group includes four species, i.e., *mansoni, rodhaini, edwardiense*, and *hippopotami*; all possess lateral spines on the eggs, and all use species of *Biomphalaria* as intermediate hosts. Also in Africa there are eight species in the *haematobium* group. Included are *mattheei, margrebowiei, bovis, leiperi, curassoni, guineensis*, and *intercalatum*; these are the terminal-spined schistosomes and require *Bulinus* spp. as intermediate hosts. In Asia, along with *japonicum*, there is a group of four species, including *maylayensis, sinensium, ovuncatum*, and *mekongi*; these have a vestigial lateral spine, or none at all, and are transmitted by species of *Oncomelania* and *Tricula*. Finally, there are four more species that comprise the *indicum* group (*spindale, nasale,* and *incognitum*) located in southeastern Asia and India; they possess various egg morphologies and are transmitted via several types of pulmonate snails. Interestingly, although *S. spindale* is an Asian species, it is more closely related to the African, *S. haematobium.*

In reading several of David's papers, I was strongly reminded of something I knew, but about which I had really not given much thought. The schistosomes are all dioecious and that means genetic recombination will occur. In turn, this leads to increased genetic diversity. Another feature of recombinant biology is the potential for hybridization. I think one of the things that impressed me as David and I talked was the complexity of schistosome systematics in certain parts of Africa, for example in Senegal, and how hybridization exacerbates this complexity. He related, "It is an interesting thing that we have morphologically different parasites, which we can always identify by biochemical means, occurring in cattle that form viable hybrids. For example, you have *haematobium, bovis,* and *curassoni* in a certain area of Africa. *Schistosoma curassoni* and *S. haematobium* look, from the standpoint of egg morphology, very similar. *Schistosoma haematobium* occurs in man, but doesn't occur in cattle or sheep. We found *curassoni* and *bovis* in cattle and sheep. When they co-occur in Senegal and Mali, they form viable hybrids, with no mating preferences. Any male and female will come together. Then, other people started reporting it from different parts of West Africa and very similar situations were found in Niger, that is, the hybridization of *S. bovis* and *S. curassoni* occurring in cattle. Then, we had someone working in northern Nigeria on cattle and, lo and behold, *S. curassoni* and S. bovis were there as well. It's very rare in these situations to find *S. curassoni* by itself. So, we thought, what is happening here with these two species? This is a very strange situation. However, cattle move a lot in West Africa. They cross boundaries and they move from one area to another before they actually end up in an abattoir. So you can never actually pinpoint their origin, or where transmission might have occurred." But, I interjected, "You've got to have the snails around, don't you?" He responded immediately, "Absolutely, you've got to have the snails around. But this is an example of how, when you bring schistosomes together, they form hybrids of all sorts. And, we also know that when hybridization occurs that there will be a change in the intermediate host specificity. In studies that have been done in the laboratory, for the most part, if you have a schistosome that will develop in snail A and another that develops in snail

B, and they are restricted to A and B, respectively, the F_1 hybrid will be able to develop in snail A and B. So, it makes you wonder whether being fairly promiscuous, if you like, gives the parasite an advantage in its ability to adapt, or switch, to available strains, or even species, of intermediate hosts."

After listening to this account, I remarked, "Holy cow! I didn't realize things were this complicated when it came to schistosome reproduction and systematics." David responded, "It is extremely complicated! Within what we call the terminal-spine egg group, the *haematobium* group, the relationships between the different species is absolutely fascinating. We get hybridization between parasites that infect humans, for example, between *S. haematobium* and *S. intercalatum*, and that's another fascinating story."

I then told David about my interview with Darwin Murrell (Essay 13) and how the new evidence suggested the original source of *Trichinella* spp. was very likely a reptile. I asked him the same question for the schistosomes of birds and mammals. He replied, "Most likely a reptile as well. We have a strange sort of schistosome in a crocodile in Australia. My guess is that's where we should be looking, but that's a very rare occurrence."

He continued, "There's a lot more work to be done and we are doing some of it here, outside the biomedical group, in relation to determination of life cycles and identification of the primary hosts. We are using molecular phylogenies in attempt to better understand life cycles. New phylogenies are being created at a fast pace, and there is a lot of sorting out to be done. The question is, and I think this is what you are implying, how can we use these phylogenies and what questions can we help answer with these phylogenies? What can this information tell us about host switching, coevolution, etc.? Can information on species diversity help us distinguish even further, for example, between *S. haematobium* and *S. mansoni*? There are implications in looking for vaccines and that sort of thing. Why is *japonicum* so different than *mansoni* and *haematobium*?" When he made the remark concerning vaccines, I couldn't help thinking about the difficulty of using immunotherapeutic measures for the control of schistosomiasis in view of all the potential hybridization and genetic recombination problems.

The mention of vaccines by David brought me to another question raised in one of David's earlier papers. In 1986, he had predicted that GST 26 (Glutathione-S Transferase) and GST 28 antigens were the most likely antigens for the development of a schistosome vaccine. I asked if he still felt that way. He responded, "Unfortunately, no. It's not something that we've worked on here at all. But Andre Capron's group had made tremendous progress with GST-based vaccines. They even went as far as trials in Senegal. But they ran into financial problems. The final phase is going to be an extremely expensive exercise, and I'm not sure they had the backing or if the results warranted their moving further forward. Interestingly, from a biological perspective though was the discovery in Zambia working on *S. matthei* that the GST vaccine was more of an antifecundity vaccine. It seemed that it reduced egg laying by the female worms. I believe there is still some work going on in China with the GST vaccine on animal schistosomiasis."

I mentioned Peter Hotez's work with a hookworm vaccine and that part of his strategy was to target the L_3 and the adult stage. I asked, "Wouldn't it be logical to target the skin stage of the schistosome?" He responded, "Yes, it would be very logical. The same thing is true for the schistosome vaccine. There are different groups working on different stages. There is a very good group at the University of York, for example, that is primarily concerned with the lung stage. Of course, the most successful approach has been to use irradiated cercariae, which gives very good protection but unfortunately cannot be used in humans. So, at the moment control is based around chemotherapy. One hope is, I suppose, information regarding the schistosome genome that will be published shortly will inspire people to take on some of the ideas and look at some of the data and actually begin to identify genes worthy of closer study."

At this point, I asked him if he was working with the World Bank and he replied, no, not directly, but they acted as a World Health Organization Collaborating Center. I then told him about my discussion with Don Bundy regarding the World Bank's use of teachers in schools as a way of delivering drugs to school-aged children in their deworming programs in several parts of the world, primarily sub-Saharan Africa. He laughed and said, "I actually had this discussion last week because that is exactly what we are doing in Zanzibar, using teachers in the schools to deliver the drugs to the children. We treat for intestinal worms as well as schisto, so we are giving out albendazole as well as praziquantel. The campaign is called 'Kick Out Kichocho' [kichocho is Swahili for schistosomiasis]. It is primarily based on chemotherapy in the first year, bringing in health education, but also bringing in this question of understanding transmission, because the reason this program got off the ground to start with was the work we did with the intermediate snail host. It was a Ph.D. dissertation on two species of *Bulinus*, *B.globosus* and *B. nasutus*, which brought us to this point in our effort to eliminate schistosomiasis on Zanzibar. Interestingly, you only find schisto on parts of the island where you find *B. globosus*. Contrary to what we thought in the past, the other *Bulinus* species is not involved with the transmission of schisto on the island. So, what you are actually doing is giving the teacher a health care responsibility and another job to do. We've had workshops and trained teachers to be aware of what they are doing, to be aware of what the different drugs are. For example, you give only one albendazole treatment, but you give praziquantel according to height or weight., actually height in this case, using a dose pole." This is the same technique described to me by Don Bundy for programs sponsored by the World Bank. He said, "This is the best method because scales are not always available and not always accurate. In these kinds of communities you can get a fairly good idea of the drug required according to a child's height." I had thought when Bundy first described the 'stick', "how ingenious!" My thinking was re-enforced when David told me they used the same 'stick'.

David continued, "This is a local program. It's a country-to-country orientated program. We have, based in London, another Gates Foundation funded program, called the Schistosomiasis Control Initiative, which is actually hoping to control schisto in six African countries, mainly by the delivery of praziquantel, and also intestinal worms as well." I asked if there is a research component to this effort. He replied, "A small program that was run through Harvard, but this was recently changed. There is money there, through small grants, mainly for people in the countries being treated. The countries included are Uganda, Tanzania, Zambia, Niger, Mali, and Burkina Faso. The program is having a big impact on morbidity caused by schistosomiasis and is leading to new integrated efforts on the neglected diseases."

David has been a systematist and a phylogeneticist throughout his professional career, and he still is, but now, even though he continues with his basic research, he has turned some of his attention to treatment and the control of schistosomiasis in many of the same countries that he has done some of his best basic research. Has it been rewarding? "It's been an eye opener for me," David explained. "It's been great actually because I'm having a different impact. It's a different perspective. You also get a feeling that things are working when you have been involved in treating 132 000 school children, as in Zanzibar. And, so, something has happened because of the project." I asked, "When will you see how much of an impact you have had?" David responded, "It's a four year project. We've done a six month follow-up. We've got 24 schools involved right from the beginning as a baseline, so we are monitoring those schools every year. I sense already that we are having an impact in these schools. These schools are operating in communities that are not at the moment being touched by treatment, so we can actually lower prevalence and intensity of infection in school children. We can begin to reduce what happens on the transmission side. At the same time, we've got village communities that are not being treated. So, to have a long-term sustainable impact, these communities will have to be looked at as well, unless we do other things. And one of the ideas here from a biological control point of view is where we have a snail that is not susceptible and we have a snail that is susceptible, and they occur in different parts of the island, can we do any sort of transfer so that we can replace the susceptible snail? Let's see if a little competition will work."

At this point, I wanted to shift direction and head in another way. I had read some of David's work and in one paper (Rollinson *et al.*, 1997), the authors said, "The schistosome genome will contain a number of conserved genes which show a high degree of homology to those of their mammalian hosts. This may reflect not only the conserved nature of the genes in question, but also the importance of key molecules in maintaining the host–parasite relationship." I then said to David, "For me, this almost sounds like a fait accompli when it comes to horizontal gene transfer between host and parasite. Would you comment and do you agree?" He responded, "Well, I think the jury is still out on that one. There is some very interesting work coming out from Japan being published in relation to horizontal gene transfer between the definitive host and the parasite. The genome database now is actually enormous and growing. It's a job that needs to be done. There will be an opportunity for better analysis between the parasites and the mammalian host. We're still studying the molluscan genome and when that comes out, and I'm sure it will, and you've got the genomes of all three players, you'll be able to make better comparisons of what's what. But gene transfer at the moment, well, that's a long way to go as far as what's what. However, you are still left with this parasite that's got this amazing ability to live for a very long period of time in a most hostile environment, masking itself with either host antigens, or by mimicking this, that, and the other. There's got to be things going on." He repeated, "There's got to be things going on." I came away from this answer that "things are going on", and it sure sounds an awful lot like gene transfer – I said that, not David.

I shifted target again by asking, "Since some of these schistosome species are zoonotic, doesn't this present a special problem for their eradication in endemic areas?" His response was an emphatic, "Yes, a huge problem, particularly with *S. japonicum*, which has a *large* [his emphasis] number of mammalian definitive hosts, and domesticated animals, particularly water buffalo. With *haematobium*, it is not a problem and with *mansoni*, you do get areas where there are some small problems with other primates, with baboons being involved."

Continuing, David then said, "Changing the emphasis, the biggest problem I think is that we have one very effective drug and that's praziquantel, but we only have one." I responded, "It creates a problem with resistance, doesn't it?" He responded in the affirmative and I asked if there was any evidence for resistance to praziquantel yet. He replied, "Yes, there is some evidence for changes in tolerance. It's very difficult to talk about true resistance yet. So far, it just means that a parasite can take more of the drug without being killed. There are examples of this in Egypt and Senegal, but it's very difficult to say much about it because the drug only kills adult worms. It doesn't kill developing worms and it works better in conjunction with the immune response. In areas of high transmission, like in Senegal, egg production seems to continue sometimes even with treatment with praziquantel, probably because the child is returning to water and being reinfected, and also because developing worms are not being killed by praziquantel treatment. So, very quickly, the egg production and excretion pattern becomes re-established. However, even though we have seen relatively rapid development of resistance among worms of veterinary importance, there is a general feeling that the same thing will not happen with schisto and praziquantel because it is not blanket coverage. There are pockets of schisto scattered around that are not being treated all the time. It's not like treating a herd of cows all at the same time. In other words, there are refugia in which the drug is not being used."

The last question I asked David that day in May was again related to resistance and praziquantel. I told him of a story I had heard about the use of atabrine during WWII and that I was not certain of its veracity. It seems that both the Japanese and the Americans were using it in the southwest Pacific to prevent the acquisition of malaria, except that the Japanese were using it in suboptimal doses, which led to resistance to the drug in certain localities. I asked first if this scenario was plausible and, second, was there any risk of the same thing happening with praziquantel and schisto? David replied, "Yes, definitely, it could happen, because you can actually select for resistance under laboratory conditions. It is a definite possibility, but there are lots of other sorts of things involved. There is no reason, no theoretical reason, why we shouldn't see the evolution of resistance to praziquantel. As I said earlier, we have seen one or two examples of increased tolerance to the drug already. You have to balance that with the practicalities of going forward with treatment, because one argument in its favor is that we possess a very powerful tool, so we should use it. That's where the practitioners for control are moving ahead. Those of us more on the research side would say, sure there is this possibility and we have to monitor it. We have to make sure that control programs have adequate surveillance going on. As importantly, we must continue with work on the schistosome genome project, which might lead to the next generation of control."

Then, I think I threw him for a loop, because when I asked him what the next generation of controls would be, he paused and sighed before he answered. His response, "Who knows? I think the best way forward would be an effective vaccine." I then offered, "In combination with an effective drug?" "No," he said, "ideally, the best way would be with a vaccine alone. That would be so much better. However, the chances at the moment seem very slim. The research community is not there." I then asked, "Do you think an effective vaccine is really practicable with all of the strain variation that you are seeing?" He replied, "I am not an immunologist, but it amazes me that when you are killing the worm with chemotherapy and releasing all those antigens into the system there is little effect on the development of immunity. And then, why does it take up until an adolescent age for any form of immunity? And you are still left with these very skilful worms masking themselves. So vaccines don't seem particularly likely at the moment, but in terms of an effective control tool, I still think a vaccine is the best way to go. Of course, the long-term solution remains the provision of safe and clean water, and general improvement of the quality of life in endemic areas."

This ended our afternoon. Several things struck me as I reflected back on our conversation. One of them was the incredible variation among the schistosomes in terms of their ability not only to colonize an individual host, but in their ability to colonize new territories and new snail hosts in both ecological and evolutionary time frames. Of course, I was aware of these features in their life cycles since the first time I took a parasitology course nearly fifty years ago. However, hearing it from an expert like David made everything I already knew seem so much more real, and the things I hadn't known so very much more fascinating. Another thing I was interested in was the contrast in approach taken for the control of schistosomiasis and hookworm disease. In the first case, a vaccine seems like it may be intractable, making chemotherapy the appropriate route, while for hookworms the approach employed was to develop a vaccine and avoid chemotherapy as the sole weapon of choice (at least this is the Hotez strategy for hookworms). In each case, it seems like the parasite's biological strategy has dictated the nature of the control used. This feature is one of the best examples I can think of to illustrate the incredible adaptations made by parasites to their absolutely unique way of making a living. It reminds me of the reason for my own choice in wanting to spend my life studying the nature of host–parasite relationships. As I listen to more and more of these parasitologists tell their life stories and of their 'addiction' for parasites and parasitism, I become more and more impressed with their incredible dedication and devotion to these animals and their lifestyles.

REFERENCES

Grove, D. I. 1990. *A History of Human Helminthology*. Wallingford: CAB International.

Katsurada, F. 1904. *Schistosoma japonicum*, ein neuer menschlicher Parasit, durch welchen eine endemische in verschiedenen Gegenden Japand verusacht wird. *Anotationes Zoologicae Japonenses* **5**: 146–160.

Katsurada, F. 1905. *Schistosoma japonicum*, a new human parasite which gives rise to an endemic disease in different parts of Japan. *Journal of Tropical Medicine and Hygiene* **8**: 108–111. (Translation of Katsurada, 1904).

Looss, A. 1909. What is *Schistosoma mansoni* Sambon 1907? *Annals of Tropical Medicine and Parasitology* **2**: 153–192.

Miyairi, K. and M. Suzuki. 1913. (On the development of *Schistosoma japonicum*.) *Tokyo Iji Shinshi* No. 1836, 1–5. (In Japanese.)

Rollinson, D. L., A. Kaukas, D. A. Johnston, A. J. G. Simpso, and M. Tanaka. 1997. Some molecular insights into schistosome evolution. *International Journal for Parasitology* **27**: 11–28.

Sambon, L. W. 1909. What is *Schistosoma mansoni* Sambon 1907? *Journal of Tropical Medicine and Hygiene* **12**: 1–11.

12

Dicrocoelium dendriticum and Halipegus occidualis: their life cycles and a genius at work

Earth knows no desolation.
She smells regeneration.
In the moist breath of decay.
'Ode to the spirit of earth in autumn', George Meredith (1828–1913)

One of the best parasitologists in the early part of the twentieth century for making discoveries involving parasite life cycles was Wendell Krull. The late, and great, Miriam Rothschild referred to him as "a genius who hid his light under a bushel" (Ewing, 2001). Part of her admiration for Krull stemmed from his research on the life cycle of *Halipegus occidualis*, the hemiurid fluke that lives under the tongues of North American ranid frogs. Others might consider him as a genius for his contribution in resolving the life cycle of *Dicrocoelium dendriticum*.

Wendell Krull is now dead, so I had to rely on other resources for special information regarding his career and some of the research he accomplished. In addition to a number of Krull's papers, one of the important biographical sources for this effort was a book written by Sidney Ewing, entitled *Wendell Krull: Trematodes and Naturalists*, published in 2001. It really is a delightful read. Over a period of eleven years, including his three years as a graduate student, Sidney had many conversations with Krull regarding his life's work. I have unashamedly quoted from Ewing's book throughout the essay. I also had the wonderful opportunity of interviewing Sidney Ewing and his wife, Margaret, who provided me with even more insights regarding Krull, including Margaret's experience in Krull's veterinary parasitology course at Oklahoma State University.

Wendell Krull plied his trade mostly in the first half of the twentieth century. He was born in Tripoli, Iowa, in December 1897. After service as a corpsman in the U.S. Navy from 1917 to 1920, he completed an undergraduate degree at Upper Iowa University in Fayette in 1921. He had a couple of years as an undergraduate student at Cornell College in Iowa before entering the Navy, which is why he graduated in just one year after returning from the service. He then obtained a Master's degree at the University of Iowa in 1924. While pursuing graduate work at Iowa, he was a teaching assistant in invertebrate zoology. Apparently, this is where, and when, he developed an interest in parasites. However, after obtaining his Master's degree, he delayed going further in graduate school, apparently preferring instead to gain some experience in academia, I guess to see if this was his 'calling' (it wasn't, at least during the first half of his career).

His first 'real' job was as Head of the Zoology Department at Northwestern College in Naperville, Illinois, with a 'grand' salary of $2300 per annum. The next year, he taught at Kansas Wesleyan in Salina, Kansas, where, incidentally, he met his future wife, Nellie Godard. After one year at Kansas Wesleyan, he and his new bride headed to the University of Michigan and the irascible George R. LaRue, with whom Krull began work on his Ph.D. degree.

When Krull finished his Ph.D., he did not return to academia, and wouldn't for several years. He instead opted to do research for the government. He left for Washington, D.C. and the Division of Zoology in the Bureau of Animal Industry (BAI), then part of the U.S. Department of Agriculture. His boss and major influence at BAI was Maurice C. Hall, who had studied parasitology with Henry Baldwin Ward at the University of Nebraska before Ward left for the University of Illinois. Another parasitologist, actually a British veterinarian, who greatly influenced Krull during those early years, was the very exacting Albert Hassall. Among other things, Hassall had teamed with Charles Wardell Stiles to found the *Index-Catalogue of Medical and Veterinary Zoology* and the U.S. National Parasite Collection, now one of the world's largest repositories of parasitological material, and a superb working collection. These times also were among the most productive of Krull's career. Examination of just those papers in his curriculum vitae that deal with trematodes and their life cycles shows 62 publications between 1929 and 1937. From a professional perspective, it is clear that Krull made a wise decision when he opted for government service. When I asked the Ewings why he chose this course rather than an academic one, neither could say for sure, except possibly it had to do with the depression years.

In 1938, he and Nellie left Washington, D.C. and headed west. Krull had been promoted to Chief of a BAI unit in Logan, Utah, where he was to stay until 1942. It was while in Utah he decided that a career change was in order. He made up his mind that he wanted to study veterinary medicine and work toward a D.V.M. degree. Ewing said in his book, "he did tell me that he never regretted the decision and that his colleagues in government service were long disbelieving." He added, "It would not be the last time that Krull was to make a bold professional move." During our interview, I asked the Ewings why he chose to go back to school after such a successful career as a government researcher, another very important career decision. Sidney replied, "He told me that he had really become interested in domestic animal parasites working for the BAI, and somehow he thought that if he was a veterinarian he could work more with them."

The tough decision to enter the Auburn University School of Veterinary Medicine in 1942 was easy compared with actually obtaining the degree. It was not that Krull could not handle the work; after all, he was an exceptionally bright man. One of the more serious obstacles he faced was in financially supporting himself and his wife, Nellie, while he was in school full time. Because of his background and experience, however, he managed to secure a position teaching parasitology to his fellow students at the same time he was participating in the ordinary class work of a full-time veterinary medicine student. It was part of the quid pro quo for matriculation, and he actually accepted the responsibility with considerable alacrity. Fortunately, that first year, Will Bailey, who had just completed the D.V.M. program at Auburn, assisted him in the laboratory portion of the parasitology course. Bailey was to spend his entire career at Auburn, teaching parasitology, developing a very good research program, and eventually becoming one of the University's top administrators in his later years.

Although Krull enjoyed the teaching requirement and the classroom, he did not care for the southeastern part of the country, having grown fond of the wide-open spaces of the western U.S. during his four years in Utah. Sidney also said, "he didn't like the heat and humidity." So, after a year at Auburn, he and Nellie moved to Fort Collins and the Colorado State University (CSU) School of Veterinary Medicine, again becoming a part-time instructor in parasitology while continuing his studies. He graduated in 1945. He chose not to practice veterinary medicine, but instead turned to academia. As Ewing described it, Krull advanced "from part-time instructor to Professor and Department Head (Zoology) and Parasitologist (Agricultural Experiment Station) at CSU in a single academic year", really quite an achievement!

Krull remained at CSU for three years, then took his talents to what was then Oklahoma A & M. (now Oklahoma State University, OSU) in Stillwater, where he became Professor and Founding Head of the Department of Veterinary Parasitology in their fledgling veterinary school. However, to justify the salary paid because of his stature and experience, he also received a joint appointment as Professor and Head-Designate of the Zoology Department in the School of Arts and Sciences. The latter responsibility he held for just a few years, but he remained as Head of Veterinary Parasitology until he was forced into retirement in 1964 at the age of 67.

The early forced retirement greatly irked Krull because he had been persuaded when hired that he could continue working until he reached 70. However, for some unknown reason, the Board of Regents at OSU drafted a new rule regarding retirement age and Krull was caught. Even though he was bitterly disappointed, he immediately secured a position at the Animal Medical Center (AMC) in New York City where he planned to launch a research effort on "the unknown aspects of the dog heartworm (*Dirofilaria immitis*) life cycle." The plan was abandoned when he was informed that officials at the AMC would require him to use naturally infected animals to conduct his research. This edict contravened his projected research protocol, so he resigned.

Soon, however, he was 'happily' hired by the veterinary school at Kansas State University (KSU) to teach parasitology (at the behest of Sidney Ewing who was by then a faculty member at KSU), which he did until he and his wife Nellie retired to the warm, dry deserts of Arizona at the age of 70. Following retirement, he was called back into service by OSU to teach one more semester of parasitology before he died of cancer in 1971 in his 74th year.

On reading Sidney Ewing's biographical sketch, I was greatly struck by Krull's superb observational skills. He was a very gifted field biologist, a real naturalist in every sense of the word. This conclusion is easily reached by examination of his curriculum vitae. His first publication was in 1929 and the second in the following year. Both of these papers came from his dissertation research. Some of this work was conducted at the University of Michigan Biological Station, situated on Douglas Lake near Pellston, in the northern part of Lower Michigan, and part of it in the environs of Ann Arbor. His first publication described a new technique for the rearing of adult dragonflies from the egg stage in the laboratory. In no small way, this procedure allowed him to then successfully pursue research for the second paper in which he described the life cycles of *Pneumonoeces* (= *Haematoloechus*) *medioplexus*

and *P.* (=*Haematoloechus*) *paraplexus*, both of which required dragonflies as the second intermediate host. To achieve success with these
life cycles, he was required to expose uninfected dragonfly nymphs to
the cercariae of both trematode species. Since he could not be certain
of obtaining 'clean' dragonflies in the field, it was essential that he
develop the method for rearing them from eggs in the lab, so, this is
what he did.

At first glance, the rearing of a dragonfly in the lab may seem
like a relatively easy and almost trivial task, but this is certainly not
the case. Anyone who has raised, or maintained, invertebrates in the
lab knows this sort of effort is in the same 'ballpark' as growing parasites in culture. In vitro culture and rearing invertebrates in the lab
are not easily accomplished. Since I have also attempted to raise certain invertebrates in the lab over the years, my feelings about growing,
culturing, or maintaining certain invertebrates are almost parallel to
the way I think about in vitro culture. I have thus always believed
that studies on parasite life cycles border on sorcery, or alchemy. Some
would disagree. For example, Miriam Rothschild was once asked if she
believed the solving of "trematode life cycles was faintly supernatural,
tinged with an element akin to black magic, or monstrously favored
by a breakdown in the laws of chance?" She responded, "Wendell Krull
published his paper on the life cycle of the two frog lung flukes – and
I knew the baffling crowded space was certainly not filled with luck."

I have read a great many parasitology papers, as well as most of
the past presidential addresses and Ward Medalist lectures published in
the *Journal of Parasitology*. As I implied above, I had previously concluded
that a bit of luck must be required to successfully determine the life
cycle mechanics of a digenetic trematode, or any other parasite for that
matter. When one thinks about the resolution of helminth life cycles in
North America during the twentieth century, a small cadre of successful
pioneers comes to mind immediately, i.e., Horace Stunkard, Will Cort,
George LaRue, Lyell Thomas, Wendell Krull, etc. Each of these men was,
beyond any other talent they may have possessed, a great naturalist. I
have concluded since I began reading and writing about some of these
men and the research they accomplished that being a great naturalist
is much more important than being very lucky.

However, after a substantial amount of cogitation on the 'discovery' of parasite life cycles and those who focus on their resolution, I have
also decided that there were at least two more factors that encouraged
success by these great parasitologists. The first was tenure. The second was an ample supply of patience. In the first instance, the almost

archaic academic practice of giving tenure is essential for those who tread on the 'thin ice' of trying to define a parasite's life cycle from 'scratch'. The reason is simple. Too many things can go wrong in the lab when simultaneously dealing with any of the several potential intermediate and definitive hosts and the various life-cycle stages of a parasite. In addition, it may take a long period of time for success to accrue, if at all. Without success in life-cycle work, there will be no publications, and without publications, there will be no promotion, and without promotion, tenure is highly unlikely. This is a simple fact of life in the academic profession. A local wag once told me that the only thing academic deans are really good at is counting (an accurate statement in my experience!).

The second requirement for success in life-cycle studies is patience, something that is certainly not part of the psyche of all parasitologists. In my opinion, endurance is crucial to what ultimately breeds success for the naturalist. For example, many biologists will not spend the time necessary to understand the biology of an invertebrate on which they may be working. Many others, simply put, do not have the 'feel' of a naturalist, i.e., they do not understand the need for treating their invertebrate subjects with 'tender loving care'. For example, Krull noticed during one of his studies that some of his snails kept crawling out of the water in the aquaria in which they were being kept, then dying. He attacked the problem and, after much careful observation, he discovered that the quality of the food supply was inadequate. He was able to make adjustments in their feeding regime and the snails lived. Ewing quotes Krull, "In order to do accurate work on the larval phases [of trematodes] it is necessary in many cases to have snails of various ages which are free from infection. A dearth of information involving the snail as a host is apparent from the life history work which has been done on snails. I dare say that this situation is owing to the lack of information which exists in the field of conchology and the lack of patience and time on the part of the helminthologist." Some of these snail problems have been resolved over the years through the contributions of great malacologists such as Henry VanderSchalie, Jack Burch, Elmer Berry, Eli Chernin, and many others. But many difficulties persist. These await the skill and perseverance of a naturalist like Wendell Krull.

If I were guiding the dissertation work of a Ph.D. student, would I suggest the life cycle of a digenetic trematode as an appropriate research problem? If I were a young Assistant Professor pursuing tenure, would I risk pursuing success in resolving a parasite's life cycle? Would I risk it? As intimated previously, such a project is fraught with potential difficulties, so I am not sure that I would, in either case for that matter. The question surely must have occurred to both LaRue and Krull, but according to the latter's biographer, Sidney Ewing, there was no personal communication from Krull to him (Ewing) that there was any doubt about whether to proceed. He just did it, and with consummate success.

Krull's second paper, the one describing the life cycles of the two trematode species, was clearly a harbinger of things to come. From 1929 to 1960, Krull published some eighty papers dealing in one way or another with the life cycles of digeneans and their hosts, a clear indicator of his skill in this area of parasitology. It is of interest to note that of these eighty publications, only five had coauthors, and only once was another person the senior author. From a personal perspective, this observation tells me several things about Krull. First, he had to be an independent thinker and capable of working alone. Several times during our conversation, both Sidney and Margaret Ewing referred to Krull as "a loner," in fact, I think you could say that both of them *emphasized* this personality trait. Second, he had to be a superb naturalist. Both Sidney and Margaret stressed this skill in their interview. Finally, it is also possible that Krull was somewhat cantankerous and found it difficult to work in collaboration with others. Miriam Rothschild alluded to this possibility in her chapter in Ewing's book. But for her, and for me, this last facet of his personality is irrelevant. In science, success is what really counts.

The variety of digeneans on which Krull worked over the years was enormous. It included at least 28 species. There were two, however, that stand out among all the others. The first of these is *Halipegus occidualis*, a hemiurid fluke that occurs under the tongues of ranid frogs, and that involves a complicated, four-host life cycle. The second is the lancet fluke, *Dicrocoelium dendriticum*, in the livers of sheep. Whereas my students and I have spent a great deal of time in a small North Carolina farm pond working with *H. occidualis* since 1983, I was quite fascinated with a chapter in Sidney Ewing's book, written by Miriam Rothschild, and entitled, 'Homage to Wendell Krull'. Under this title is an epigram that reads "*a genius who hid his light under a bushel*[*]". At the bottom of the page, the footnote for the asterisk says, "The container capable of holding a bushel (eight gallons)", presumably an annotation for those who have no knowledge of the nonmetric system used in North American agriculture for quantifying crop production. The late Miriam Rothschild was also one of the really great naturalists of the twentieth century, a Fellow in the Royal Society, and, incidentally, an Honorary Member of the American Society of Parasitologists. Rather than describing my own interest in *H. occidualis*, or that of my students, I would rather excerpt some of the comments made by Rothschild in her tribute to Krull. Then, I wanted an answer to the following question. Was his success with *H. occidualis* and *Dicrocoelium dendriticum* serendipitous, or even partly so? After carefully reading Ewing's book and Krull's papers, I think I have an answer.

   Halipegus occidualis was first described by Stafford in 1905. In 1935, Krull published its life cycle. In doing so, he produced a truly fascinating picture of the way in which a digenetic trematode can circumvent just about all of the possible odds against its chances for survival. Most digenean life cycles involve three hosts, although a few have two. However, not only does *H. occidualis* require four hosts to complete its cycle, one of them is a 'paratenic' host, an organism used to bridge an ecological, or trophic, gap between two other hosts in the life cycle.

   Not being present at the time, and not having spoken with Krull about his thinking process, it is not possible to place myself in his mindset as a way of understanding how he approached the problem. If you read his paper carefully though, the answer is very clear. First, he knew that *H. occidualis* is a hemiurid fluke, and that most hemiurids are parasites in the stomachs of marine fishes. I am certain that no other hemiurid life cycle was known at the time, so Krull was indeed breaking new ground when he pursued *H. occidualis*. However, there were several other pieces of exceedingly useful information about which he was aware. It was this information and how he used it that allowed him to unlock the secret of this life cycle and, I suspect, with relative ease. Among other things, he knew that the parasite's definitive hosts were green frogs, and that green frogs eat dragonflies. He was also aware that these arthropods had to be involved in the parasite's cycle since he had previously isolated metacercariae of *H. occidualis* from naturally infected dragonflies. Then, in an abstract published in 1932 as part of the ASP's annual meeting program in the *Journal of Parasitology*, Lyell J. Thomas reported observing *Cercaria sphaerula n. sp.* infecting *Cyclops vulgaris*. Thomas correctly identified these cercariae as those produced by a hemiurid fluke and that they had emerged from *Helisoma trivolvis* isolated in the Douglas Lake region of Michigan. Krull was aware of this abstract because he cited it in the 1935 publication in which he described the life cycle of *H. occidualis*. I am certain that Krull made the connection between Thomas's cercariae and those of *H. occidualis*. So, even before he began experimental work on the parasite's life cycle,

Krull had all the 'dots' on the parasite's life cycle map. What remained was for him to connect them.

The 'tack' he used was typical 'Krullian'. He raised his snails, *Helisoma antrosa* (= *anceps*), from eggs in the laboratory. His snails were, accordingly, free of infection. Ostracods were also reared in the lab and, therefore, also free of infection. Green frogs, *Rana clamitans*, were collected as tadpoles and maintained in the lab until they had undergone metamorphosis to the adult stage. Before starting any feeding experiments, 45 of these frogs were necropsied to confirm they were without parasites, and, naturally, being reared in the lab they were all uninfected.

He began his lab experiments by obtaining eggs of *H. occidualis* from one of five adult parasites collected from under the tongue of a green frog collected in the area of Beltsville, Maryland. These eggs were used to expose six small *H. anceps* (he described them as one fourth to one third grown) in a covered stendor dish containing filtered water and calcium carbonate. Five of six snails died before shedding cercariae, but all contained intramolluscan stages of a trematode, including both sporocysts and rediae. Based on his description, it can be assumed that the snails died because they had been exposed to too many parasite eggs. Thus, the hepatopancreas of each snail was ravaged by larval stages of a trematode. The organ was virtually nonexistent on necropsy, being replaced almost entirely by developing parasites. However, the sixth snail did not die and began shedding typical hemiurid cystophorous-type cercariae 94 days postexposure.

Krull also found that the snail must eat the eggs of *H. occidualis*; they do not hatch and release miracidia as with so many other trematodes. Exposure of ostracods to cercariae was a simple task and successful infection in the lab was easily accomplished. In fact, he discovered that, when given the opportunity, his ostracods gorged themselves with cercariae. Finally, he obtained a metacercaria from a naturally infected *Libellula incesta* in the field, fed it to a frog, and 22 days later, an adult *H. occidualis* was present under the tongue of the green frog. Except for one step that I will mention a bit later, Krull had successfully completed the life cycle of *H. occidualis* in the lab!

I can understand why Miriam Rothschild was so impressed with Krull's skill in deciphering the parasite's life cycle. He made it seem easy. In reality, it was easy. Why? As pointed out before, it was because, a priori, he had first made several key observations regarding the biology of the parasite and the hosts. He had done some very careful groundwork before he began the study. Krull was not guessing about the steps involved. He already had a very good notion of what he was going to find.

As an aside, I was interested that he made no effort to expose dragonflies to infected ostracods in the laboratory. Neither Ewing, nor Krull, offered an explanation for this omission. One of my former graduate students, Derek Zelmer, was successful in achieving success with this step. I am sure others had accomplished it before Derek, but he also made a rather interesting observation regarding development of the parasite in the ostracod, one that I am sure would have pleased Krull. Thus, Derek was able to show that if infected ostracods were maintained in the lab for a long enough period of time, the parasite would mature to the metacercaria stage, which he then used to successfully infect green frogs, thus bypassing the paratenic dragonfly host. I should add that Derek believed, and I agree, this process never occurs in the field because ostracods are not a normal prey item for green frogs. In fact, we both doubt green frogs can see ostracods, let alone eat them.

In reading Krull's 1935 paper, one of the most fascinating aspects of the parasite's biology had to do with the cercaria morphology and the manner in which the parasite accessed the ostracod's body cavity. This process also held great fascination for Miriam Rothschild, in part, I think, because during her years at the Plymouth Marine Laboratory in England, she had described a new hemiurid, *Cercaria sinitzini n. sp.*, in Novitates Zoologicae (Rothschild, 1938). Even though hemiurid cercariae are without tails, their complexity is significant, and I think this is what really grabbed Rothschild's attention.

Cercariae of *H. occidualis* are of the cystophorous type. They are spherical in shape, without a tail, but with a tiny appendage that resembles a handle. An elongated cercaria body and a delivery tube are folded inside a "double-walled transparent cyst, enveloped by a thin membrane which follows [the] cyst wall rather closely on [the] side opposite [the] handle, with space between [the] membrane and cyst wall increasing towards the region of the handle." (Krull, 1935). In effect, the body of the cercaria is buried inside the tail, meaning we have a trematode larva that does not swim. Swimming cercariae normally remain infective to the next host in the life cycle for 24–36 hours, but use the time to distribute themselves spatially in such a way as to increase their chances of locating the next host. In contrast, the nonswimming cystophorous cercariae of *H. occidualis* remain infective much longer, increasing their chances of transmission to ostracods. Thus, some cercariae remain infective to their potential ostracod hosts for up to

18 days after emerging from the snail. In both swimming and non-swimming cercariae, the limiting factor in survival is probably their supply of glycogen, since both larval types are nonfeeders.

On emerging from the snail, the cystophorous cercariae drop to the substratum and lie motionless. When they are eaten by an ostracod, the caudal appendage (handle) is manipulated by the ostracod's mouthparts. Within a few moments, the delivery tube is explosively extruded and the microcrustacean's gut wall is penetrated immediately. Literally, in a fraction of a second, the body of the cercaria is hurled through the delivery tube and into the ostracod's hemocoel. Within the body cavity, the anhydrobiotic body of the parasite immediately imbibes water and grows rapidly in size.

There are several interesting features about how this happens. First, entry into the body cavity is apparently a rather traumatic experience for the ostracod. In his 1935 paper, Krull describes the process as follows: "occasionally when the cystoid cercaria is discharged into the mouth of the cyclops the latter makes a terrific spurt for a moment, then lies motionless with the appendages widely separated, as if dead on the bottom of the container for as long as a minute in some cases. While it is in this condition a part of the delivery tube with the attached cyst of the cercaria usually projects from the mouth. The ostracod suddenly regains its equilibrium and appears normal except that it now contains an active larval fluke which is almost always in the body cavity and only occasionally in the intestine." Based on Krull's description, it almost resembles the way someone might be expected to respond to a hard blow to the solar plexus, e.g., in having their 'wind' knocked out.

Second, it seems that smaller ostracods are more vulnerable to successful transmission than large ones. Krull indicated that both large and smaller ostracods feed voraciously on the motionless cercariae. However, in large ostracods, the discharged body of the cercaria does not enter the gut wall, apparently because the delivery tube is not long enough to reach its target. Instead, these cercariae enter the intestine and are digested. Krull counted the bodies of as many as sixteen cercariae in the distended intestine of a single ostracod. He also noted that if one of the motionless cercariae is missed by the ostracod on the first foraging foray, it would surely be picked up on the second. It almost makes one wonder if the parasite is not releasing some sort of chemical agent to attract the ostracod.

Third, the expulsion process can be easily stimulated using a probe and watched with an ordinary dissecting microscope. Superficially, manipulation of the handle appears to cause the 'explosion'.

However, a very elegant study by Derek Zelmer indicates that the explosive force is produced by a rapid osmotic change inside the cyst when the outer membrane is breeched. The only other biological phenomenon of which I am aware that is even remotely similar to the release of the cercariae body from the membranes comprising the cystophorous tail of *H. occidualis*, is ejection of the poisonous barb of a cnidarian's nematocyst (= cnida). In the latter case, Brusca and Brusca (2003) indicated the velocity of release by the nematocyst was estimated at 2 meters/second, with an acceleration of 40,000 g, very remarkable indeed. I suspect that both the velocity and the *g*-force for the exploding cercariae of *H. occidualis* are very comparable to those of the nematocyst.

Finally, one can only wonder about the evolution of such a curious phenomenon. In her 1938 *Novitates Zoologicae* paper, Miriam Rothschild said, "The extraordinary degree of specialization shown by this group of cercariae is unique, and it is difficult to conceive how the delivery apparatus, with its peculiar function in the life-history of the cercaria, can have arisen." She continued, "[since] this was probably the first stage in the evolution of the present peculiar method of excystment of the cystophorous cercariae, it gives us no hint how the extremely complicated and delicately adjusted delivery system first came into being." I agree entirely with her assessment. The elegant words, "extraordinary", "unique", "peculiar", "complicated", and "delicately adjusted" are all appropriate for the description of an unusual, but highly successful, transmission step in the life cycle of this most curious trematode.

Most academic administrators, from chairs, to deans and vice presidents, have abandoned the research bench. This is a natural pattern. Krull was no different; well, almost. After going to Stillwater and OSU in 1948, his research productivity was squelched, greatly curtailed, by a combination of heavy teaching loads, advising students (and faculty), and dealing with small budgets as Chair of his Department of Veterinary Parasitology. However, one summer around 1950, he was summoned by John H. Whitlock of Cornell University School of Veterinary Medicine in Ithaca, New York, to help one of his students, Courtland Mapes, resolve the life cycle of another trematode. Krull responded to the challenge with great excitement and enthusiasm. The research over the next several summers was to yield nine papers. For many, this research represented one of Krull's best pieces of work. I asked Sidney if Krull and Whitlock were friends and he responded, "No, but they became friends." So why did Whitlock go to Krull for help? He replied that, "Whitlock and Mapes were hung up. They were at a dead end.

He [Whitlock] went to someone and asked for help. The response was, 'Krull is your man'."

Dicrocoelium dendriticum is not native to North America. It was introduced from Europe, apparently in the early twentieth century, although I am not sure of the precise time. For several years, the life cycle of *D. dendriticum*, the so-called lancet fluke, was said to be similar to that of *Fasciola hepatica*. This assertion was based on publications emanating from several English, German, and Canadian parasitologists. So this had to be correct, right? After all, more than one prominent parasitologist had published a paper saying the cycle included just two hosts, a snail and a sheep, with free metacercariae encysted on grass as the transport mechanism between the two hosts. We now know that this was an error, and that ants are second intermediate hosts for the parasite. The story behind this discovery is a fascinating one, and is told eloquently by Sidney Ewing in his biography of Krull.

Wendell Krull and Cortland Mapes, the graduate student at Cornell University, were to resolve the life cycle of *D. dendriticum*. Once again, just as he did for *H. occidualis*, Krull made good use of his skill as a naturalist, and that of an experienced parasitologist, to develop several important insights for understanding the parasite's life cycle. For example, as pointed out by Ewing, Krull knew that *F. hepatica* transmission occurred in the wettest portions of pastures where its molluscan intermediate host is typically found. I do not know if Krull had read Thomas's account of his work with *F. hepatica*, or not. If he had, he would have known the importance of water in the habitat distribution of the snail host to *F. hepatica* and for that parasite's transmission to sheep. In the first year of Thomas's study, water was abundant, and so were the snails and infected sheep. In the second year, though, there was drought. The snails disappeared and so did the infection in sheep. In fact, Thomas had to completely abandon his research efforts in the dry year. In the case of dicrocoeliasis, transmission apparently occurred "on hillside pastures with no such habitat for snails"(Ewing, 2001). This observation was made by Krull in his first summer at Cornell. In 1960, Krull and Ewing had a long conversation regarding his work at Cornell. Ewing relates, "Krull told me the day he arrived in New York he dissected one of Mapes' snails and when he saw the cercaria, he knew that something was terribly awry in the literature. He knew, purely on the basis of cercarial morphology, that there had to be a second intermediate host", but, he had to prove it beyond any shadow of a doubt.

The molluscan host in New York was already known to be *Cionella lubrica*, definitely not an aquatic snail, but a terrestrial species instead.

Moreover, sixteen years previous to Krull and Mapes (1952a, b), a pair of German parasitologist, W. Neuhaus and O. Mattes had described how *D. dendriticum* used "schleimballs" produced by snails as vehicles for the direct transport of cercariae from the snail to the definitive host (or at least they thought it was snail to sheep transmission). Krull and Mapes were required to first demonstrate that *C. lubrica* also produced slimeballs in North America, which they reported in their 1952 paper. They also discovered that the stimulus for slimeball production and release from the snail through the respiratory pore was a reduction in ambient temperature. However, their conclusion regarding the use of slimeballs for carrying cercariae to the definitive host was at odds with those of the Germans. The two Americans stated, "it can safely be assumed that the function of the slimeball is to carry the cercariae to another host. The actual fate of the slimeball and the method of infection of the final host in this cycle are not clear to us at this time. Our attempts to infect definitive hosts, to be reported in a subsequent publication, have raised some question as to the validity of the premise that the slimeball is the transfer agent from the snail to the definitive host" (Krull and Mapes, 1952a).

A subsequent paper, only one page in length, was a bombshell with respect to transmission from the snail to the definitive host (Krull and Mapes, 1952b). In it they said, "We have observed that ants mistake these [slimeballs] as choice food items and carry them back to the colony. We have determined by controlled feeding experiments with sheep that the ant, *Formica fusca*, is the second intermediate host of *D. dendriticum*." They then stated, "Inasmuch as we have been unable to obtain an infection without this second intermediate host, it is suggested that previous accounts of the life cycle of this important trematode are probably in error." This last sentence is about the greatest understatement I have ever seen in the parasitological literature. With this finding, Krull and Mapes established that the two-host pattern as propounded by the German, English, and Canadian parasitologists was, as put by Krull several years later, a "false" life cycle (Ewing, 2001), and that it was a three-host cycle instead. Ewing quotes a personal communication from Roger J. Panciera, an OSU veterinary school graduate who was at the field study site when Krull was making his observations regarding the relationship between slimeballs and ants. Panciera says that Krull spent many hours, lying prone, or crawling around the pasture, observing ants and how they would acquire the slimeballs, then carry them to their nests. We should expect this sort of behavior from a naturalist 'on the rampage'.

One cannot help but be impressed with the way in which Krull approached this life cycle problem. He and Mapes did not accept the printed literature, probably for several reasons. First, they were unable to confirm that slimeballs could be used to infect sheep directly. They tried, but it would not work. Guesswork would suggest that the English, Germans, and Canadians did not think about an infected ant with wanderlust and access to a barn where uninfected control sheep were being held (as suggested by T. W. M. Cameron to Bill Campbell in a personal letter). Second, *C. lubrica* was clearly a terrestrial snail, not aquatic like those that transmit *F. hepatica*, another example of why the naturalist was successful and the previous investigators were not. Third, when Krull had his first look at the cercaria of *D. dendriticum*, he knew the parasite had to have a second intermediate host. Sidney Ewing reminded me that Krull was an expert on snails, both their biology and culture. Some folks are blessed with great skills at growing plants and are frequently described as possessing 'green thumbs'. Others, like Wendell Krull, for whatever reason, have an uncanny capacity to grow invertebrates in the laboratory. They are also the true naturalists. With this information in hand, Krull and Mapes, one a veteran parasitologist and the other a 'rookie' student, went at the problem and solved it.

Toward the end of my interview with the Ewings, I asked them, "What was it about Krull that made him the consummate naturalist? If you could name any one thing, what would it be?" Margaret, who had taken both of his veterinary parasitology courses, responded first, "You mean any one 'complex' of things," obviously implying, at least to me, that Krull was not a 'simple' man. Then, Sidney said, "You know, he was an exceedingly keen observer, and he was a 'loner'." Sidney continued, "Another thing, Krull never really taught and did research at the same time," a point echoed by Margaret almost immediately. She added, "When he did research, he devoted his whole being to it, and when he taught, he would do the same thing. He was very focused. He was meticulous. And the rigor in being a tough guy hangs together with his other personality characteristics." I asked, "When you say he was a tough guy, you don't mean he was not a fair man, do you?" Margaret quickly responded, "Oh no, it was just that his personal standard was very high, with a great deal of rigor in what he was doing. This contributes to his success as a naturalist. He doesn't make just one observation, which would be wonderful in itself. For example, look at the range of observations that was needed for him to figure out the *D. dendriticum* life cycle." She also said that he was skillful in "ranking the quality of the information and that helped him enormously in problem solving. Another thing about him that helped was that he was a very enthusiastic person. When he was interested in something, he was *very* interested!"

Sidney then remarked, "He was also an amazingly enthusiastic teacher. The veterinary students feared him because he was so damnably demanding, but they also loved him." Margaret said that he was known to give 'pop quizzes', a classic way of getting the students' attention. At the beginning of a semester, students would offer a visible sigh of relief when he came into a lecture room and saw he was not carrying quizzes. They learned quickly, however, that this did not mean they were to be without one because he would frequently stash them in a desk drawer before class and then calmly walk to the desk and withdraw them.

After reviewing all of the procedures followed by Krull in resolving the life cycles of *H. occidualis* and *D. dendriticum*, the final question was, was serendipity involved? The answer is a resounding, and very emphatic, no! In the course of his research, he used plain common sense, coupled with intimate knowledge of the biology of parasites and hosts. This knowledge he gained through personal observation, and by deduction on reading the available literature. His discoveries were made with a prepared mind, a conclusion also emphasized by both Margaret and Sidney Ewing during our conversation. That was how he did it. Was he a genius, as inscribed by Miriam Rothschild? I suspect he was, but he also was a very patient naturalist (and he had tenure, even though *he* surely didn't need it)!

REFERENCES

Brusca, R. C. and G. J. Brusca. 2003. *Invertebrates*. Sunderland, Massachusetts: Sinauer.
Ewing, S. A. 2001. *Wendell Krull: Trematodes and Naturalists*. Stillwater, Oklahoma: College of Veterinary Medicine, Oklahoma State University.
Krull, W. H. 1929. The rearing of dragonflies from eggs. *Annals of the Entomological Society of America* **22**: 651–658.
Krull, W. H. 1930. The life history of two frog lung flukes. *Journal of Parasitology* **16**: 207–212.
Krull, W. H. 1935. Studies on the life history of *Halipegus occidualis* Stafford, 1905. *American Midland Naturalist* **16**: 129–143.
Krull, W. H. and C. R. Mapes. 1952a. Studies on the biology of *Dicrocoelium dendriticum* (Rudolphi, 1819) Looss, 1899 (Trematoda: Dicrocoeliidae), including its relation to the intermediate host, *Cionella lubrica* (Muller). III. Observations on slimeballs of *Dicrocoelium dendriticum*. *Cornell Veterinarian* **42**: 253–276.

Krull, W. H. and C. R. Mapes. 1952b. Studies on the biology of *Dicrocoelium dendriticum* (Rudolphi, 1819) Looss, 1899 (Trematoda: Dicrocoelidae), including its relation to the intermediate host, *Cionella lubrica* (Muller). VII. The second intermediate host of *Dicrocoelium dendriticum*. *Cornell Veterinarian* **42**: 603–604.

Rothschild, M. 1938. *Cercaria sinitzini n. sp.*, a cystophorous cercaria from *Peringia ulvae* (Pennant 1777). *Novitates Zoologicae*, **41**: 42–57.

13

Trichinosis and *Trichinella* spp. (all eight of them, or is it nine?)

Like a hog, or a dog in the manger, he doth only keep it because it shall do nobody else good, hurting himself and others.

Anatomy of Melancholy, Robert Burton (1577–1640)

The first time I saw live *Trichinella spiralis* was when I traveled to Chapel Hill, North Carolina, and began an NIH postdoctoral fellowship in the laboratory of Professor John E. Larsh. It was there that I observed probably Jim Hendricks or Norm Weatherly cervically dislocate a mouse, then skin and gut it. I watched as they took a pair of heavy scissors, and cut the mouse into small pieces before dumping everything into a 500-ml Erlenmeyer flask, containing a foul-smelling concoction of hydrochloric acid and pepsin. The flask was placed on a hot plate equipped with a magnetic stirrer, and allowed to stand for a few hours. The flask was then removed and the contents were poured through several layers of cheesecloth before gently centrifuging it and pouring off everything except the pellet at the bottom. A Pasteur pipette was then placed into the pellet and a small amount was removed to a microscope slide to which was added a cover slip. When I gazed through the microscope, I was astonished. There were hundreds of live, first-stage larvae of *T. spiralis*. It was an amazing site, one that I have not forgotten. As a student, I had seen these larvae in *in situ*-tissue sections stained with hematoxylin and eosin. Anyone who has taken an introductory parasitology course has had the latter experience, but, the live worms – they were impressive.

I remember learning about *T. spiralis* as an undergraduate student in an introductory parasitology course taught by Robert M. Stabler at Colorado College during my senior year, but I don't recall anything special, except the tissue sections with the embedded larvae. It was not until I was doing graduate work at the University of Oklahoma with J. Teague Self that I learned very much that really stayed with me. I recall that he and McWilson (Mac) Warren taught a seminar in which all their graduate students were required to participate. Mac was a young Assistant Professor at the University of Oklahoma's medical school in Oklahoma City, before he left for the Centers for Disease Control and Prevention (CDC) in Atlanta where he spent the rest of his career. At the time, Jim McDaniel, John Janovy, Jr., Fred Hopper, Henry Buscher, Horace Bailey, and I were all working as graduate students in J. Teague Self's lab, who, with Mac Warren, would parcel out the parasites. We graduate students would prepare and present oral reports on one of them each Friday at a bag lunch, alternating between the Oklahoma City and Norman campuses. One of my assignments was *T. spiralis*. I can remember it well. In fact, I still have the notes I used for that seminar squirreled away somewhere in one of my ancient filing cabinets. One of the things etched in my memory of the parasite was the enormous range of hosts from which it had been reported. At the time, I was puzzled because it was known even then to have an almost cosmopolitan distribution and that it had been found in everything from crows to walruses, and from mice to bears. Of course, we all knew at that time that the parasite normally cycled through rats and pigs, and that humans were infected by eating poorly cooked pork, and even bear meat. It is really amazing to see how much the systematics of the old *T. spiralis* has changed since the early 1960s.

When I began thinking about this book, I knew I wanted to include an essay on *T. spiralis*, in part because of the parasite's intriguing biology, my own long-standing interest in the parasite, and the parasite's absolutely fascinating history, especially in the nineteenth century when its life cycle was being debated. Two of the people I initially contacted as resources for this parasite, were William (Bill) C. Campbell and K. Darwin Murrell. Both of them provided me with reprints, but I interviewed Darwin for this essay. I saved Bill for the ivermectin story that was told earlier.

As I inferred above, one of the remarkable things regarding the *T. spiralis* I first knew was the huge diversity of hosts in which the parasite had been reported, everything from carnivores to herbivores and from mammals to birds. An explanation for this diversity in hosts rests with the fact that there are now eight different species, including the two most recently described, i.e., *Trichinella papuae* and *T. zimbabwensis*. Not only are these latter two species the most recently described, but they were isolated from reptiles, i.e., caimans, crocodiles, pythons, and turtles. Pozio *et al.* (2004) claim, based on these findings, that *Trichinella* spp. represent an ancient group that was present in poikilothermic reptiles before the evolution of homoeothermic birds or mammals, an interesting hypothesis.

Another aspect of most species of *Trichinella* that is of great interest is the manner in which the larval stage (L_1) is capable of transcriptionally altering and then controlling the biology of the cell that it penetrates after being freed by its mother and migrating to specific skeletal muscle sites in the thorax, neck, and face. The 'nurse cell', or modified skeletal muscle cell, first loses all of its contractile fibers; the nucleus of the nurse cell divides twice, creating four nuclei, all of which are 2N with respect to chromosome number. The cell's mitochondria become vacuolated and dysfunctional, producing an anaerobic environment that reflects the type of anaerobic metabolism employed by the larva. Surrounding the cell, a circulatory rete is formed, stimulated by the release of EVGH (Endothelial Vascular Growth Hormone) from the parasite, presumably to supply additional nutrients for the nurse cell and its occupant, and to remove metabolic waste. Eventually, in most species of *Trichinella*, a collagenous cyst, or capsule, surrounds the nurse cell and the L_1 stage, effectively sequestering both. There are, however, three species that do not encapsulate. According to Darwin, it is unclear if they cause the formation of a 'nurse cell'; he suggested that there might be evidence that these L_1 trichinae move from cell to cell.

Before venturing into any discussion regarding the new species discoveries, I want to first revisit some of the historical aspects of the parasite's discovery, description, and life cycle. The historical accounts are many and replete with tales of fierce competition, misjudgments, errors of assumption, etc. The parasitologists involved include a veritable 'who's who' of that era, e.g., Kuchenmeister, Leuckart, Virchow, and Zenker (among others) in Europe, and the great Joseph Leidy in North America.

One of the best historical renditions of early *Trichinella spiralis* research is that of Campbell (1983), and it is mostly from this account that I will summarize the early work. Very little is known regarding the antiquity of the disease caused by *T. spiralis*, but chances are it has been with the human species since we first appeared. It was probably seen many times before 1835, but it was not recognized for what it really is until 2 February 1835. On that day, James Paget, a first year medical student, observed trichinae while performing an autopsy at St. Bartholomew's Hospital in London. He did not know what they were and the Hospital was without a microscopist, or, for that matter, a microscope. So, Paget trundled over to the Zoology Department of the Natural History Museum, but was again stymied by the absence of a microscope. He was referred to Robert Brown, Head of the Botany Department. Brown was asked by Paget if he knew anything about worms and, he replied, "No, thank God."

However, in the meantime, Thomas Wormald had sent specimens of trichinae from the same cadaver (worked on by Paget) to Richard Owen (who was later to become Head of the Museum and an archrival of Charles Darwin after the latter published *On the Origin of Species*. In fact, Owen was the advisor to 'Soapy' Sam Wilberforce, Bishop of Oxford, before and during the famous debate that occurred at Oxford University between the latter and Thomas Huxley, Darwin's 'bulldog' and combative supporter). Owen desperately wanted to publish a description of this new parasite species and entered into an agreement with Paget that he, Paget, would receive priority for the discovery, since it was, after all, the student who made the discovery. However, when the description was published later in 1835, Paget was to receive but minor credit from Owen, "as was his nature", according to Campbell (1983). The original generic name was *Trichina*, but this was occupied by a fly description published in 1830; subsequently, Railliet in 1895 renamed the parasite as *Trichinella spiralis*, and it stuck.

From 1835 to 1846, several other investigators in Europe contributed further commentary regarding the parasite's description. An important event of serious consequence for the parasite's description and its life cycle, however, was to occur in Philadelphia in October of 1846. Joseph Leidy, referred to as the 'Father' of American parasitology by none other than Henry Baldwin Ward (1923), sat down to a dinner of pork one evening (I don't know if it was a roast or chops). While cutting the meat, he noted the presence of a large number of white specks, which immediately reminded him of similar white specks he had observed several times in the flesh of cadavers at autopsy. A curious man, he saved some of the pork and took it with him the next day to his laboratory at the Philadelphia Academy of Sciences where he examined it microscopically. He was amazed to see that they were tiny, encysted worms that reminded him of *Trichina (Trichinella) spiralis*, so recently described by Owen in London. The problem he thought, however, was that Owen's parasite was from human flesh and the one he was examining came from a hog. His larvae were dead, killed, Leidy concluded, by cooking (an important contribution to understanding the public health importance of such an act). Not long after, Deising referred to Leidy's parasite (incorrectly) as *Trichina affinis*, thinking in fact of Leidy's parasite as a microfilaria. According to Ward, Leidy's friends urged him to enter the 'priority' battle, but he demurred, saying (cited by Ward, 1923), "The important thing is that the discovery or fact should be made known. It is of little consequence who made it."

According to Campbell (1983), it was Kuchenmeister who had the idea that the parasite could be transferred from hogs to humans, but "he failed to make the others pay attention." Interestingly, the first one to complete the parasite's life cycle was Professor M. Herbst in Gottingen, Germany, but his report was "largely ignored" (Campbell, 1983). It seems that Herbst kept a pet badger to which he fed meat from animals he dissected in the lab. When the badger died, it too was dissected and found to be heavily infected with trichinae. Herbst then fed flesh from the infected badger to three young puppies. Several months later, they were killed and were found to have extensively parasitized flesh. Again though, his report, published in 1851, was not in the least influential with regard to research dealing with the parasite's life cycle.

According to Reinhard (1958), Dujardin in France and von Siebold in Germany reached similar conclusions in the early 1850s that trichinae in the flesh of animals were in fact the offspring of an already known adult nematode, namely *Trichuris* sp. Then, in 1855, Kuchenmeister dredged up Leidy's 1846 observation of trichinae in the flesh of hogs and suggested this was the vehicle for infecting humans. In 1859, both Rudolph Leuckart and Rudolph Virchow joined the 'hunt', and the competition was keen.

However, several errors were made in the process, most of them because each person wanted to be first into the literature. For example, Kuchenmeister suggested that *Trichuris* developed from the tiny trichinae found in the flesh. Leuckart also championed this hypothesis. In 1857, he fed infected meat to several mice and was dazzled when at necropsy a few days later he found many tiny worms of the same kind, only larger than trichinae. He had thus learned that trichinae would excyst in the gut of mice and that they would grow there. What he did not realize, however, was that he was actually seeing adult *Trichinella spiralis*. Virchow, in the middle of 1859, found trichinae in a cadaver at necropsy. He fed flesh from the cadaver to a sickly dog. According to Campbell (1983), "In the intestine of that wretched dog, Virchow found innumerable tiny nematodes that he recognized as adult worms." He also knew they were not *Trichuris*, thereby refuting both Kuchenmeister and Leuckart's hypothesis that *Trichuris* developed from the small trichinae in the flesh of humans. Humorously though, in his attempt to beat

Leuckart to the punch, Virchow sent a letter to the Paris Academy of Sciences describing his discovery, but the writing was so poor it took two months for the German to be translated into French! In the meantime, Leuckart was not waiting in the wings. He was attempting another experiment. His approach was correct, but he, like Virchow, was too hasty. Leuckart took trichinae from a cadaver and fed them to a hog. Four weeks later, he killed the animal and found *Trichuris*, confirming, he thought, Kuchenmeister's hypothesis that *Trichuris* developed from trichinae. Like Virchow, he wanted to see his results published immediately, so he quickly sent a letter to Professor von Beneden in Belgium "asking him to present the new information at the next academy meeting." Von Beneden was so accommodating that he had it inserted in the printed record of September 1859. Two mistakes were made, however. First, Leuckart believed that he had confirmed Kuchemeister's hypothesis that *Trichuris* developed from the trichinae. Second, von Beneden misread Leuckart's handwritten note and indicated that a "duizend" worms had been recovered, not a "dutzend". In other words, von Beneden said there were a thousand *Trichuris*, not a dozen, the latter being the more accurate enumeration – by far!

It was about that time that Freiderich Zenker made a huge discovery, one that would ultimately provide the definitive link between *T. spiralis* (trichinae) in hogs and trichinellosis in humans. In December of 1859, a young woman was admitted to a hospital in Dresden with exaggerated muscle pain and fever. The diagnosis was typhoid fever. About two weeks later, she died. Zenker immediately removed muscle tissue and examined it microscopically. He was astounded to see dozens of tiny worms moving in the tissue. In fact, virtually every piece of muscle he examined was 'crawling' with these same tiny worms. Zenker recognized them immediately as trichinae. As noted by Campbell (1983), "The first clinical case of trichinosis ever recognized was also the first known fatal case."

A few weeks later, Zenker examined the mucous tissues in the woman's gut and found tiny adults, with larvae of the same kind he had seen in the woman's flesh. He also took flesh from the dead woman and fed it to a dog. He gave some muscle tissue from the same woman to Virchow who also fed it to a dog. Leuckart's gift of the same flesh from Zenker was also fed to a dog and a hog. All of the animals developed infections. Zenker's curiosity was further aroused, but not yet satisfied. He now believed that trichinae in flesh became adults in the gut when infected flesh was consumed. He recalled the woman worked on a farm some eighty miles distant from Dresden. He felt compelled to go there and he did. The journey was to conclude the second part of the *Trichinella spiralis* story, i.e., the parasite's life cycle, and begin a third phase, the one having to do with its control. On arriving at the farm, he discovered that the farmer's wife had been severely ill with the same symptoms suffered by the dead woman and that the butcher who had slaughtered the hogs and prepared the meat for consumption had also been violently ill. Zenker discovered that ham and sausage from the Christmas celebration were still on site. When he examined it, the flesh was full of trichinae. As stated by Campbell (1983), "Zenker had broken the mystery of trichinosis."

Most of the trichinosis reported in Europe was confined to Germany, although there were reports in countries as far away as the U.K. and Ireland. The epidemic proportion of the disease is rather misleading. Unlike measles, flu, etc., outbreaks of trichinosis were much more confined, but when it struck, it did so with great vengeance. Mortality ranged as high as 18–20%. There was indeed a problem.

But how was the control problem resolved? Not without a lot of money, some ugly political confrontation, economic sanctions placed on importation of pork from one country to another, and time. There were two ways to solve the control issue. The first was to prevent hogs from being infected and the second was to prevent infected hogs from being consumed by an unsuspecting and vulnerable public. In Germany, where the trichinosis problem was greatest, laws against the sale of infected pork were passed in many localities and armies of inspectors examined freshly butchered pork. In the U.S.A., however, pork inspection was never really attempted on a large scale. It was just too expensive. The result was an embargo of pork from the U.S.A. in the late nineteenth century by several European countries. Because so much pork was being embargoed, there was great economic pressure placed on pork producers and exporters. The U.S. response was varied. Some wanted counterembargoes, while others wanted to proceed with meat inspection. Things got pretty hot for a while. Initially, the U.S. retaliation included a tax on the importation of sugar from Germany. Finally, in 1890, inspection of export pork was approved by Congress and signed into law by President Benjamin Harrison. This cooled things down a bit, but did not resolve the export/import problems because difficulties with trichinosis in Germany persisted. Then, France imposed an embargo on U.S. pork and the Germans passed laws requiring pork from the U.S.A. to be reinspected on its arrival, thereby increasing the cost of American pork and making locally grown pork much less expensive. One of the serious problems in Germany was the poor quality of inspectors, and of outright graft in the process of inspection, e.g., bribery, etc.

Resolution of the problem came with the appointment of Charles Wardell Stiles as a scientific attaché to the American embassy in Berlin. Stiles' credentials as a scientist were of the highest quality; for one, his doctorate had been secured in the laboratory of Rudolph Leuckart. His efforts were focused in several directions and he was successful. One of the things he pushed hard was trichinoscopy by trained inspectors. Gradually, the prevalence of trichinosis in Germany declined to levels far lower than in the U.S.A., the latter country having abandoned meat inspection for trichinae completely. In fact, the levels of trichinosis in the U.S.A. in 1970 were higher than in Germany in 1870 (Campbell, 1983). In the first half of the twentieth century in the U.S.A., the prevalence of trichinosis was relatively high. The passage of laws banning the feeding of uncooked garbage to hogs to control hog cholera reduced the prevalence of trichinosis somewhat, but the disease remained. In no small part, its persistence is due to the zoonotic character of species in the genus. As noted by Campbell (1983), "*Trichinella spiralis* is destined to remain with us, both in nature and in the laboratory. It is not an endangered species."

I wrote earlier that when I first started my career as a parasitologist, there was just one species of *Trichinella*, i.e., *T. spiralis*. However, Garkavi (1972) described a nonencysting form in birds, *T. pseudospiralis*, doubling the number of species in the Trichinellidae. Now, there are eight. The story of the elucidation of the next six species is fascinating. For the telling of the story, I went to K. Darwin Murrell, a leader in the research on this parasite over the past several decades, and a past president of the International Commission of Trichinellosis.

By the late 1950s, at least forty species of mammals were known hosts for *T. spiralis*, about a hundred years after Leuckart, Leidy, Kuchenmeister, Virchow, and Zenker were engaged in discovering the biology and life cycle of the parasite. Darwin said that by the 1980s, the host list for *T. spiralis* had expanded to at least a hundred species. Presently, most specialists for this parasite believe that there is not a single mammalian species that it cannot infect. I do not think, and most others concur, there is another species of parasitic helminth that has as wide a range of hosts as *T. spiralis*. Superimposed on this broad spectrum of host–parasite interactions and interrelationships, are seven other species of *Trichinella*, each with a unique pattern of epizootiology in wild animals and epidemiology in humans.

One of the most widely used textbooks (*Introduction to Parasitology*) in North America during the middle of the twentieth century was written by Asa Chandler and Clark Read (1961). In it, they refer to the Trichinellidae as having but one species, i.e., *Trichinella spiralis*. However, as Darwin said, "people began to notice some rather strange things involving the biology of the parasite." He pointed out that the first person to call attention to any sort of notion regarding the possibility of an anomalous strain of *T. spiralis* was Bob Rausch (1980; p. 353), who indicated that specimens of *T. spiralis* from Africa and central Europe were morphologically identical. But, he went on, their "biological characteristics, on the other hand, may be distinctive. Besides being relatively unable to infect laboratory rats, as we have observed in Alaska, the northern strain may not be infective for swine. We failed to establish infection in a pig by feeding it partially dried meat from an arctic fox, but were successful in infecting two young black bears by similar means."

Although Darwin credits Rausch with being the first to suggest the possibility of *T. spiralis* strains, i.e., in the Arctic versus everywhere else, he also believes that George Nelson "finally, and in a formal and systematic way, documented that not all *Trichinella* are the same." There was an outbreak of trichinellosis in Kenya, one of several since the great endemics of the mid and late nineteenth and early twentieth centuries, like those seen by many of the nineteenth century parasitologists who worked out the biology of trichinosis for the first time. However, in this case, the source of infection was definitely not domesticated swine, but instead was the bush pig. The study by Forrester *et al.* (1961) documented the clinical features of the disease. As part of that research, George Nelson removed a piece of gastrocnemius muscle from an infected human and experimentally attempted passage through laboratory rats, but without success. Further investigation revealed the parasite also was not infective for domesticated swine. They were, however, able to isolate larvae from lions, serval cats, spotted and striped hyenas, side-striped jackals, and domesticated dogs. Nelson went back to the London School of Hygiene and Tropical Medicine and obtained the strain of *T. spiralis* being maintained there. He found that it would infect both rats and pigs with ease. He was familiar with Rausch *et al.*'s (1956) work in which they had reported the results of a survey of Arctic mammals, finding *T. spiralis* in everything from whales to hares, and that a strain isolated from a fox had been isolated and maintained by Bob Rausch since that time. Nelson obtained the strain from Rausch, but he could not get that one to go in rats or pigs either. At this point, Nelson began to realize that *T. spiralis* was quite variable.

Most of Nelson's research was conducted in East Africa. In the late 1960s, Gretillat and Vassiliades (1968), however, examined a number of wild and domesticated mammals in West Africa. Basically, their work confirmed that of Nelson on the other side of Africa. They isolated the parasite in both domesticated cats and dogs, but the domesticated pig was not susceptible. When they did obtain an infection in a pig, the larvae were killed by the host and calcified within a matter of days. Laboratory infections with wild rodents and hedgehogs were also unsuccessful. Primates, however, such as baboons and other monkeys, were susceptible to infection, along with felids.

In our conversation, Darwin then indicated that about this time "The Russian parasitologists were becoming active with *T. spiralis* in a big way, but it was difficult to tell what they were doing because their work was being published in obscure Russian journals." One of these investigators was V. A. Britov, "who is your picture of a 'Russian bear', a great big guy, with a huge beard, . . . [who] has been more or less restricted to working in Vladivostok his whole career." It was Britov (1971) who initially decided that these *T. spiralis* variants "should be raised to at least the status of sub-species." Britov initially recognized *T. s. nativa*, *T. s. nelsoni*, and *T. s. domestica*, with the latter being the most widely distributed subspecies. (It should be noted, however, that the first reference [at least that I have seen] to a subspecies, i.e., *T. spiralis arctica*, was made by Everett Schiller and Clark Read in 1960 [cited in Chandler and Read, 1961]. They said that *T. spiralis arctica* "is essentially incapable of infecting rats but readily infects deer mice, and certain carnivores.")

Subsequently, however, Britov (1980) and several other Russians argued that, in fact, there were four species that constitute the genus *Trichinella*, i.e., *T. spiralis, T. psuedospiralis, T. nelsoni*, and *T. nativa*. Britov used several criteria for making this judgment. First, he stated that each of the species was genetically isolated from the other. Second, each species has its own set of transovarially transferred endosymbionts. Third, he argued that each species exhibits a certain degree of host specificity. For example, he noted that *T. spiralis* was "well adapted to pigs, rats, mice, and other laboratory animals, but very little to adult dogs." He continued, "*T. nelsoni* and *T. nativa* infect mainly canines, but do not survive well in pigs or rats, and *T. pseudospiralis* is the only species adapted to birds [as well as mammals]. He also pointed out that the geographic distribution of the mammalian species is important. *Trichinella nativa* is Holarctic, above the 40th parallel, whereas *T. nelsoni* is a southern species, restricted to "45° of the north geographical latitude to the south tip of Africa." *Trichinella psuedospiralis* does not occur north of the 45th parallel, but its distribution has not been determined otherwise. Finally, he noted that trichinae of *T. nativa* are able to survive freezing temperatures of −20 °C to −25 °C, while those of the other species die within a few hours. At this point, Darwin pointed out that some sort of 'systematic' order was beginning to take place with respect to species in the genus.

Not long after Britov's (1980) publication, an international meeting was held in London to deal with problems relating to nematode systematics. Terry Dick of the University of Manitoba was a participant and raised several serious questions regarding the taxonomy of *Trichinella* spp. (Dick, 1983). In this paper, Terry suggested that there are four factors that should be included in species recognition. These included: "(1) intraspecific variability; (2) influence of the host; (3) genetic variability; and (4) gene flow between geographical isolates of *Trichinella*." After careful consideration of the data generated through interbreeding experiments elsewhere and those in his own laboratory (using mostly Arctic isolates), Dick (1983) concluded that, "it is possible that we are dealing with a series of semi-species or incipient species in various stages of speciation. Most will probably disappear or become part of the normal variability for *T. spiralis*, while others such as *T. spiralis* var. *pseudospiralis*, may progress to species status particularly if it stays isolated geographically and circulates through bird hosts. The African isolates, although having many features similar to arctic isolates, may, if separated long enough, attain species status also." He concluded his paper by saying, "Based on our current knowledge, it appears that *T. spiralis* is the only bona fide species in the genus *Trichinella*." This was obviously a conservative position at the time Terry reached his conclusions. However, as noted by Darwin in our interview, there were still a number of other North American parasitologists who had done experimental work on various isolates, i.e., Read, Schiller, Todd, McKerrow, and had pointed to what they referred to as 'strain' differences.

Again, in the early 1980s, several parasitology groups, primarily in Europe, turned to the morphology of the various isolates as a way of determining differences and clarifying the question of species, subspecies, or strain, within *Trichinella*, but with mixed results. Lichtenfels *et al.* (1983), however, using scanning electron microscopy (SEM) and without any a priori or intuitive notions, concluded that there were no morphological differences between three subspecies, i.e., *T. s. spiralis*, *T. s. nativa*, or *T. s. psuedospiralis*. In other words, according to Darwin in our interview, "no morphological markers" were associated with these subspecies that would allow the separation of one from another.

The pork producers kept asking another question, "Are we wasting our time? So, if we clean up the farms, is this thing [*Trichinella* sp.]

going to jump from nature back in again?" The answer was, according to Darwin, "We had no idea, because there were open questions such as, if you find it in a wild animal, then is it capable of infecting a pig?" So, he began to accumulate a bank of isolates in rats and pigs from as many different sorts of wild animals as he could possibly obtain. Gerry Schad was his partner in much of this important epidemiological research. He initially believed that the *T. spiralis* that infected pigs was of one sort (a domestic strain, subspecies, or species?) and that the sylvatic kind was just that, another kind (a strain, subspecies, or species?). But, as he went along, he began to "find that some of these isolates from wild animals would also infect pigs and rats." He wondered about the explanation. He said that he would obtain "bear isolates from the Pocono Mountains of Pennsylvania, some of which would infect pigs and rats, and other bear isolates from the Poconos that would not." Because of these findings, Darwin and Ralph Lichtenfels began to rethink their position regarding species differences in *Trichinella*.

Clarification of the systematics problem for *Trichinella* spp. however, became the product of the rapid advances that were then occurring in molecular systematics in general, and was made using biochemical/molecular procedures that were initiated by several *Trichinella* research groups in North America and Europe. Information regarding the genetic variability would, it was hoped, finally resolve a number of basic epidemiologic/epizootiologic and taxonomic questions that had plagued workers in this area for several decades. John Dame, a molecular biologist and Darwin (Dame *et al.*, 1987) produced genetic evidence for the presence of synanthropic (subspecies) in sylvatic hosts. In other words, they were able to show that domestic *T. s. spiralis* was able to cycle into wild animals. They concluded, "The sources of pork for wild animals which can harbor *Trichinella spiralis spiralis* may be plentiful and, as man encroaches on the sylvatic biotope, there is increased risk of exposure of wild animals through refuse and garbage."

Five parasitologists in Italy and the U.S.A. took the final step in the systematics analysis of *Trichinella* and created a new taxonomic scheme for the genus. These included (in alphabetic order) Guiseppe La Rosa, Ralph Lichtenfels, K. Darwin Murrell, Edoardo Pozio, and Patrizia Rossi. According to Darwin, by the early 1990s, Pozio's group in Italy had begun doing "a lot of isozyme analysis of *Trichinella*, aided by some sophisticated statistics." Murrell said that he had a "lot of biological data and isolates that I sent to Edoardo Pozio, who had just established the International *Trichinella* Reference Centre." The groups began collaborating and finally published three landmark papers in the *Journal of Parasitology* (La Rosa, *et al.*, 1992; Pozio, *et al.*, 1992a, b). In these papers, they analyzed and correlated "the biological characterizations, the biochemical characterizations, and then a summary paper on the taxonomic revision of *Trichinella*." Up to this point in time, most groups recognized only four subspecies, or strains, depending on whoever was doing the taxonomy. However, based on the isozyme analysis and the biological data, the Anglo-Italian group concluded there were eight distinct gene pools, designated T1–T8; according to Murrell *et al.* (2000), there are now nine gene pools, with the last being designated as T9. Using Pozio's scheme, they were identified as T1 = *T. spiralis sensu stricto*; T2 = *T. nativa*; T7 = *T. nelsoni*; and T4 = *T. pseudospiralis*. In the last of the three papers, they added T3 = *T. britovi*. To quote from La Rosa, *et al.* (1992), "The absence of evidence of gene flow among the gene pools and the high level of allozymic differentiation between the groups support the concept that the genus *Trichinella* is composed of several sibling species." Among the biological characters used to establish the groups included were the number newborn larvae produced per female worm under in vitro culture conditions, nurse cell development time, resistance of trichinae to freezing, and environmental characteristics of the location where the larvae were isolated.

Pozio *et al.* (1992b) included a complete (up to that point in time) taxonomic revision of the genus *Trichinella*, plus a description of *Trichinella britovi n. sp.* This was the landmark paper and is the one most frequently cited for those referring to the systematics of the *Trichinella*. Since then, three new species have been described, including, appropriately, *Trichinella murrelli*. The other two are *Trichinella papuae*, from a wild pig (and salt-water crocodiles) and *T. zimbabwensis*, from crocodiles (and probably mammals); both are nonencapsulating forms, similar to *T. pseudospiralis*. One of the things I like about the summary paper of the group (Murrell *et al.*, 2000) is a statement made in the last paragraph. They say, "In the interest of stable nomenclature, we believe the vast amount of information accumulating about *Trichinella* can be best handled by the scientific community if the well studied phenotypes are given names. Furthermore, the use of scientific names and T-designations as used by La Rosa *et al.* (1992) and Pozio (1992b) for isolates not fully characterized provide a system less likely to result in premature assignment of scientific names to isolates than would possibly occur if subspecies were commonly used." Eight years later, Murrell *et al.* (2000) provided another update on the *Trichinella* spp. systematics situation.

Research on *Trichinella* spp. continues unabated. For example, the International Reference Centre in Rome acts as a repository of all new isolates discovered from around the world. Darwin Murrell, in collaboration with Eric Hoberg at the ARS (USDA) in Beltsville, Maryland, have undertaken to identify the geographic location (longitude and latitude) and hosts of all *T. spiralis* reports in North America, including Mexico, with a view toward understanding the biogeography of the parasites. Included will be any information dealing with intensity of infection, prevalence of the parasite in the population from which it was taken, and then a reference for it. Such a database will allow risk assessment studies, for example. Darwin said that he raised this question regarding *T. spiralis* and wild animals because he wanted to know if the parasite occurs by itself, in a completely sylvatic cycle, or does it only occur in animals where and when there was a previous record of *T. spiralis*? One of the problems is that the records of *Trichinella* in North America are scattered over fifty years in so many different journals, and in some that are very obscure, e.g., *The Canadian Field Naturalist Newsletter*. Once these data are assembled, Darwin and Eric have plans to place this information online so that anyone with an interest in trichinellosis can access it. Darwin said he had about 260 citations already entered, but many more remain.

Beginning with Paget and Owen, and then continuing with Zenker, Leuckart, Leidy, Virchow, and Kuchenmeister, some of the world's earliest and best helminthologists have been hard on the trail of *Trichinella spiralis*. We now know there is not one species, but at least eight, partly easing my early frustration in trying to understand the epidemiology and epizootiology of a parasite known (at the time) to infect at least fifty species of hosts. Despite the fact that we now believe there are eight species of *Trichinella*, it is our present understanding that *T. spiralis* is still the reigning helminth parasite with respect to the breadth of host specificity. The story involved with understanding the biology and systematics of these parasites has been absolutely fascinating to learn. My suspicion, and something shared by Darwin, is that research on this parasite will be even more revealing as time goes forward.

As mentioned elsewhere, one of the aims of these essays is to determine the extent of serendipity in discovery. If one goes back and reads the early history of trichinellosis, I suppose that the original identification of the parasite in cadavers by Paget and then Leidy was serendipitous. However, in the latter case, Leidy was also very astute in that he made a connection between the pork he was eating for dinner one evening and trichinae in the cadaver he had observed several days earlier. He was apparently very good at 'connecting the dots'. The later discoveries by Leuckart, Virchow, Kuchenmeister, and Zenker were not serendipity. In most cases, they were actually generating hypotheses and testing them. They were definitely following a trail. The same thing can be said for Nelson, Dick, Britov, Pozio, Murrell, Lichtenfels, and others of the modern era. These investigators were also following trails. They first built a strong database, then generated hypotheses and tested them. However, most of the systematics research up to that point must be considered as pure guesswork.

The systematics of *Trichinella* remained largely as opinion until the molecular/biochemical studies of Terry Dick, Edoardo Pozio, Darwin Murrell, and their colleagues. When they began using these technologies, the opinion approach took a back seat (where it belonged!). The population genetics that emerged in the work of these and other investigators was pivotal in taking us to the point we now occupy. The identification of nine distinct gene pools based on isolates from around the world was the significant breakthrough. Much remains to be done, as described by Darwin in our conversation, but we are getting there.

REFERENCES

Britov, V. A. 1971. Biologic methods of determining *Trichinella spiralis* (Owen, 1835) varieties. (In Russian.) *Wiad Parazytol* **17**: 477–480.

Britov, V. A. 1980. The species of *Trichinella*, their specificity and their role in initiating disease in humans and animals. *Helminthologia* **17**: 63–66.

Campbell, W. C. 1983. Historical introduction. In *Trichinella and Trichinosis*, ed. W. C. Campbell pp. 1–30. New York: Plenum.

Chandler, A. C. and C. P. Read. 1961. *Introduction to Parasitology*. 10th edition. New York: Wiley.

Dame, J. B., K. D. Murrell, D. E. Worley, and G. A. Schad. 1987. *Trichinella spiralis*: Genetic evidence for synanthropic subspecies in sylvatic hosts. *Experimental Parasitology* **64**: 195–203.

Dick, T. A. 1983. The species problem in *Trichinella*. In *Concepts in Nematode Systematics*, ed. A. R. Stone, H. M. Platt, and L. F. Khalil, pp. 351–360. New York: Academic Press.

Forrester, A. T. T., G. S. Nelson, and G. Sander. 1961. The first record of an outbreak of trichinosis in Africa south of the Sahara. *Transactions of the Royal Society of Tropical Medicine and Hygiene* **55**: 503–513.

Garkavi, B. L. 1972. Species of *Trichinella* isolated from wild animals. *Veterianariya* **10**: 90–91.

Gretillat, S. and G. Vassiliades. 1968. Particularités biologiques de la sud-ouest-africaine de *Trichinella spiralis* (Owen, 1835). *Revue d'Elevage et de Médicine Vétérinare des Pays Tropicaux* **21**: 85–99.

La Rosa, G., E. Pozio, P. Rossi, and K. D. Murrell. 1992. Allozyme analysis of *Trichinella* isolates from various host species and geographical regions. *Journal of Parasitology* **78**: 641–646.

Lichtenfels, J. R., K. D. Murrell, and P. A. Pillit. 1983. Comparison of three subspecies of *Trichinella spiralis* by scanning electron microscopy. *Journal of Parasitology* **69**: 1131–1140.

Murrell, K. D., R. J. Lichtenfels, D. S. Zarlenga, and E. Pozio. 2000. The systematics of the genus *Trichinella* with a key to the species. *Veterinary Parasitology* **93**: 293–307.

Pozio, E., G. La Rosa, P. Rossi, and K. Darwin Murrell. 1992a. Biological characterization of *Trichinella* isolates from various host species and geographical regions. *Journal of Parasitology* **78**: 647–653.

Pozio, E., G. La Rosa, K. D. Murrell, and J. R. Lichtenfels. 1992b. Taxonomic revision of the genus *Trichinella*. *Journal of Parasitology* **78**: 654–659.

Pozio, E., I. L. Owen, G. Marucci, *et al.* 2004. *Trichinella papua* and *Trichinella zimbabwensis* induce infection in experimentally infected varans, caimans, pythons, and turtles. *Parasitology* **128**: 333–342.

Rausch, R. L. 1980. Trichinosis in the Arctic. In *Trichinosis in Man and Animals*, ed. S. E. Gould, pp. 348–378. Springfield, Illinois: C. Thomas.

Rausch, R. L., B. B. Babero, R. V. Rausch, and E. L. Schiller. 1956. Studies on the helminth fauna of Alaska. XXVII. The occurrence of larvae of *Trichinella spiralis* in Alaskan mammals. *Journal of Parasitology* **42**: 259–271.

Reinhard, X. 1958. Landmarks in parasitology. II. Demonstration of the life cycle and pathogenicity of the spiral threadworm. *Experimental Parasitology* **7**: 108–123.

Ward, H. B. 1923. The founder of American parasitology, Joseph Leidy. *Journal of Parasitology* **10**: 1–21.

14

Phylogenetics: a contentious discipline

There never was in the world two opinions alike, no more than two hairs, or two grains; the most universal quality is diversity.

'Of the Resemblance of Children to their Fathers', *Essays*,
Book II, Michel de Montaigne (1533–1592)

Whenever my wife and I go to London, we always stay at the same hotel in Kensington. Our favorite pub, The Goat, is nearby on Kensington High Street. Not too long ago, I recall sitting with Pete Olson and one of his postdocs in our pub enjoying a pint of lager. Pete, his postdoc, and I were talking about the *Journal of Parasitology* and some of what goes on in terms of editing. During the conversation, I mentioned that the only really contentious issues that have arisen over the years generally involved confrontations between referees and authors concerned with systematics/phylogenetics papers. I said to Pete, "Aside from natural personality clashes between certain folks, why do you suppose this is the case?" Pete responded immediately, like he knew the question was coming and had prepared an answer in advance. He said, "This is the only area of parasitology where opinion is acceptable in print form."

When Pete said this, I knew exactly what he meant, because when one thinks about it, a genus and species is really nothing more than a hypothetical construct. It is an opinion, well considered and documented in most cases, but still an opinion. When dealing with hypotheses, people will often have different ideas about them. The new molecular technologies are helping to alleviate some of the contentiousness in systematics, but certainly not all of it. I asked Steve Nadler if he would agree with Pete's conclusion, or not. He responded, "Yes, in the sense that, as an evolutionary biologist, or a systematist, you are dealing with history. It is historical reconstruction when you create a hypothesis and then try to test that hypothesis, which is very different from experimental work that you can repeat in the laboratory. Any time you are talking about trying to reconstruct the past, given whatever data we currently possess, it's obviously just a testable hypothesis because, number one, the data can change, i.e., you can gather more of it, or you can look at a different locus on a chromosome, and then you may generate a different hypothesis. And, second, it may be contingent on how you analyze the data, i.e., different methods of phylogenetic inference can give you different tree topologies and the interpretation might be different depending upon which one you might choose. So, for example, I think there is a fundamental difference in the way in which a phylogeneticist and a biochemist work. The latter can go into the laboratory and repeat the experiment over and over. History is not that way. The phylogeneticist is trying to interpret history. There is but one history that we are trying to recover and it cannot be repeated."

When I began the interviews for this book, I felt comfortable talking to the wide range of people that I had selected for inclusion. I also felt confident that I would be able to converse with each of them and ask intelligent questions even though in several cases I was way outside my area of expertise. Steve Nadler was no different. I had known Steve for many years and we had become friends over that period of time. I should have realized I would have problems when I invited Steve 'into my parlor'. I should have recognized this when I sat down to go through the papers he sent me. I was hung up immediately. Part of the reason for my problem was very simple and it rests with Steve's explanation for the way a phylogeneticist works. Most of us, as parasitologists, are empirical in our approach. We are experimentalists. But this is not the way of phylogeneticists. Yes, they generate hypotheses, but they cannot be as empirical in their approach because they cannot go into a lab and do experiments. Of course they can do protein electrophoresis, or DNA sequencing, and use cladistics to analyze data. They are able to examine an historic record, but only the one that is tied up in the manner in which DNA has come together in some unique way. Moreover, the data generated from these approaches can be used only to infer phylogenies, to infer the systematic status of a group of organisms, or even to infer the existence of a species. At best, these inferences can be no more than educated guesses, or hypotheses. They are, as Steve said, expressed within an historical context and, as such, usually cannot be tested. I must emphasize, so my evolutionary biology friends will not 'freak out', I am not denigrating or demeaning the value or the reality of this approach in parasitology. Quoting from Nadler (2002), and I agree with him entirely, "Species are real and discrete units in nature, and . . . they result[ed] from descent with modification. This evolutionary perspective provides a conceptual framework . . . to view species as independent evolutionary lineages, and provides approaches for their delimitation." However, the phylogenies that these species reflect can only be inferred; they cannot be tested, certainly not in the same way an empiricist can test a hypothesis.

In beginning this essay in this manner, I guess I am trying to explain why I chose to go in several directions rather than tell a story like I have in most of the other essays. When Steve sent me a selection of his papers, he packaged them into several topics, or groups. I then chose a few of these to examine more fully and discuss with him when he came to Wake Forest for Joel Fellis's dissertation defense in the spring of 2005. I also noted while reading these papers that phylogenetics has a great deal of terminology not used by us ordinary, run-of-the-mill biologists or parasitologists, so I thought it would be a good thing to start with definitions of a few terms frequently used by the phylogeneticists. I began by asking him to define congruence. He said, very simply, "Matching." He continued, "There can be two kinds of congruence. One is topological, or branching congruence. Another kind refers to congruence in time. Sometimes, for example, genetic data can give you information that relates to the timing of a speciation event." I interjected at this point by asking, "To do the latter are you required to use certain genes?" He responded, "Well, one of the things you need is a relatively constant rate of evolution within a gene. Given a gene within a group of organisms, you may not have a constant rate of evolution. In other words, you are really inferring that substitutions are occurring in roughly a time-dependent fashion. It's a molecular clock sort of idea, at least to some degree. If the clock is too irregular, then you are not going to be able to use it to infer the timing of speciation events. There is no way to correct it basically. So, it's not just that some genes can be bad. In some cases, no matter what gene you look at, you may not have enough regularity. Thus, if you have two lineages and for some reason there is a speed up in one of them, then you cannot compare them." I interrupted by asking, "How do you know when this occurs?" He replied, "There are pretty good tests for it. In other words, if you have a gene and you have a group of organisms sampled, you can tell if you have constant rates of evolution by tests based on the branch lengths. But, you can't always tell how to correct it. If you have a high rate of variation, there are not a lot of ways to correct for that. So congruence in timing is much more tricky than topological congruence." I said, "when you are talking about congruence then, you are really referring to coevolutionary events?" Steve responded, "Not quite. I would distinguish between coevolution and cospeciation or cophylogeny, because I think coevolution implies some reciprocal evolution, or reciprocal adaptation to most people. Most folks think of coevolution as like predator-defense mechanisms."

I then shifted my inquiry by asking him to explain the difference between a plesiomorphic character and homoplasy? Steve said, "A plesiomorphic character simply represents the ancestral state and homoplasy refers to recurrent similarities, like when you have the same character state in two organisms, but they develop it independently." Then, I asked about the difference between an apomorphic character and a synapomorphic character? He responded, "An apomorphic character is a derived character state in a single lineage. It would be like a group of organisms that share a common ancestry and maybe a unique morphology develops in just one of them, independently of the others. A synapomorphy is a shared derived state. In other words, if two species have a morphological variance and they inherit it from a common ancestor, it is a shared derived state."

I then turned to another topic of interest, namely, the idea of cospeciation. At this point, I read a quote from one of Steve's papers (Hafner and Nadler, 1990) and asked for his reaction. "You say in your *Systematic Zoology* paper that documented cases of widespread cospeciation are rare. Maybe I'm thinking about cospeciation in the wrong way, but what about pinworms and primates? What about hookworms and primates? What about *Plasmodium* species and primates? Can't you think of cospeciation in each of these groups?" Steve responded by saying, "I believe your premise is inaccurate. When we wrote that, we were thinking of the phylogenies of both hosts and parasites. There are lots of potential cases out there where there could be coevolution events and cospeciation. In that particular study, we used as an example our studies on pocket gophers and chewing lice 'to illustrate how genetic distances can be used to explore relative rates of genetic change in the two groups and to investigate relative timing of cospeciation events in the assemblage'. We were trying to document the phylogeny of the host and one for the parasites in which you have a significant amount of congruence between the two. In lots of the cases, like the ones you mentioned, you don't have the phylogeny of both the hosts and the parasites. For example, you may have the systematic treatment where there are ten species of parasites. But, that really doesn't tell you that there is cospeciation until you have a phylogeny or a phylogenetic investigation for both." I then asked, "This means you can't infer anything like that can you?" Steve replied, "No, because it could be that there has been widespread colonization among those parasites."

I then asked another question, "If you are going to see cospeciation, wouldn't something like 'host capture' sometimes go hand-in-hand with it? In other words, when humans came out of Africa many thousands of years ago, presumably they came out with a set of geohelminths. Pinworms were probably among these nematodes as well. But, pinworms were not dependent necessarily on our culture, our lifestyle, if you will. To acquire hookworms, as well as *Ascaris* and whipworm, for example, we had to switch from a hunter-gatherer culture to an agricultural one. Couldn't this be an example of cospeciation? Am I on the right track in my thinking about this?" Steve said, "The idea of cospeciation in this case is that basically you have a parallel phylogenesis between primate parasites and their hosts. If you are acquiring these parasites from other hosts as a result of changes in host lifestyle, adapting an agricultural way of living, then this is not a cospeciation event. Of course, in your example, parasites acquired in this way could subsequently undergo speciation events, given time and the development of suitable host specificity."

At this point, I asked him to explain what he meant by the idea of genetic distance. He said, "Genetic distances are a summary measure of the overall amount of similarity, or lack thereof, between two taxa. It kind of goes back to the idea that when people were doing protein electrophoretic work, the early emphases in molecular systematics were not cladistics. Let me explain it in another way using DNA as an example. Imagine that you compare just two species. You take two sequences and you line them up. You look at the number of differences at a single locus and count the number of differences between them. Let's say in one comparison there are two differences. Then, you do another comparison of two different species and you find that there are 50 differences. One of the things we know about molecular evolution is that the more different things are, the more hidden substitutions there are. In other words, you have a finite amount of sequence. Over time, some of the same sites that changed previously will have further substitutions, so that the sequence that has two differences is probably a pretty close estimate to the true number of substitutions that occurred and that the two species have shared a common ancestor more recently. The two that have 50 differences are probably not a very good representation of the true number of differences between those species because over a longer period of time, some of those sites have probably had multiple changes. So a distance measure will try to correct for that. At the nucleotide level, the pair that has two differences is still going to have a low distance. The one that has 50 is going to be greater, because we know that is an underestimate of the true number of differences." Thus, according to Sober (1988), "Distance measures take overall dissimilarities between pairs of taxa, rather than the character states of each, as data. They have been," as Steve then said, "especially popular because certain kinds of molecular data [like those for protein immunology or DNA hybridization] come in this form. For example, techniques are now available that will allow one to take strands of DNA obtained from different species and compute the fraction of sites on those strands at which they differ; one can do this without knowing the actual sequence of molecular characters found at any site on either strand. Instead of inferring phylogenies from characters, one infers them from distances." Steve summarized, "Genetic distance is an estimate of the true number of changes that have occurred since the two species shared a common ancestor."

I am not a phylogeneticist and certainly I am not 'up to snuff' when it comes to the molecular techniques employed by those working in this area. I do know, however, that as time has passed, molecular techniques have become more sophisticated. A question that occurred to me as I examined some of Steve's earlier work on biting lice and pocket gophers was that he and his colleagues switched methodologies between 1988 and 1994, even though these two papers were dealing with a very similar question. Were their conclusions any different using different techniques to answer basically the same question regarding cospeciation and relative rates of evolution? In reality, the issue revolves around the choice of techniques used in answering questions regarding amounts of cospeciation and rates of evolution. In the first study, they used starch gel electrophoresis to compare 31 protein loci in pocket gophers and 14 from biting lice. In a subsequent study, they chose to examine DNA sequences of the gene encoding mitochondrial cytochrome oxidase I. Sandwiched between these two investigations was an effort to explore "relative rates of genetic change" in pocket gophers and biting lice using a nonparametric test of association between genetic distance matrices for the hosts and their parasites (Hafner and Nadler, 1990). During our conversation, Steve remarked, "If you are trying to summarize something about evolutionary history, then basing it on multiple loci is a definite advantage, rather than one locus. The 1994 paper is based on one mitochondrial gene, which presumably represents the entire mitochondrial history. In some ways, I think the technique for the earlier work was, in theory, of greater utility because it was multilocus. On the other hand, the technique of electrophoresis doesn't pick up all the variation that might be there. So, when you compare these alleles on a gel, if they have different mobility then you know you have a real difference. If they migrate differently, then you know they have a net difference in charges of the migrating molecules. If you have a difference, then you know the underlying proteins are different, and, of course, then the underlying nucleotides are different. If two of these molecules migrate in the same fashion, then it doesn't necessarily mean that they are the same because there could be a lot of hidden variation both at the DNA level, and even at the protein level. You can imagine that if you have one charged amino acid substituted for another similarly charged amino acid, you are not going to see a difference necessarily." He continued, "You asked earlier about a plesiomorphic case. If you were to do a cladistic analysis of protein electrophoretic data, you might be assuming that there are two species that share this ancestral 'state', the plesiomorphic state. But in fact, it may turn out that this one and the other are not really alike, but they happen to have migrated in the same way. So, there has always been that drawback. I think the one paper, with fourteen loci in it, is near the lower end of the boundary of what one would like to publish using electrophoretic data. It would have been better to have a larger suite of electrophoretic data, even better yet would be to have sequence data from multiple loci, more than just the mitochondrial information. But, you can still infer trees from those protein electrophoretic data. The difficulty with them in terms of inferring trees is that some of the approaches that are used and that are favored today, like cladistic approaches, are somewhat problematic to protein electrophoretic data for reasons that I won't get into. But coding the data can be difficult and even controversial for protein electrophoretic data. I think from the more analytical standpoint, sequence data are so much better, because of things that you can do with the data. It comes down to two things, pattern, i.e., you can certainly get a reasonable tree using protein electrophoresis data, and distance, i.e., if you want to compare distances among nodes in a tree, and we did that successfully in our louse pocket gopher studies. The ideas that we used in the 1994 paper, for the most part, were developed from protein electrophoretic work."

I then asked, "Did you see major changes in the conclusions you drew from 1988 to 1994 using the different approaches?" Steve replied,

"No, we didn't. But, if you can get more precise DNA data, then that would be the route to follow. And today, it is easier to get DNA data than it is to do protein electrophoresis. The latter approach is a real pain, i.e., preparing buffers and the like takes a lot of time. Nowadays, it's just more practical to work with DNA. I don't think there is anything wrong with protein electrophoretic data, except there could be cases in which taxa could actually share the same state, but you can't resolve it. I think when you look at that many loci, I think what happens is, it all comes out in the wash."

I have asked several of my interviewees, i.e., Roy Anderson, Don Bundy, and Peter Hotez, their views on overdispersion and genetic predisposition to infection. I wanted the perspective of an evolutionary biologist about the same thing, so I asked Steve, "Do you have any feelings about this idea one way or the other?" He replied, "Yes, I have to believe that there has to be some genetic predisposition to infection. How much of the overdispersion is due to genetic predisposition versus behavior and environment, I don't know. But, in a lot of these cases, I just feel that genetics plays a role." Well, at least I got my question asked and I was satisfied with his answer, not quite as forcefully given as Anderson, Bundy, and Hotez, but, he agrees, there is a genetic component to the epidemiology of these geohelminths.

During Steve's career, he has worked variously with ascaridoid nematodes, with hookworms in marine mammals, and several other systems, and I wanted to get into some of this work. To do so, I wanted to ask him about something he published in the *Journal of Molecular Evolution*. The idea that rotifers and acanthocephalans are related in some way is not a new one. Rotifers, nematodes, acanthocephalans, and several other taxa are frequently grouped together as the Aschelminthes. In fact, if memory serves me, this idea appears prominently in one of the classic invertebrate books of my early days by Libby Hyman (1951). I asked Steve about this, and he replied, "The Aschelminthes is not a valid higher taxon because it is not monophyletic." To re-emphasize, the idea that rotifers and the acanths are related is definitely not a new one. However, the Garey *et al.* (1996) paper, according to Steve, "Is a pretty interesting finding, previously supported just by morphological observations, but now by molecular data as well. The story here is that Tom Near, who was an undergraduate in my lab at the Northern Illinois University, decided to do a Master's in my lab and was looking for a project. I suggested that he do some work on acanth phylogeny. So, he was doing some sequencing of acanths and came across this paper describing the possible connection between rotifers and acanths, and he wrote Jim

Garey who was working at Duquesne University in Pittsburgh. Jim was
big on invertebrate relationships and a good one to contact. He was very
helpful." Why were the results of such great interest? The answer is simple. As stated in Garey *et al.* (1996), "Until now a free-living sister group
of a major obligate parasitic taxon has not been identified, hampering
comparative studies of character change associated with the evolution
of parasitism. Comparisons between acanthocephalans and free-living
rotifers should prove instrumental in solving long-standing problems
such as the relative importance of secondary character loss vs. character innovation in the evolution of a parasite life history (Brooks and
McClennan, 1993)." Unfortunately, this insightful prediction is, as yet,
unfulfilled, primarily I believe because no one has pursued an answer.
The results generated by the Garey *et al.* (1996) paper were apparently
quite controversial when the paper was first published. As Steve said,
"At the time, the rotifer people were really upset. There are a lot of
rotifer biologists and they didn't like the idea of the acanth-rotifer relationship, but it is now pretty strongly supported. The question then
became, which rotifer group is sister to the acanthocephalans? That's a
little trickier. However, in collaboration with Martin Garcia-Varela, who
worked as a postdoc in my lab at Davis, we have obtained data from
multiple loci indicating that bdelloid rotifers are the sister group to
the acanthocephalans."

I asked Steve, "How can these results be used to examine the
question dealing with free-living versus parasitic systems?" He replied,
"I think one of the things that you might want to ponder would be,
what sort of genetic differences would accompany the transition from a
free-living state to a parasitic one? What genes, for example, are unique
to acanths versus rotifers? You could have genome comparisons, for
example, or you could look at expressed genes. It's a fundamentally
interesting question. We get this in the nematodes as well, where there
are free-living taxa that are closely related to parasitic ones. Some of
the most important plant parasites are this way. In the SSU [small sub-
unit RNA] data, at least, and one other locus, it has been shown that
they are most closely related to some free-living nematodes, which are
microbial feeders. It's another case where, how did you get from the
free-living microbial feeding state, with a stoma that is pretty much
unadorned, to a parasitic mode of life in which the mouth is elaborately
modified? The model organisms that we should be using to look at the
comparative aspects have really changed now that we know more about
evolutionary history. For example, when root-knot [nematode] females
migrate into the roots of their host plant, they stay there and induce

the formation of a multinucleate nurse cell complex, like *Trichinella*. It's basically a feeding site for the female nematodes. She then secretes eggs into a gelatin-like matrix. There is another group of plant parasites, the *Heterodera*, or cyst nematodes, which are very similar to the root-knot worms. Classically, in plant parasitic nematology, these two groups of worms were considered to be members of the same family, the Heteroderidae. However, using phylogenetic analysis, we have discovered that this is not true at all. It's a case of convergent evolution, with two nematodes that both end up as parasites, both feeding at the same site in the root, but they are not sister groups. For a long time, people were thinking that they were very similar. On the other hand, there was some evidence for differences before the phylogenetics research based on TEM analysis of the respective nurse cell structure. It's amazing about the convergence that can occur among some of these nematodes."

Perhaps one of the most interesting areas of research pursued by Steve Nadler and his colleagues deals with the hookworms of pinniped species on the west coast of North America, then down the western coast of South America, and around the tip to the Falkland Islands on the eastern coast. The earliest described species in this group is *Uncinaria lucasi*, from the Pribilof Islands in the Bering Sea. In 1958, O. Wilford Olsen made the initial headway in describing the life cycle of this parasite, which was finally resolved by Olsen and Lyons in 1965. In the northern fur seal, the cycle is quite different from the typical hookworm of terrestrial mammals. Infections occur in seal pups when they suckle milk and simultaneously consume L_3 larvae from the mother's mammary glands. After about a three-month life span, adult hookworms are spontaneously shed from the pups. Hookworm eggs released from the adult female worms develop in the rookery sand to the L_3 stage, then penetrate the skin of all seals, whether males or females. L_3s infecting male seals are at a dead-end. In females, however, the L_3s head for the females' mammary glands, then lie in wait for lactation and suckling to begin during the next birthing cycle. The hookworms are particularly dangerous if they occur in large numbers, causing death in heavily infected hosts.

The species in the northern fur seal of Alaska is definitely *U. lucasi*. However, another species, *Uncinaria hamiltoni*, in South American sea lions, has been described using morphological characters and it too looks valid. Nadler *et al.* (2000) examined a hookworm from California sea lions and the one in the northern fur seal (*U. lucasi*) using nuclear ribosomal DNA and concluded that they were different species. Since the California sea lion hookworm was thought at the time to be a variant of South American *U. hamiltoni*, this would mean that they were dealing with three species, not two, confirming their multivariate analyses data. They concluded, "These hookworms represent two species that are not distributed indiscriminately between these host species, but instead exhibit host fidelity, evolving independently with each respective host species."

In our interview, Steve said, "The original paper [Nadler *et al.*, 2000] dealing with these hookworms was based on a single locus. We have not published some of the new results, but the work on mitochondrial data show the same thing. I'm now using nucleotide sequence data and analyzing it phylogenetically to look for evidence that these taxa are independent evolutionarily. When you do a phylogenetic analysis using ribosomal DNA and you include lots of individual hookworms from California sea lions, there may be unique substitutions that circumscribe all of the California sea lion parasites, and other ones circumscribe parasites from the northern fur seals making them different in this regard. Does this mean they are evolving independently? There is sort of a population level part of this question, i.e., when we say that here is a change, a substitution, that appears to be derived and is unique for this group, could it be a polymorphism? Let's say you only surveyed one or two nematodes and you said, okay, those are going to be my representatives for the California sea lion nematodes and then a couple more for the fur seal nematodes. This approach is often taken in phylogenetic studies when you assume that you know with which species you are dealing. However, you might make a mistake because it could be that some of these substitutions that you are using to demarcate a group actually are polymorphic. In other words, there could be multiple states at that site and you think they are fixed in a population genetics sense, but they are not. So, basically, it's an evolutionary distinction on what species are and on recognizing species as being independent evolutionary units. The way I'm advocating to approach this problem at present is not to use one locus alone, but to look at multiple gene loci. I am now using the nematodes' mitochondrial locus and a nuclear locus. If you see the same pattern, then it's one of those cases where 'two is more than twice one'. You've got two completely different genomes, a mitochondrial genome and a nuclear genome, and in both you see evidence that these things are evolving, and have been evolving, independently, through time. So, we are now extending this analysis to include a variety of hookworms from other hosts. What we hope to do is go back and probably redescribe those species morphologically and then, whenever other species are found in other hosts, to describe them as well."

Steve continued, "This is a fascinating system." I agreed enthusiastically. He then said, "The collaborator (Bob DeLong) we have has access to San Miguel and San Nicolas Islands off the California coast where we did some of the 2000 study. And they have found something new and unusual. On San Miguel, in California sea lions, there is even more pathology in the last two years. And what they have found is that adult hookworms are crossing the gut wall and ending up in the peritoneal cavity." I interrupted at this point and asked, "What on earth are they doing in the peritoneal cavity?" Steve replied, "Well, they are causing peritonitis is what they are doing. Lots of pups are now dying. You know, we also get animals that are migrating and dying (from other causes) during that migration. A few years ago, Gene [Lyons, another collaborator] was able to obtain tissue from some of these dead animals. He removed the blubber and placed it into a Baerman funnel. The larvae, which are in arrested development, crawled out. I have taken individual larvae and done PCR on them, and either sequenced them or done a diagnostic RFLP analysis on them. It turns out that so far we have only been able to get California sea lions that are migrating but, in a few of these animals, some of the larvae that were in arrested development in the tissues were the northern fur seal parasites, not a high percentage, but they were in arrested development mainly in the blubber. So we are continuing these studies. We are collecting hookworms from the peritoneal cavity and from the intestine. We think it is the wrong parasite because it would not be migrating that way if was in the right host species." I could tell from his demeanor and voice that he was really excited about this particular line of research. I too think it sounds very exciting and that it is not an intractable problem. It will take some time to resolve, but the wait will be worth it once he and his colleagues have met the challenge.

It was getting close to the end of our discussion and I really hated to stop, but I felt it was time (I have only so much space to fill in the book). So, I asked him, "Where are you going to be ten years from now?" I think I caught him off guard as he responded, "Hmmm, that's a good question. Well, you know, we're doing a lot of phylogenetic work on nematodes and, now don't get me wrong, but some of it is boring. There is such a tremendous diversity among the nematodes and I have really become interested in certain aspects of nematode biology and natural history. It's an entirely new area for me. You know, I've talked to a number of people about this and producing a tree isn't very much fun if there isn't much biology to go with it. It gets kind of old. And I have people call me up all the time and invite me to work with them on this nematode and that, and could easily produce a tree if there is enough material. And, you could look at stuff like character evolution, which is almost routine and map those things on a tree, and develop hypotheses based on the tree. But it seems to me that it would be more interesting to get back to some of the biology of the organisms, maybe even some experimental things, to try to tie together evolutionary hypotheses with the real biology of what's going on. Ten years from now, I'm not sure exactly what I'll be doing, but it will be along those lines, i.e., what do these animals do, and do close relatives do it the same way? Another thing I have thought about is something I said in the review I wrote for the *Journal of Parasitology* (Nadler, 1995), and that would be to do a good comparative study on population genetics structure by looking at a couple of sister species."

Before Steve leaves phylogenetics for what he hopes to be 'greener pastures', I am confident that he will finish the hookworm work. This should be a very interesting and revealing piece of research. He also has embarked on some anisakid phylogenetics that will also be productive. I don't know what Steve will be doing ten years from now, but I would bet a lot of real 'bucks' that he will become good at it and that he will become a leader in that field, just as he is in the realm of phylogenetics and systematics at the present time.

REFERENCES

Brooks, D. R. and D. A. McClennan. 1993. *Parascript.* Washington, D.C.: Smithsonian Press.

Garey, J. R., T. J. Near, M. R. Nonnemacher, and S. A. Nadler. 1996. Molecular evidence for Acanthocephala as a subtaxon of Rotifera. *Journal of Molecular Evolution* **43**: 287–292.

Hafner, M. S. and Nadler, S. A. 1990. Cospeciation in host-parasite assemblages: Comparative analysis of rates of evolution and timing of cospeciation events. *Systematic Zoology* **39**:192–204.

Hyman, L. 1951. *The Invertebrates.* Vol. III: *The Pseudocoelomate Bilateria*. New York: McGraw-Hill.

Nadler, S. A. 1995. Microevolution and the genetic structure of parasite opulations. *Journal of Parasitology* **81**: 395–403.

Nadler, S. A. 2002. Species delimitation and nematode biodiversity: phylogenies rule. *Nematology* **4**: 615–625.

Nadler, S. A., B. J. Adams, E. T. Lyons, R. L. DeLong, and S. R. Melin. 2000. Molecular and morphometric evidence for separate species of *Uncinaria* (Nematoda: Ancylostomatidae) in California sea lions and northern fur seals: hypothesis testing supplants verification. *Journal of Parasitology* **86**: 1099–1106.

Sober, E. 1988. *Reconstructing the Past: Parsimony, Evolution, and Inference.* Cambridge, Massachusetts: MIT Press.

15

Toxoplasma gondii, Sarcocystis neurona, and *Neospora caninum*: the worst of the coccidians?

The sick are the greatest danger for the healthy; it is not from the strongest that harm comes to the strong, but from the weakest.

Genealogy of Morals, Friedrich Nietzsche (1844–1900)

Without question, the most speciose group of eukaryotic parasites are the coccidians. For me, the three most significant members are *Toxoplasma gondii, Sarcocystis neurona,* and *Neospora caninum*. One way or another, J. P. Dubey has had a direct hand, literally, in discovering and naming two of these organisms, in resolving one of the parasites' life cycle, and in developing a significantly deeper understanding regarding the biology of all three. Two of these three coccidians, namely *T. gondii* and *N. caninum*, infect more people, and cause more abortions in sheep and cattle, than any other protozoan parasite. The third, *S. neurona*, inflicts serious neurological damage to horses within the U.S.A. Because of the significant medical, veterinary, and economic importance of these parasites, I wanted to talk with J. P. and learn what I could about his interest in these organisms and the diseases they cause. Having known J. P. for probably thirty years, I thought I knew the answer, but I wanted to satisfy my curiosity and see if my guesswork was correct, so I headed for Beltsville, Maryland, in early May of 2006 to find out.

His 'love affair' for protozoan parasites, especially *Toxoplasma gondii*, goes back a long way. It actually began with the plagiarism case involving a prominent researcher at the Institute of Veterinary Research in India, even before he started research for his Master's degree. My first question for him in our interview was directed at why he became interested in protozoans, and not parasitic helminths? What was it about protozoans that drew his early research interest? He responded, "Actually, helminths in India are more important economically, or at least they were at the time I began my studies. In domestic livestock, helminth parasites are much more important, causing much debilitation and even mortality. However, the role of protozoans in some of these situations back when I started was not known or fully appreciated. After all, for example, the role of *Toxoplasma* in sheep abortion had only just been discovered." I came away with the feeling that he believed these organisms represented a real intellectual challenge. More importantly though, they embodied a serious medical and veterinary menace in J. P.'s mind and, simply put, he wanted to diminish the problem they presented.

On completion of his Master's degree and receipt of the prestigious Commonwealth Scholarship, he was off to the University of Sheffield in England for the expressed purpose of working on *Toxoplasma* in the lab of Colin Beattie and Jack Beverley. He said, "I went to Sheffield to work with *Toxoplasma* in sheep, because I was a veterinarian. I knew that Beverley had done some remarkable work in discovering the abortion storms in sheep caused by *Toxoplasma* and that thousands and thousands of lambs are lost every year." He told me that during the second week in Sheffield, he and Jack Beverley sat down and talked about what he wanted to do. J. P. told him he wanted to work with *Toxoplasma* in sheep. However, he said that Beverley was not really very responsive initially and that he wanted to think about it overnight. He explained, "The next day, Beverley told me that there was a lot of new interest in working with *Toxoplasma* in cats . . . It was then that I found out about Bill Hutchison at the University of Strathclyde in Glasgow. Not long before I arrived, Bill had come down to Sheffield to consult with Beverley and Beattie about a paper he was going to submit for publication in *Nature*. Jack told me that Hutchison said he had just discovered that *Toxoplasma* is transmitted through cat feces via the eggs of a nematode, *Toxocara cati*. Hutchison informed him what he had done in the experiment, but Beverley was a very cautious person. Hutchison should be very thankful he consulted Beverley because he urged Hutchison to insert a sentence in the last paragraph of the paper that reported all of what was present in cat feces in addition to *Toxocara cati* eggs. In the concluding paragraph, Beverley had him add that there were nematode eggs, along with bacteria and *Isospora* oocysts, and that it was uncertain if fecal transmission was by nematode eggs, or some other means. Retrospectively, the *Isospora* oocysts turned out to be those of *Toxoplasma*!"

After Hutchison left, Beverley told J. P. about part of his conversation with Hutchison, but not all of it. Hutchison's paper appeared in *Nature* about a year later, so Dubey knew the story about a year ahead of everyone else. Since there were only a few people in the world who knew about Hutchison's results, J. P. was able to begin working with cats immediately. His initial idea of working with sheep "went out the door".

J. P. then said, "Let me tell you the whole story of Bill Hutchison's research. Beverley had gone to Glasgow to give a lecture about the transmission of *Toxoplasma*. Jack had been wondering whether defecating crows might be important in the transmission of *Toxoplasma* to sheep in pastures where they grazed. Hutchison was in the audience when Beverley gave this lecture and he then became interested in the problem. One of the things that Beverley had said during his lecture was that this protozoan was very fragile, that placing it in water could even kill it. This caused Hutchison to begin thinking about other ways in which transmission could occur rather than via crows. It led him to hypothesize that maybe *Toxoplasma* could be physically transported by another parasite, similar to the way *Heterakis gallinae* carries the protozoan, *Histomonas meleagridis*, which causes 'Turkey blackhead' in chickens. He began looking for a nematode that could infect both cats and humans. It was at this point that Hutchison came up with the idea. He thought that perhaps the eggs of *Toxocara* could transmit *Toxoplasma*. At first, however, Hutchison wanted to do an experimental test using *Toxocara canis* in dogs because this helminth was known to be zoonotic. However, because the University of Strathclyde was so new, there were no facilities to keep dogs. So, he had to use a cat, which he initially maintained in a cage at his home before moving the animal to the University where the cage was kept on one of his lab benches. He fed *T. gondii* tissue cysts to the cat he had infected with *Toxocara cati* two months earlier. He isolated eggs of *Toxocara* from the cat's feces and placed them in beakers so they could embryonate in water. He then fed the eggs to mice. The mice became infected with both *Toxocara* and *Toxoplasma*."

The results clearly supported his hypothesis. J. P. remarked, "If Hutchison had chosen the dog instead of the cat for his experiment, he would never have discovered the oocysts. It was just chance [pure serendipity!] that it happened that way. He did the experiment with a cat, and that is why and how I got started with *Toxoplasma* and cats."

I then said to J. P., "But you knew that Hutchison was on the track of something, didn't you?" He responded in the affirmative. I then interjected, "Did you believe his story? I mean you persisted in your work with cats. How did you separate his error from the truth?" He replied, "Well, you know, when the paper was first published, I believed it. It was Jack Beverley who was very suspicious of the whole thing. He did not believe in it. So, he encouraged me to continue with my work. He reminded me that very little was known about *Toxoplasma* in the cat, so he asked me to do many things. One of the things I pursued was infection in naturally infected cats. He wanted me to isolate *Toxoplasma* to confirm that the parasite was really there. If it is there, then we should find it in the cat, and we did. Something else I discovered in my three years at Sheffield was that *Toxoplasma* could be transmitted from a cat that was worm free. At the time though, we thought that our mice must have been contaminated with something. We thought that there must be something else in the lab that was contaminating our animals. Another thing that happened was I isolated *Toxocara* eggs from a naturally infected cat, but was never successful in getting *Toxoplasma* transmission using its eggs. In that case, we filtered the *Toxocara* eggs through a sieve. In hindsight, I now know that we had collected the eggs of the nematode, but had filtered out the *Toxoplasma* oocysts. Those who know about *Toxocara* eggs also know they are very sticky. We were working with coccidian oocysts that were sticking to the eggs of *Toxocara*," which is why Bill Hutchison was able to make the connection between the nematode and *Toxoplasma*. However, he and Jack Beverley did not realize this at the time. So, during his stay in Sheffield, the issue of *Toxoplasma* transmission using *Toxocara* eggs was never really resolved. In other words, the results they generated confirmed Hutchison's hypothesis in some experiments, but not in others.

J. P. told me that, "In the meantime, Hutchison had done his wonderful experiment again. He repeated his first experiment using two cats this time, instead of one. One of the cats he infected with *Toxocara cati*, and the second was worm free. Then, he fed *Toxoplasma* to both cats. He collected the eggs and embryonated them as he had done before. Then, he dewormed the cat infected with *Toxocara* and infected the second cat with *T. cati*. He then fed *Toxoplasma* cysts to both cats again and, after isolating and embryonating eggs of *T. cati* from the second cat, he fed the eggs to mice. Again, only the inocula with *T. cati* eggs produced toxoplasmosis in mice. The results of this experiment were identical to the first one he had tried with just one cat. In other words, if there was no *Toxocara*, there was no transmission of *Toxoplasma*. In hindsight, we now know that the chance of this experiment working successfully is one in ten million, or even a hundred million. It was only an act of God that it happened the way it did. There is no truth to the association between *Toxocara cati* and *Toxoplasma*. I still believe that Bill Hutchison was a very honest man, and that he reported [the results of] his experiments as he saw them. But the chances of this happening [are extremely remote]."

I then asked why Hutchison repeated the experiment? J. P. responded that it was because he had used just one cat the first time. I then asked if Hutchison's results were challenged in any way at the time? He said that they were not challenged because he was so far ahead of everyone else. But, I then said, "He was not ahead of anybody. His results were an aberration." J. P. nodded and agreed. "If you run this experiment a thousand times, it will fail 999 times. And the reason it will fail is that we now know that there is a very strong immunity to *Toxoplasma* in the cat. This is a classic example of why we should always repeat experiments in biology."

Jack Frenkel at the University of Kansas Medical Center had obtained a grant to work on the idea of nematode transmission of *Toxoplasma gondii*, and invited J. P. to join him after his hiatus to India following completion of his Ph.D. at Sheffield. He was anxious to continue his work on *Toxoplasma*, so he eagerly accepted the offer. He continued the story. "I was in a good position to contribute, because I was ahead of everyone and I knew the whole history of what had already been done. In addition, because of my Master's degree, I was already familiar with coccidian biology. The first month I was in Jack's lab, we infected cats with *Toxocara* and *Toxoplasma*. We knew that *Toxocara* eggs were about 80 microns in diameter, so we took a 44-micron sieve and filtered cat feces that were infected with both parasites. We then microscopically examined the filtrate to ensure there were no *Toxocara* eggs present, and they were gone. We then took the filtrate, gave it to mice, and they died from toxoplasmosis. We had proved that the nematode had nothing to do with the transmission of *Toxoplasma*."

I then asked J. P. if he had any idea that *Toxocara* was not really involved with the transmission of *Toxoplasma* while he was still in Sheffield working on his Ph.D. He said at first he was not suspicious, but that changed with time. And, he reminded me, Jack Beverley was very suspicious from the beginning. Part of his reasoning was related to the huge diversity and numbers of hosts that had been identified for *Toxoplasma*. "So," I said, "you took this idea with you to Frenkel's lab in Kansas City, where you and Jack designed an experiment to test the hypothesis." He replied, "Yes, that is correct, and, you know, Jack was very good at designing experiments. Together we did this. It was an exciting period in my life. I remember not sleeping many nights because I was so thrilled. We also employed a new approach in that we used newborn kittens to do the experiments. Everybody in the past who had worked on the problem had used older animals, whether they were sheep, or goats, or cats. The idea was that kittens would not be infected with *Toxocara*. And, if they were newborn kittens, they should be free of everything else too. When we fed *T. gondii* tissue cysts to one-day-old kittens and found the oocysts three days later, we knew we were on the right track."

"One thing that delayed publishing the real life cycle story, however, was we were sporulating *Toxoplasma* oocysts in potassium dichromate, like everybody else. We would sporulate them in potassium dichromate, then wash them, and feed them to mice. When we did this, the infectivity of the oocysts was always lower than the counted number of oocysts. If we counted 100 000 oocysts in an inoculum, the infectivity was always tenfold, or a hundredfold lower. Jack kept asking me, 'How can we deal with *Toxoplasma* if the infectivity and the numbers don't match?' He said that we would be making the same mistake that Hutchison had made. So, we came up with this idea of trying 2% sulfuric acid rather than potassium dichromate for sporulating oocysts of *Toxoplasma*. We then would use sodium hydroxide to neutralize the sulfuric acid inoculum. The sodium sulfate that is produced does not harm the mice. When we tried this approach, there was a hundred percent match. When we inoculated 100 000 oocysts, that is how many became established. Later on, we discovered that the dichromate affects the oocyst wall in some way, and this is what prevents 100% excystation. We also discovered that we could inoculate this mixture of sodium hydroxide, sulfuric acid, and oocysts, intraperitoneally and subcutaneously, and obtain excellent results, even better than oral intubation. In doing this, we discovered that coccidia are infective to their hosts parenterically."

I asked J. P. if Bill Hutchison's publication of the *Toxoplasma* life cycle in 1965 had much scientific impact, even though it was later shown to be incorrect by you and Jack Frenkel. J. P. responded that it did. "Yes," he said. "There was an explosion of ideas. Many labs wanted to work on this transmission. Harley Sheffield and Leon Jacobs at NIH were both working with *Toxoplasma*. In fact, our paper and one by Harley Sheffield that proved the identity of *Toxoplasma* in cats by electron microscopy appeared in the same issue of *Science*."

J. P. then told me an interesting story regarding priority in resolving the *Toxoplasma* life cycle. It seems that the Frenkel/Dubey/Miller paper was supposed to have been published in December 1969, but was delayed until February 1970 because the Editor of *Science* wanted to devote an entire issue to the moon landing and walk by Neil Armstrong on 20 July 1969. Hutchison and Siim in Denmark wanted to claim priority for the *Toxoplasma* life cycle since they had also discovered the oocyst of the parasite and its significance to the parasite's transmission. (Bill Hutchison collaborated with Siim's lab in Denmark because the facilities he had in Scotland were so poor.) J. P. said, "there was a lot of rivalry at that time and they wanted to publish their work sooner than us. So, Hutchison and Siim got a one-paragraph note into the last issue of the *British Medical Journal* in 1969 in order to claim priority. Their main paper followed in January. Our paper was published in February 1970, like theirs, but it would have made it into December 1969, except Armstrong landed and walked on the moon. There was a lot of controversy about the issue." I asked him how priority was actually determined? He replied, "The priority has never really been established. There have been many papers published about the conflict between the two sides. But, in the end, the fact remains that much of the definitive proof was provided in Jack Frenkel's lab. I was a young investigator at the time and was excited about these sorts of things. However, Jack always told me, 'look, at the end of the day, what remains is, what proof did you provide to establish an hypothesis?' And I think that people who will read about the discovery of the life cycle, how it was done, what methods were used, will make the right decision" about who did what and when.

A few years ago, J. P. contacted me about some new research he was undertaking and asked if the *Journal of Parasitology* would be willing to publish a series of what might appear to be repetitive Research Notes. After he explained his objective, I enthusiastically agreed. During our interview, I asked him to elaborate on the new focus so I could include it as part of this essay.

He told me that after *Toxoplasma gondii* was described and research began, it was soon thought that there were several species, one in the dog, another in birds, and so on, and they were all given separate species names. In 1948, however, a reliable serological test for *Toxoplasma* was developed and it was soon realized that all of these different 'species' were not different after all. They were not only morphologically and serologically identical, they were likewise cross-protected. Using these three criteria, it was finally concluded that there was just one species, i.e., *Toxoplasma gondii*, and that it infected almost all warm-blooded animals. Then, it was found that the isolates of *Toxoplasma* were very similar genetically to each other, "so much so that a clonal theory was developed for *Toxoplasma* by Sibley and Boothryoid." In other words, it was proposed that all isolates of *Toxoplasma* were ultimately derived from a single source. He told me that, "Until recently, it was believed that all *Toxoplasma* isolates were different from each other in only two alleles. Dan Howe and David Sibley began to analyze all the isolates they could find from around the world. They proposed that you could divide *Toxoplasma* into three genetic groups, which they termed Type I, Type II, and Type III. They also proposed that, phenotypically, Type I *Toxoplasma* was mouse virulent and that Type II and Type III *Toxoplasma* were relatively avirulent. In fact, they found that 100% of the Type I isolates were lethal for mice. However, many people did not believe in this genetic clonal theory. Still, most people thought they were all alike."

About seven years ago, J. P. was approached by Tovi Lehmann, a biologist working out of the Centers for Disease Control in Atlanta, Georgia, and who is now at NIH in Bethesda, Maryland. He asked if J. P. would share with him the *Toxoplasma* isolates that he was maintaining in his lab. J. P. remarked, "You know, of all the people working on *Toxoplasma*, I had the greatest number of isolates from food animals. I told Lehmann that Howe and Sibley tested isolates that had been maintained in the lab either in mice or in cell culture for very long periods of time. In addition, the history of many of these isolates was not known. I was really adamant that you can change *Toxoplasma* biologically by keeping it in cell culture or in mice. So, I told Tovi that in nature, we do not have highly virulent *Toxoplasma* like the RH strain. The latter is a highly virulent, man-made strain that was isolated in 1939 by Albert Sabin from the brain of a young boy infected with *Toxoplasma*. It has been in mice since that time, but is still 100% lethal to mice. However, we later found that it is so 'mouse-adapted', it does not even produce oocysts when introduced into cats. We have since found that if you pass any *Toxoplasma* isolate in the lab thirty times that it will not produce oocysts any more. I told Tovi that many of these strains of *Toxoplasma* are really useless in a way because they do not behave like the wild type any longer. In addition, we don't know the histories of many of them."

Based on this kind of information, Tovi and J. P. conceived an experimental design to get around the problems associated with lab-maintained *Toxoplasma*. What they decided to do was go into the field and obtain *Toxoplasma* from just one kind of host, and bring it to a single lab where it could be handled using a standardized series of protocols. The host they selected for the isolation of *Toxoplasma* in the field was the free-range chicken. J. P. said, "I thought of the free-range chicken because it does not move around very much. Most of them will not go more than a few hundred yards from their home. So, we came up with this idea that we are going to examine the population genetics of *Toxoplasma* based on isolates from the free-range chicken. Many people feel that I am crazy for doing this study in chickens from so many different countries, but it has an objective to it. We started out in Brazil, where we found excellent collaboration with Solange Gennari and Lilian Bahia-Oliveira. One of Lilian's graduate students, Daniel Silva, went door-to-door on the outskirts of Rio de Janeiro. He bought the chickens, killed them, and brought selected tissues and organs back to Beltsville. We inoculated parts all of those chickens into mice." Although J. P. did not mention it, another reason for using chickens is that they are 'ground feeders' and likely to ingest oocysts if they are present in the soil.

J. P. continued, "We are early into the study, but I was so surprised when I found that when you inoculated these isolates from Brazil into mice, they started dying the second week. I had worked with *Toxoplasma* for thirty years and I had never seen something like this happen, i.e., that mice started dying from toxoplasmosis the second week after they were inoculated. In fact, I became seriously worried that a virus had invaded my lab, my mouse colony, or something like that. So, I examined them more carefully and confirmed that we were dealing with virulent *Toxoplasma* that was killing the mice. As a result, each of the isolates was cryopreserved after its primary isolation. We not only cryopreserved each isolate, we proceeded to inoculate five outbred Swiss mice from each isolate and recorded the mortality data. We collected *Toxoplasma* DNA from each infected mouse. I am glad that the *Journal of Parasitology* is publishing some of this research. It may seem repetitive, but it is not. So far, we have collected strains from most of the South and Central American countries. We also have isolates from Asia, Africa, and North America, including Mexico, for comparative purposes. All of these isolates are being genetically analyzed. The initial analysis was based on one gene. There was initial criticism that we were repetitive and that we were using only one gene, but our main motive was in documenting the origin of each isolate, where it came from, the geographic location, and give a designation to each isolate. We intentionally obtained chickens from houses that were at least 500 meters apart, so we know that our isolates are independent of each other. Amazingly, the isolates are extremely different. For example, Type II *Toxoplasma* is completely absent from Brazil. We have also found that the isolates from Chile are entirely different than from Brazil, although they are adjoining countries. The isolates from Peru are also different than from Brazil." Based on what J. P. told me about this effort, I am absolutely confident that their results will be very useful and that they will be able to ultimately tell a fascinating story regarding the world-wide distribution and diversity of *Toxoplasma* as determined by solid genetic evidence.

I knew we would get to *Sarcocystis neurona* and *Neospora caninum* before ending the interview, but there was another question I wanted to ask first. The end of the *Toxoplasma* discovery story offered me an opportunity to 'use' one of the world's experts to answer it. I told J. P. that whenever I teach my parasitology course, I always include the coccidians. When I do, one of the things I always mention is the fact that there are at least 30 000 species of *Eimeria*. I asked him to first explain why are there so many species in the group and, second, are all of them really valid taxa?

He sort of smiled and replied, "It is a very puzzling situation, isn't it? I believe that all of the coccidians were originally transmitted fecally, with one-host life cycles. They mutated rapidly and, over time, became adapted to different life-cycle patterns. The eimerians retained their one-host life cycles. There is simply a lot more mutation going on in that group, but we don't know why. They are very host specific, and they are genuine species. You cannot cross-transmit them. For example, the eimerian species that occur in sheep and goats are morphologically indistinguishable, but you can't take the *Eimeria* from a sheep and put it into a goat, or vice versa. I think it is logical to assume that these two species had a common ancestor and, as animal husbandry progressed, they became adapted to either sheep or goats. But it remains a mystery as to why eimerians are so host specific, and site specific within their hosts. Then, in contrast, look at *Toxoplasma* at the other end of the spectrum. It has a heteroxenous life cycle. It infects virtually every warm-blooded animal in the world, and it is not really site specific. However, it has retained specificity with respect to completion of its life cycle in felines. Many people refer to it as being neurotropic, but I personally don't believe that it is, although in humans and in mice it is more often in the brain. In contrast, the parasite is more often localized in musculature of food animals. It looks more reasonable that musculature is the right place to be for these animals since the life cycle depends on carnivory for success." He also believes that *Toxoplasma* originally was a coccidian of cats, and that it gradually became adapted to nonfeline hosts. It depended on domestication of the feline for it to have been acquired by humans. So, he did not have a direct answer for my question as to why there are so many species of *Eimeria*, except that there was a great deal of mutation involved, which was always accompanied by natural selection.

I then turned to a second coccidian parasite with which J. P. is closely identified, *Sarcocystis neurona*. After five years with Jack Frenkel, J. P. left for an academic position at Ohio State University in Columbus, Ohio. The very first week he was there, he was shown slides from the brain of a horse that had died with a toxoplasmasmosis-like illness. When Ohio State experts first viewed the slides, the parasite was identified as *Toxoplasma*. He told me that there was a particular interest in the dead animal because it had belonged to a millionaire. Interestingly, it was being transported in a trailer when the vehicle was struck and wrecked by a car. The horse was injured in the wreck and, while hospitalized at Ohio State, it was given large doses of cortisone to reduce some of the inflammation. Eventually, the horse died with an overwhelming *Toxoplasma*-like disease (the condition later became known as equine protozoal meningeoencephalitis, or EPM). At necropsy, tissue was taken from the horse's brain and spinal cord, and examined. This is when they discovered the *Toxoplasma*-like organisms.

J. P. said that when he examined the slides, he was amazed to see hundreds of these organisms in the intact neurons, something that he had never observed previously. He looked carefully for *Toxoplasma* cysts, but, oddly, none was to be found. There were, however, cyst-like structures, but when he used a *Toxoplasma*-silver stain, the cysts would not take it. They also were not PAS-positive as they should have been if they were those of *Toxoplasma*. So, J. P. wrote a paper with other pathologists, describing these aberrant characteristics (Dubey *et al.*, 1974). The paper concluded by saying that they were not dealing with *Toxoplasma*, but something different.

In the same year, investigators at the University of Pensylvania and Illinois State University published papers in which they claimed they had found *Toxoplasma* in horses. J. P. in fact obtained the slides from the group at Penn. When he viewed them, he concluded that they were *Toxoplasma*-like, but not *Toxoplasma*, and very similar to the organisms he had seen in the millionaire's horse.

Between 1974 and 1990, he said he had seen the same parasite in the brains of horses on numerous occasions. From 1974, when the first three papers were published, until 1990, J. P. said that he tried to grow the unknown organism in cell culture, but was consistently unable to achieve success. Finally, in 1990, he had an opportunity to collaborate with Dwight Bowman at Cornell University in an effort to solve the long-standing problem. Dwight sent a spinal cord from a horse that had died due to EPM to J. P. at the Agricultural Research Service center in Beltsville. Stan Davis was a postdoc who came to Dubey's lab from the vet school at North Carolina State University to work on *Toxoplasma*. J. P. asked him to try and grow the organism in cell culture. He said that they had two cell lines at the time, bovine monocytes and cardiopulmonary endothelial cells. Surprisingly, Davis got it to grow in bovine monocytes, but they still did not know what it was. Before they were successful with the in vitro culture, J. P. said he had made a decision with respect to naming the organism. He had decided to call it *Sarcocystis neurona* based on histology and structure of the organism. Because of their success in growing the new parasite in cell culture, it made his previous decision faster and more definitive. The paper describing the new species was ultimately published in the *Journal of Parasitology* in 1991 (Duby *et al.*, 1991). Oddly, the appearance of the paper was met with a great deal of consternation by a number of investigators. One German group even wrote the Editor of the *Journal of Parasitology*, Brent Nickol, protesting publication of the new species description, and virtually demanding that it, and J. P., be summarily condemned for their lack of credibility. Of course, this did not happen, mainly because of the wise judgment of Brent Nickol, who was supported by a couple of very knowledgeable and astute referees.

I asked J. P. if he could ascribe any serendipity to the discovery of *Sarcocystis neurona*. He responded, emphatically, "Yes, the discovery of the sarcocyst was pure luck. You see, the parasite was at first known only in the horse. We then tried to induce encephalitis in the domestic cat by inoculating *Sarcocytis neurona*. At that time, the sarcocyst was not known. I had given cortisone to the cats, and waited for something to happen for about two months, but they remained clinically normal. I then killed and necropsied them. I was trying to see if *Sarcocystis neurona* grew in them at all. I checked the muscles, nerves, and brains, and was amazed to see sarcocysts in the musculature. I had never expected it and no one at that time would have predicted their presence in that site. This was definitely an example of serendipity!" I concurred.

He then felt that he had to prove they were sarcocysts of *S. neurona*. He said, "I'm not a very patient person, someone who can just sit and wait for things to happen. In this case, I did though. I called Bill Saville at Ohio State to see if he had any parasite-free opossums. He had been successfully rearing these animals and I wanted to know if he had any available for what I wanted to do. He said that he didn't have any, and that I would have to wait for another year to get some." He continued, "It was one of the most difficult years of my life, but I made it through.

In the meantime, I had repeated the experiment. I inoculated another group of cats and waited the whole year for my opossums. We killed these cats and fed muscles from them to the parasite-free opossums. Within a short period of time, the opossums shed oocysts. We fed them to interferon gamma gene knockout mice and horses, and proved they were sarcocysts. This is when I contacted you [GWE] and said, look, we have something important for the *Journal of Parasitology*, where the paper describing the experiments was published in 1997 (Dubey *et al.*, 1991). This is how the life cycle of *Sarcocytis neurona* was completed in 2000."

The discovery of the opossum as the true definitive host was also based in part on the research of a graduate student, Clara Fenger, at the University of Kentucky in Lexington, Kentucky. She was amplifying a gene that she had discovered in cat and opossum feces when she had found something in opossum feces that matched a gene cloned from cell cultures of *S. neurona* I had given to her supervisor, David Granstrom. She then hypothesized that opossums were the definitive hosts for *S. neurona*. She took the opossum feces containing sporocysts and fed it to horses. The horses developed EPM. I am pleased to say that some of this work was also published in the *Journal of Parasitology* in 1996 (Fenger *et al.*, 1996).

The third parasite for which Dubey's name is frequently associated is *Neospora caninum*, and there is an interesting story behind it as well. Inge Bjerkas was graduate student in the Pathology Department at the Norwegian College of Veterinary Medicine, Oslo, Norway, where he found a toxoplasmosis-like illness, i.e., encephalomyelitis and myositis in a litter of dogs. He and his colleagues published their findings in *Zeitschrift fur Parasitenkunde* in 1984 (Bjerkas *et al.*, 1984). They described the organism and the pathology with which it was associated, but were unsure of the identity of the parasite.

As it turns out, J. P. was writing a book on toxoplasmosis at the time and sent a letter to Bjerkas, asking if he would provide a slide so he could examine the parasite, but Bjerkas never responded to his request. So, J. P. contacted James Carpenter, a pathologist at the Angell Memorial Animal Hospital in Boston, Massachusetts, a facility that deals exclusively with small animals. J. P. told him about the Norwegian graduate student who had described the *Toxoplasma*-like illness in the dog and asked if he would share any canine tissues or other information with him. Carpenter sent him all the slides he had for dogs and cats that had been necropsied at the Hospital since 1948. There were literally thousands of them. Carpenter had also meticulously kept case histories for all of the animals on which he had done the necropsies. J. P. kept the treasure in the basement of his home for three years. He told me, "And I examined every one of them in minute detail, and I read their case histories. I found that there was something different about some of these dogs. This is when I discovered the thick-walled cysts in the dogs that had suffered from toxoplasmosis-like disease. Then, I remembered when I was teaching at Ohio State University that I had read a graduate student thesis done in 1957 on toxoplasmosis in dogs. The student had written that in eyes of these dogs, there were thick-walled cysts." He told me that he went through all of these dog slides that had thick-walled cysts and separated them from everything else.

Subsequently, James Carpenter in Boston called J. P. and told him that he had a dog with ulcers. Carpenter said that a resident at the Hospital had prepared a smear from one of the ulcers and that he wanted J. P. to look at it. After he received the slide and examined it, he told me, "I was convinced that we were looking at a different organism than *Toxoplasma*. It was the smear that made my decision. When I looked at it, virtually every organism in the smear was in division by endodyogeny. When they are like this, superficially they look like *Leishmania*. By the time I returned Carpenter's call to tell him of the news, the dog had been euthanized because the resident thought it had *Leishmania*. This cemented my decision in naming the organism *Neospoa caninum*. I called the Editor of the *Journal of the American Veterinary Association* (JAVMA) and told him that I had something very important and I would like to publish it in the *Journal*. He agreed."

At USDA, there was a rule that all publications must go through an external review before it can be submitted for publication. Darwin Murrell, who was a Director at the lab at the time, decided he would send J. P.'s paper describing the new species to Norman Levine, one of the real giants in protozoology of that era. Levine read the paper and wrote to both Darwin and J. P. that he disagreed with the latter's determination regarding a new genus and species. J. P. soon met with Darwin and told him that he was convinced it was a new organism, despite his great respect for Levine. J. P. reminded Darwin that it was clinically and morphologically different from *Toxoplasma* and that he was still of the very strong opinion that it was a new species. J. P. said, "I told him that if I named it *Toxoplasma canis*, there would be a lot of confusion because there had already been something that was named *Toxoplasma canis* or *caninum*. I was afraid that if I used the latter name, it would not bring attention to the clinical aspects of the parasite that I wanted to make. That's the reason for giving it a new generic name. The decision was made to go ahead and send it to JAVMA. It was then that the Editor of JAVMA told me that they could not publish new names and descriptions of new taxa. I said, look, that's not what I want to do. I really want to publish this as an unrecognized new clinical entity and then he agreed (Dubey *et al.*, 1988)."

Neospora caninum is now known as one of the major causes of spontaneous abortion in livestock. I asked J. P. why it took a relatively long time to link the parasite to this problem? He explained, "I always had these thick-walled cysts on my mind. I had gone to Australia to visit a colleague named Bill Hartley, who showed me the slide of the spinal cord from a one-day-old calf that had died, and there were thick-walled cysts present. I returned home with a slide that Bill Harley had given me, and the presence of the thick-walled cyst in the calf stuck in my mind. A veterinary pathologist, John Thilsted, who worked at New Mexico's Veterinary Diagnostic Laboratory, soon contacted me. He had a herd of cattle that was aborting. Thilsted was terribly frustrated by the situation because he had examined many aborted fetuses and could not find a cause. He had sent slides for me to inspect. Much to my surprise, I found some small cysts in the fetuses and, although the cysts were very small, they had thick walls! Right away, a connection was made in my mind. John Thilsted told me that he was going to be associated with a new journal, called the *Journal of Veterinary Diagnostic Investigation*, and would it be okay if we publish a clinical report describing the new find? I told John that this was not *Toxoplasma*. It was very much like this new *Neospora* organism that we had recently discovered in dogs. So, when we published it (Thilsted and Dubey, 1989), we referred to it as neosporosis-like organism, and this is how the connection between *Neospora caninum* and bovine abortion was first made."

I then asked J. P. how long these spontaneous abortions had been occurring in cattle. He responded, "Oh, forever!" I reacted by asking, "Why did it take so long to figure out their etiology?" He said, "The reason was that people had always been looking for *Toxoplasma* in aborted fetuses, and they never found it. Part of the reason they had gone down the wrong road was that *Toxoplasma* causes abortion in sheep, so they made a natural mistake in assuming that it caused the same problem in cattle, but they were wrong." But, why hadn't they seen these cysts in cattle? He said it was because there are so few of them and, in most cases, brains were not examined. He then told me that he even had a grant to explore the cause of cattle abortion and that he had introduced *Toxoplasma* in cows, but had never induced an abortion.

J. P. said that, after these cysts were found in aborted fetuses, Brad Barr and Michael Anderson at the vet school of University of California-Davis deserved much of the credit in shedding further light on the etiology of the abortion problem in cattle. To his credit, the Dean at the University of California-Davis vet school decided that every veterinary pathologist who worked in the diagnostic lab there had to be involved in research. According to J. P., "These two young pathologists, Brad Barr and Mike Anderson, took it upon themselves to investigate the cattle abortion problem. They systematically examined hundreds of aborted fetuses and they consistently found *Neospora caninum* in them using antibodies that we provided for specific diagnosis." They concluded that *Neospora caninum* was the major, single cause of spontaneous abortion in North American cattle. The work, started by Dubey and completed by Barr and Anderson, has saved, and will save, the cattle industry millions of dollars on an annual basis.

I asked J. P. if the life cycle of *Neospora caninum* was known. He said, "It took another ten years to find the oocyst of the parasite. Milton McAllister at the University of Wyoming (now at the University of Illinois) made the initial discovery. He had fed *Neospora* cysts to dogs, but could not find oocysts in the feces. However, about a third of the mice fed feces from dogs became seropositive to *N. caninum*. He contacted me and asked if I would participate in a definitive experiment. Four dogs were fed numerous *N. caninum* cysts in his lab. He sent me feces from these dogs that we examined microscopically. I called him in the middle of the night and told him to open a bottle of champagne to celebrate his success because his dogs were found to be shedding oocysts." Infecting gamma interferon gene knockout mice that we had in my lab proved the identity of the oocysts; these mice died of neosporosis when fed canine cysts. Pita Godim, McAllister's graduate student, found the coyote as another definitive host for *N. caninum*, which makes perfect sense since coyotes are sympatric with cattle on the pasture where cattle normally graze. J. P. told me, however, that postnatal transmission of *N. caninum* is still a mystery. However, transmission in a herd is vertical and most efficient. I asked why the parasite had not been seen before in aborted fetuses. He responded that, in the first place, the parasites are rare and not abundant in the issues. Second, until Barr and Anderson started looking, no one ever checked the brain. It just was not done. J. P. said, "Jack Frenkel always told me, 'always fix the whole mouse, formalin is cheap'. How right he was."

There was one other important player in the *Neospora caninum* story that must be identified. This person is David Lindsay, presently a Professor in the vet school in Blacksburg, Virginia, but back then a young postdoc out of Auburn University who was working in J.P's lab in Beltsville. J. P. remarked that, "David has these magic hands," because he was the first to culture *Neospora caninum* using an in vitro cell culture process. He not only cultured the parasite, he was involved in preparing antibodies to it, and then in developing a serodiagnostic test. It is easy to see why David has been so successful as a scientist. He has what we might term a 'knack' for this sort of thing. J. P. calls it "magic hands", but I have always referred to it as having a 'green thumb'.

Early in our interview, I asked J. P. if he started his career with the aim of simply doing good science, or if he had a 'humanitarian' motive. Without hesitating, he replied, "Humanitarian." He went on to explain that he was concerned about the young child with irreparable hydrocephaly, and the thousands of spontaneously aborted lambs or calves each year around the world. I believe that his contributions to the fields of parasitology, public health, and veterinary medicine are not modest, but massive. He is approaching 68, and he could easily and justifiably retire if he chose, but he wants to contribute even more. By continuing, I suspect that he will!

REFERENCES

Bjerkas, I., S. F. Mohn, and J. Presthus. 1984. Unidentified cyst-forming sporozoon causing encephalomyelitis and myositis in dogs. *Zeitschrift fur Parasitenkunde* **70**: 46–54.

Dubey, J. P., G. E. Davis, A. Koestner, and, K. Kiryu. 1974. Equine encephalomyelitis due to a protozoan parasite resembling *Toxoplasma gondii*. *Journal of the American Veterinary Medicine Association* **165**: 249–255.

Dubey, J. P., A. L. Hattel, D. S. Lindsay, and M. J. Topper. 1988. Neonatal *Neospora caninum* infection in dogs: isolation of the causative agent and experimental transmission. *Journal of the American Veterinary Association* **193**: 1259–1263.

Dubey, J. P., S. W. Davis, C. A. Speer, *et al.*, 1991. *Sarcocystis neurona n. sp.* (Protozoa: Apicomplexa), the causative agent of equine protozoal myeloencephalitis. *Journal of Parasitology* **77**: 212–218.

Fenger, C. K., D. E. Granstrom, J. L. Langemeir *et al.* 1996. Identification of opposums (*Didelphis virginianus*) as the putative definitive host of *Sarcocystis neurona*. *Journal of Parasitology* **81**: 916–919.

Frenkel, J. K., J. P. Dubey, and N. L. Miller. 1970. *Toxoplasma gondii* in cats: fecal stages identified as coccidian cysts. *Science* **167**: 893–896.

Hutchison, W. M. 1965. Experimental transmission of *Toxoplasma gondii*. *Nature* **206**: 961–962.

Thilsted, J. P. and J. P. Dubey. 1989. Neosporosis-like abortions in a herd of dairy cattle. *Journal of Veterinary Diagnostic Investigations* **1**: 205–209.

Summary

There is endless merit in a man's knowing when to have done.

Francia, Thomas Carlyle (1795–1881)

As it turns out, the title for this book is hugely misleading. When I began writing about two years ago, I really thought that many of the discoveries in parasitology were of a serendipitous nature. However, after talking with my group of scholars/parasitologists and seriously evaluating some of the historical aspects of parasitology, I discovered this is simply not the case. I would say, in fact, that of all the major discoveries made in the field of parasitology over the last 150 years, no more than 5–10% can be considered as truly serendipitous. Yes, there were occasions when somebody found something, or saw something, for which they were not looking. For example, in one of my first interviews, Dick Seed pointed out that very early in the twentieth century, there were serendipitous discoveries made regarding antigenic changes in the African trypanosomes. Keith Vickerman observed, quite by accident, some sort of 'coat' on trypomastigotes of *T. brucei* in the salivary gland of a tsetse fly. He put 'two and two' together and came up with a hypothesis that, when tested, generated positive evidence for the presence of VSGs. Sidney Ewing discovered erhlichiosis in a North American dog by accident. Arthur Looss, while working in Egypt, accidentally dropped culture medium containing L_3 larvae of hookworm on his own skin and this led him to identify the entry route for these parasites into humans. This discovery was paralleled by Will Cort's work when he accidentally infected himself with cercariae that cause swimmer's itch. Bill Campbell's discovery that ivermectin would kill preadult heartworms and microfilariae, but not adults, was pure serendipity. In fact, as Bill said, "a Hollywood scriptwriter could not have done a better job in terms of a story line." There were several other discoveries that seemed to have been serendipitous, but, in most of these cases, the person was looking for something; they just did not know what it was. In these cases, however, i.e., whenever luck, or chance, intervened, the person recognized the observation as being significant. Like Sidney Ewing said when he saw *Erhlichia*, and recognized it for what it was, "I was ready." This confirms that Pasteur was most certainly correct regarding observation and the prepared mind.

In these men and women I interviewed, there were several qualities that I noticed as I went along. In the first place, they are all dedicated, and I mean, really dedicated, to their field of parasitology. In fact, if I had to choose a word to describe them, each of them, it would be pugnacious. I do not mean that they are bullies or belligerent. All of them had a goal, or goals, that they set for themselves, and then they stuck to their mission. They did not quit. I should note too that they were all successful, whether they achieved their goals or not. If they did not meet their objective, it was not because they quit. It was, in a few cases, because they were working on an intractable problem or, in a few others, because the technology necessary to achieve success had not been developed by the time they retired. Another notable feature of several was their absolute humanitarian commitment. They were not in it because they had a selfish interest in themselves, or for themselves, or for fame or fortune. They possessed a genuine concern for others. Finally, all of them were mentors of one sort or another. They were all fond of helping others, especially young folks. Some of them never set foot in a classroom, but stayed in the laboratory throughout their careers. One of those interviewed was a newspaper reporter who wrote a beautiful story about a South American tree, the bark of which could cure malaria – and, he did not even "particularly like plants." Whether in a classroom, a laboratory, the field, or on a 'beat', all of them were/are important purveyors of information for consumption by those not 'in the know'. In my mind, this makes them mentors.

Another thing that I picked up during my interaction with these people was their humility. It was remarkable. To be sure, all of them had rather large egos. But in each case, their self-confidence, self-assurance, and self-security completely suppressed their egos. In other words, they all knew they were good, but they were not concerned about broadcasting it. They did not feel the need.

I could have picked some others to talk with, but I am satisfied with the topics I chose and the people whom I interviewed. The time these people gave to me was something I will never forget. I hope that I have been able to pass along something about their huge passion for learning and their deep commitment to conveying this information and knowledge to humankind. These are truly extraordinary people! And the real bottom line – so were their discoveries and contributions to our science.

Index

Anderson, Roy   65–76, 203–218
   and Harry Crofton   70
   and the Wellcome Trust research
      centre at Oxford University
      74–75
   Bartlett, Morris, and *Stochastic
      Processes in Ecology and
      Epidemiology*   69, 70
   Bundy, Don, a student and
      colleague   71, 74–75, 215, 216,
      219–235
   Gates Foundation   75
   Imperial College of Science and
      Technology   61, 66, 72–73, 74–75
   May, Bob, a research colleague   70
   Ministry of Defence   61–62, 66,
      75–76
   negative binomial model   70
   Oxford University   69, 73–74
   the overdispersion concept   70
   transition to epidemiology   71–72
   undergraduate and graduate
      education   62–70
   Wooten Rob, a fellow student at
      Imperial College   101–102

Beale, Geoffrey
   The Genetics of Paramecium aurelia
      4, 8, 121
Borrelia burgdorferi   164, 165, 171, 172
   *Ixodes scapularis,*a spirochete vector
      165, 170, 171
   Ixodes pacificus, a spirochete vector
      169, 170
Bundy, Don   37–41, 219–235
   Anderson, Roy, a mentor at Kings
      College and Imperial College
      40, 41
   conflicting opinions regarding
      control of geohelminths
      227–229

Booker, Simon, and GIS in
   epidemiology   232–233
Crofton, Harry   40, 41
'Focusing Resources on Effective
   School Houses' (FRESH)   233
Hall, Andrew, and the 'weight pole'
   232
life as a child in Singapore   38
May, Bob   40, 41
research in Montserrat   227
research in the West Indies
   225–226
schooling in England   38–41
the move to Oxford   231–232
to Imperial College   226–227
Whitfield, Phil, the mentor at
   King's College   39–41
World Bank   38, 41, 215, 233–235

Campbell, Bill   18–25
   and ivermectin   19, 25, 175–187
   and Trinity College, Dublin   19
   and T. W. M. Cameron   22
   Dicrocoelium dendriticum   22
   Fasciola hepatica   23, 24
   Fascioloides magna   24
   going to the United States   21–22
   member, National Academy of
      Sciences   18
   Onchocerca volvulus and ivermectin
      19
   Schistocephaus sp. and *in vitro*
      culture   20
   Smyth, J. Desmond, the mentor at
      Trinity College   19–20
   the *Dicrocoelium dendriticum* seminar
      22–23
   the Merck experience   23–25
   Todd, Arlie, the mentor at the
      University of Wisconsin   20, 22
cinchona tree   188–202

Chelsea Physic Garden, London
190–192
decline in cinchona bark
production in South America
196–197
Honigsbaum, Mark   189, 201–202
'Jesuit's bark'   189
Ledger, Charles   197, 197–198
Markham, Sir Clements   197,
198–199
Royal Botanic Gardens at Kew,
London   190, 198, 199, 200
Sloane, Dr. Hans   192
Society of Apothecaries
Spruce, Richard   197, 198–199
The Fever Trail   189, 198, 199, 200,
202
Traditional Medicinal Malaria Plants
190
transport of quinine to Europe
192–195
'War of Jenkins' Ear'   195–196

de Kruif, Paul
The Microbe Hunters   4, 7, 43, 47,
79, 121, 252
Desowitz, Robert   9–11
and an episode with malaria
128–129
entry into London School of
Hygiene and Tropical Medicine
(LSHTM)   10
Garnham, P. C. C., a mentor at
LSTHM   10
Mulligan, Hugh   11
Schaudinn, Fritz, and the malaria
life cycle error   10
Shortt, Henry, as the mentor at the
LSTHM   10, 11
The Malaria Capers   128, 140
transmission of HIV?
Who Gave Pinta to the Santa Maria
128
Dicrocoelium dendriticum   282–297
Cameron, T. W. M., an error in the
literature   296
Cionella lubrica, production of slime
balls   294–295
Formica fusca, the second
intermediate host   294, 295
Mapes, Courtland   293, 294,
295
resolving the life cycle   293–296
Whitlock, John   293
discovery of insect vectors for
protozoan parasites
Babesia bigemina   134
Boophilus annulatus   134

Bruce, Sir David   1, 2, 4, 110–113,
134
Kilbourne, F. L.   134
Manson, Sir Patrick, and malaria
135
Manson, Sir Patrick, and *Wuchereria
bancrofti*
Smith, Theobald   134
Dubey, J. P.   100–106
Agricultural Research Service,
Beltsville, Maryland   106
a job at the Indian Veterinary
Research Institute   102–103
Beattie, Colon, a mentor at the
University of Sheffield   104
Beverley, Jack, the mentor at the
University of Sheffield   104
Commonwealth Fellowship
104–105
early education in India   101–102
early graduate work at Mathura
Veterinary College, India
103–104
doctoral research in England   104,
104–105
Fayer, Ron   106
Frenkel, Jack, a postdoctoral
mentor at the University of
Kansas Medical Center   103, 105
marriage   104
Montana State University   105–106
Neospora caninum   328
Ohio State University   105
Pande, B., a mentor in vet school
103
Sarcocystis neurona   328
Toxoplasma gondii   328
transmission of *Toxoplasma gondii*
87–88, 103, 103–104, 104,
328–337
University of Sheffield   104

epidemiology   203–218
Anderson, Roy   203–218
and computers   206–207
and genetic predisposition
210–211
and HIV   211, 215
and infectious diseases   204–205
and the future   215–216
and the use of drugs in controlling
geohelminths   229–230
and treatment of infectious disease
212
and tuberculosis
and vaccines   214–215
anthelmintic drugs and resistance
214

epidemiology (*cont.*)
   Booker, Simon, and GIS in
      epidemiology   232–233
   delivery of drug treatment
      230–231
   economics of infectious disease
      control   213
   conflicting opinions regarding
      control of geohelminths
   Crofton, Harry, and the concept of
      overdispersion   207–208
   define parasitism   208–210
   'Focusing Resources on Effective
      School Houses' (FRESH)   233
   from eukaryotes to prokaryotes and
      viruses   206
   genetic predisposition and
      transmission biology   211–212
   Hall, Andrew, and the 'weight pole'
      232
   Hawdon, John, and hookworm
      research in China   212
   Hotez, Peter, and hookworm
      research in China   212
   International Development
      Association
   mass treatment   212
   Merck, and onchocerciasis control
      215
   new approaches by Anderson and
      May   205–206
   parasitism, an ecological
      perspective   203–218
   poverty and infectious disease
      214
   selective treatment   212
   seminal epidemiology papers
      204–205
   targeted treatment   212–213
   transmission biology   210–212
   United Nations Development
      Program (UNDP)   231
   World Bank and Don Bundy   215,
      216, 233–235
Ewing, Sidney and Margaret
      28–37
   Anaplasma marginale research by
      Sidney   36
   and Derek Zelmer   291
   and Wendell Krull at Oklahoma
      State University (OSU)   31–34
   Ehrlichia sp., first discovery in North
      America   33–34
   Halipegus occidualis   29, 288,
      282–293, 297
   Kocan, Dick, a student colleague at
      the field station   32
   Metastrongylus spp. in pigs   30, 31

   Murrell, Darwin, a student
      colleague at the field station   32
   off to Kansas State University
      35–36
   Parasites, People, and Places   29
   Steffens, Margaret, Sidney's future
      wife   32, 282
   the firing of Wendell Krull   35
   Todd, Arlie, Sidney's mentor at the
      University of Wisconsin   30–31
   Tylenorhynchus ewingi   30
   University of Michigan Biological
      Station (Douglas Lake)   31–57
   Wendell Krull: Trematodes and
      Naturalists   29, 282

Gates Foundation   47, 75, 215
geohelminths, and the 'Unholy
      Trinity'   219–235
   Ancylostoma duodenale   219
   and the use of drugs in treating
      geohelminths
   Ascaris lumbricoides   219, 221
   Bundy, Don   224–226
   Collandruccio, Salvatore, resolving
      the *A. lumbricoides* cycle   221
   Collandruccio, Salvatore, resolving
      the *T. trichiura* cycle   221–222
   conflicting opinions regarding
      geohelminth control   227–229
   delivery of drug treatment
      230–231
   difficulties in resolving
      geohelminth life cycles   220
   estimated infection numbers by the
      'Unholy Trinity'   219–220
   'externality' phenomenon in
      treating geohelminths   230
   'Focusing Resources on Effective
      School Houses' (FRESH)   233
   Hall, Andrew, and the 'weight pole'
      232
   Koina, Shemesu, proof of a
      monoxenous life cycle for *A.
      lumbricoides*   221
   Necator americana   219
   pathologies produced by the
      geohelminths   223
   resolving the *Strongyloides stercoralis*
      life cycle   222–223
   Stewart, Francis, and the internal
      migration of *A. lumbricoides*   221
   Trichuris trichiura   219, 222,
      225–226
   United Nations Development
      Program   231
   World Bank and Don Bundy
      233–235

Hawdon, John   53–61
  and Gerry Schad, the mentor at the
    University of Pennsylvania   55
  early education   53–55
  hookworm, genetic variation, and
    vaccine   58–59
  Hotez, Peter   53, 57–58
  Pollock, Rich, and glutathione   56
  postdoc at Yale   60
  research at Penn   31–57
  research in China   58, 59–60, 212
  Theiler, Max   158–160
HIV-AIDS   150–163
  a case history of AIDS   154
  and natural selection   160–161,
    211
  drug therapy   155–156
  problems in drug delivery
    162–163
  reverse transcriptase   152, 153
  vaccine   154–155, 215
  virus description   151–156
Honigsbaum, Mark   61–65
  discovering the cinchona tree and
    malaria   62–63
  interest in diseases   62–70
  learning about malaria   63–64
  London *Guardian*   65–76
  on becoming a journalist   61–62,
    66
  Oxford University, New College
    61, 66
  retracing the cinchona discovery
    route   64
  The Fever Trail   189, 198, 200, 202
  Valverde's Gold   64
hookworm disease   236–253
  Bentley, C. A., experiments in India
    238
  benzimidazole   243
  chemotherapy and hookworm
    control   242–243
  conflicting opinions regarding
    hookworm control   241–242
  connecting the worm and the
    disease   237–239
  definite serendipity   238
  Dubini, Angelo, and the discovery
    of hookworm   237
  establishing a field connection in
    China   59, 248–249
  genetic variability and the
    epidemiology of hookworm
    disease   252
  Hawdon, John   245
  hookworm life cycle   237–239
  human testing of the vaccine
    250–251

ionizing radiation of hookworm
    larvae for a vaccine   244–245
Looss, Arthur, and hookworm life
    cycle   237
mebendazole   243
Salk, Jonas, suggests the need for a
    murine host model   247–248
Schad, Gerry   245
selection of a vaccine testing
    location   250–251
search for the correct hookworm
    antigen(s)   246–247
sequencing the antigen   247
source of the protein antigen   247
spread of hookworm in Europe
    237
Stiles, Charles, and the Rockefeller
    Foundation   239
survey in China for hookworm
    disease   236
tackling the hookworm problem in
    the U.S.   239–240
the hunt for private financial
    support   249–250
vaccines and hookworm control
    243–252
Hotez, Peter   42–47
  Ancylostoma duodenale   46
  Ancylostoma caninum   46
  and hookworm research in China
    212
  Cerami, Tony, as the mentor at
    Rockefeller University   45
  conflicting opinions regarding
    control of geohelminths
    227–229
  Gates Foundation, and hookworm
    disease   47
  hookworm immunity   42, 46, 47
  off to Rockefeller University   46,
    44–47
  Schad, Gerry, the mentor at the
    University of Pennsylvania
    53–61
  Rockefeller Foundation's 'The Great
    Neglected Diseases'   46
  Warren, Ken, a mentor at
    Rockefeller

ivermectin   175–187
  a break with tradition in the
    discovery of ivermectin   178–180
  advantage of ivermectin   183–184
  a 'tandem assay'   179
  Campbell, Bill, and the Merck
    experience   177–186
  Cuckler, Ashton   177
  effect on *Dirofilaria immitis*   182

ivermectin (*cont.*)
  Eimeria muris   178
  empirical screening   176
  Erhlich, Paul, an early proponent of
    immuno- and chemotherapy
    176–177
  Merck Institute for Therapeutic
    Research   177
  milbemycin   175
  mode of action   184
  Nematospiroides dubius   178
  Perspectives in Biology and Medicine
    185
  research atmosphere at Merck
    177–178
  serendipity and the treatment of
    dog heartworm   182–183
  the discovery of ivermectin
    180–182
  the Kitasato Institute   180
  treatment of *Onchocerca volvulus*
    184–186
  treatment strategies for infectious
    disease   175–176

Krull, Wendell   30, 282–297
  a consummate naturalist   296–297
  and Miriam Rothschild   282, 286,
    288, 288–289, 290, 291, 293
  career   282–285
  Dicrocoelium dendriticum, resolving
    the life cycle   293–296
  Halipegus occidualis, resolution of
    the life cycle   282, 288–293
  Mapes, Courtland, and *Dicrocoelium
    dendriticum*   293, 294, 295
  Pneumonoeces (= *Haematoloechus*)
    medioplexus   285
  Pneumonoeces (= *Haematoloechus*)
    paraplexus   285
  serendipity   297
  trematode life cycles, general
    comments   286–287
  vet school experiences   284–
    285
  *Wendell Krull: Trematodes and
    Naturalists*   282, 288
  Whitlock, John, and *Dicrocoelium
    dendriticum*   293–296
  Zelmer, Derek, and *Halipegus
    occidualis*   291, 293

Lord, Pat   95–100
  a job in the Biology Department,
    Wake Forest University   99–100
  graduate work at Wake Forest
    University Medical Center
    97–98

Lord, Richard   98
  postdoctoral work   98–99
  undergraduate work at North
    Carolina State University   96–97
Lyme disease   164–173
  an emerging disease   164–173
  Borrelia afzeli, an etiological agent
    of Lyme disease   165
  Borrelia burgdorferi, an etiological
    agent of Lyme disease   164, 165,
    171, 172
  Borrelia garinii, an etiological agent
    of Lyme disease   165
  Burgdorfer, Willy, and discovery of
    the spirochete   165
  controversy regarding identity of
    Ixodes dammini, and as a vector
    168–172
  disease in the southeastern U.S.
    172
  distribution of Lyme borreliosis
    165
  Ixodes dammini, the deer tick   165,
    169, 170
  Ixodes pacificus, a spirochete vector
    165, 169, 170
  Ixodes scapularis, a spirochete vector
    165, 170
  nonhuman hosts   165, 167
  Old Lyme, Connecticut   164
  pathophysiology of the disease
    165–166, 167–168
  problems in diagnosis   172–173
  Speilman, Andy   168–172
  transmission of the spirochete
    165, 166–167

malaria   128–148
  and serendipity   147–148
  Anopheles gambiae   144
  Apicomplexa   130
  acquisition of malaria   128–129,
    133–137
  chloroquine   129
  completion of the malaria life cycle
    140–142
  Culex spp.   137
  DDT   143–144
  discovery of the etiological agent
    132–133
  discovery of *Plasmodium* spp. life
    cycle   133–137
  Hawking, Frank, 'the Bishop'
    141
  host switching   145–147
  Huff, Clay   140
  Koch, Robert   137–138
  Laveran, Charles   130, 131–132

Manson, Sir Patrick, and his role in discovery of the etiological agent 133–134, 139, 188
'omnis cellula e cellula' 131
phylogeny 145
quinine 129
Ross, Ronald 1, 2, 130, 133–137
Schaudinn, Fritz, and his error 139–142
shikamate pathway 145
Silent Spring 143
WHO's effort to eradicate malaria 143
Murrell, K. Darwin 12–18, 32
and Denmark's Centre for Experimental Parasitology 17
and the Naval Medical Institute (NAMRI) 13, 14–15
Beltsville, and ARI 15–16
Nansen, Peter, tragic death 17
Naval Medical Research Unit-2, in Formosa 13–15
Strongyloides ransomi 15, 129
Strongyloides ratti 15
Taenia crassiceps 14
Trichinella spiralis research 15–16

Nadler, Steve 77–85
ascaridoid worms 83, 83–84
back to LSU 84–85
Department of Nematology, UC-Davis 85
early education and science 78–79, 83–84, 316
Giardia spp. 83
graduate education 79–84
Northern Illinois University 85
postdoctoral work 85
undergraduate education 79
Neospora caninum 328
and J.P. Dubey 328–344
biology and life cycle 340–344
Frenkel, Jack, 'formalin is cheap' 343
Lindsay, David, and 'magic hands' 343–344
McAllister, Milton, and life cycle 343–344

Oliver, Jim 86–95
Borrelia burgdorferi and Lyme disease 90
early life in Georgia 87
Eibson, John, President Georgia Southern University 94, 95
first teaching position, University of California-Berkeley 92, 93
Florida State University 88–89

Georgia Teachers College (Georgia Southern University) 87–88, 94–95, 104
graduate work at Johns Hopkins 91
graduate work at the University of Kansas 91–92
postdoc in Australia 92–93
Short, Robert (Bob), the mentor at Florida State University 88
Shuster, Sue 90
Southeastern Society of Parasitologists presidential contest in 1975 86
University of Georgia 87–88, 93–94
U.S. army duty 89–91

phylogenetics 315–327
and Pete Olson 315
approach to phylogenetics 316–318
contagion and genetic predisposition to infection 322
contention in phylogenetics 315–316
cospeciation 318–319
evidence relating acanthocephalans to rotifers 322–323
future research 326–327
genetic convergence among nematodes 323–324
genetic distance 319–320
hookworms of pinnipeds 324–326
molecular techniques 320–322
plesiomorphic characters versus homoplasy 318
Plasmodium cynomolgi 141
Plasmodium elongatum 140
Plasmodium falciparum 128–129, 136, 140, 142, 143, 144, 147, 195
Plasmodium gallinaceum 140
Plasmodium malariae 146, 195
Plasmodium vivax 128, 140, 141, 143, 147, 195
Pseudodiplorchis americanus, an amazing monogene 254–264
an early publication regarding *P. americanus* 254–255
funding and long-term research 256
geographic distribution of *P. americanus* 261
immune response to *P. americanus* 260
life cycle of *P. americanus* 257, 258–259

Pseudodiplorchis americanus (cont.)
  pathology caused by *P. americanus*
    259–260
  seasonal changes in population
    biology of *P. americanus*   260–
    261
  serendipity leads to Arizona
    255–256
  spadefoot toad biology   254–264
  toad reproduction   256–258
  Xenopus laevis, and *Protopolystoma*
    xenopodis research in Wales
    263–264
  Xenopus spp. research in Africa
    255

quinine   188, 188–190, 202
  Cromwell, Oliver, aversion to
    quinine   195
  de Ribera, Francisca Henriquez,
    fourth Countess of Chinchon
    189, 194
  discovery as an antimalarial drug
    189
  Manson, Sir Patrick, announcement
    to BMA re Ross's discovery of the
    Plasmodium sp. life cycle   188
  pulvis comitsae, 'the powder of the
    countess'   189

Rollinson, David   48–53
  education prior to the Natural
    History Museum   49–51
  Schistosoma mansoni, circadian
    rhythms and egg excretion   49,
    50
  The Natural History Museum,
    London   48, 48–49, 50, 51–52

Sarcocystis neurona   328
  and J. P. Dubey   328
  and serendipity   339
  biology and life cycle   338–340
  Fenger, Clara, and the life cycle
    340
  publication problems   339
schistosomes, the split-bodied flukes
    265–281
  A History of Human Helminthology,
    D. I. Grove   268
  control of schistosomiasis
    277–279
  delivery of drugs   276–277
  drug resistance   279–280
  historical aspects of schistosome
    discovery   267–269
  hybridization of schistosomes
    273–275

identification of the second
  schistosome, *S. mansoni*   268,
  270
molecular phylogenies   275
resolving the life cycle of
  Schistosoma haematobium
  269–270
resolving the life cycle of *S.
  japonicum*   271–273
schistosome genome project   278
schistosome species groups
  273–275
treatment with praziquantel
  278–279
vaccines   275–276, 280
Seed, Dick   3–5, 115
  and ß-galactosidase in *Escherichia
    coli*   5
  frustration with VSG research
    126–127
serendipity   345–346
  and discovery   2, 115–118
  and discovery of ivermectin   178
  and Keith Vickerman   9, 125
  and life cycle of *Sarcocystis neurona*
    339
  and malaria research   147–148
  and Richard Tinsley   255
  and Wendell Krull   297
  Campbell, Bill, and Merck   23
  definition   2
  Ehrlichia sp., isolation by Sidney
    Ewing   34
  ivermectin and treatment of dog
    heartworm   182
  Schistosoma haematobium   268

Tinsley, Richard   25–28
  postdoc time   27–28
  Queen Mary's College   28
  University of Bristol   26, 28
  Westfield College   28
  Xenopus spp.   26, 27
Toxoplasma gondii   328
  and J. P. Dubey   328
  and population genetics   334–338
  and *Toxocara cati*   329
  and the University of Sheffield
    329–332
  Beattie, Colin, a mentor in
    Sheffield   329
  Beverly, Jack, the mentor in
    Sheffield   329
  biology and life cycle   328–337
  coccidian speciation   334–338
  Frenkel, Jack, and the life cycle
    332, 329–332, 333
  genetic groups   335

Hutchison, William 329–332
publication priority problem
333–337
sporulation, and the 'dichromate
problem 333
Trichinella spp. 299, 299–313
and politics 305–306
and systematics 306–311
Britov, A. V., elevation to subspecies
status 308–309
Dick, Terry, and taxonomy of
Trichinella spp. 309
continuing research 312
epidemiological questions from the
pork producers 309
final systematics solutions
310–311
historical aspects of *Trichinella*spp.
301–305
Lichtenfels, Ralph, no subspecies
morphological markers 309
molecular systematics 310
Paget, James, and discovery of *T.
spiralis* 48–49, 301
Owen, Sir Richard, and theft of a
discovery 302
Nelson, George, and problems in
Trichinella spp. systematics 307
serendipity 312–313
transcriptional alteration of host
cell by L$_1$ stage 301
Trichinella spiralis, early personal
experience 300–301
trichinosis 299–313
Warren, MacWilson, seminars
300
Zenker, Freiderich, link between *T.
spiralis* in hogs and trichinosis in
humans 304–305
Trypanosoma brucei brucei 4, 109,
124, 125
and antigen coat 9
and ferritin-conjugated antibody
125
Trypanosoma evansi, and transmission
by tabanid flies 110
Trypanosoma brucei gambiense 109
Dutton, J. E., and discovery of
111
Trypanosoma brucei rhodesiense 109,
114
Stephens, J. W. W., and discovery of
114
Trypanosoma vivax 124
trypanosomes 108–127
and their variant surface
glycoproteins (VSGs) 1, 4, 9,
115, 108–118, 124–126, 127

and the original hypothesis
regarding the VSG 126
Desowitz, Bob, early explanation of
VSGs 124
features in common 109
Geimsa era 121
genetic changes associated with
VSG changes 118–121
range in disease caused by
108–109
nagana 109
success of African trypanosomes
114–115
trypanosomiasis
Bruce, Sir David, and discovery of
cause of nagana and sleeping
sickness 1, 2, 4, 110–113,
134
Bruce, Sir David, and second
sleeping sickness commission
into Uganda 110, 113
Castellani, Aldo, and first sleeping
sickness commission into Uganda
112
Castellani, Aldo, and second
sleeping sickness commission
into Uganda 2, 113
control of 109
Nabarro, D. N., and second sleeping
sickness commission into Uganda
113
South Africa, Natal, and work of Sir
David Bruce 110–113

Vickerman, Keith 4, 5, 6–9, 115,
121–126
Fellow, Royal Society 4, 6, 7
frustration with VSG research
126–127
Glasgow University 6, 7, 9
into Uganda looking for *T. b. brucei*
122
into a career 7
Journal of Cell Science 125
mitochondria and metabolism
research 124
Metchnikoff, Eli 7
seeking a Ph.D. 8
the Uganda adventure 123–124
trypanosome energetics switch 8

yellow fever 150–163
development of vaccine 158–160
discovery of etiological agent
157–158
epidemiology 156–157
Theiler, Max 158–160
virus description 151